Population and Community Biology
PLANT SUCCESSION

Population and Community Biology Series

Principal Editor

M. B. Usher
Chief Scientific Advisor, Scottish Natural Heritage, UK and Reader, Department of Biology, University of York, UK

Editor

R. L. Kitching
Professor, Department of Ecosystem Management,
University of New England, Australia

The study of both populations and communities is central to the science of ecology. This series of books explores many facets of population biology and the processes that determine the structure and dynamics of communities. Although individual authors are given freedom to develop their subjects in their own way, these books are scientifically rigorous and a quantitative approach to analysing population and community phenomena is often used.

Already published

PLANT SUCCESSION
Theory and prediction

Edited by

David C. Glenn-Lewin
Department of Botany, Iowa State University, Iowa, USA

Robert K. Peet
Department of Biology, University of North Carolina, North Carolina, USA

Thomas T. Veblen
Department of Geography, University of Colorado, Colorado, USA

CHAPMAN & HALL

London · Glasgow · New York · Tokyo · Melbourne · Madras

Published by Chapman & Hall, 2–6 Boundary Row, London SE1 8HN

Chapman & Hall, 2–6 Boundary Row, London SE1 8HN, UK

Blackie Academic & Professional, Wester Cleddens Road, Bishopbriggs, Glasgow G64 2NZ, UK

Chapman & Hall, 29 West 35th Street, New York NY10001, USA

Chapman & Hall Japan, Thomson Publishing Japan, Hirakawacho Nemoto Building, 6F, 1-7-11 Hirakawa-cho, Chiyoda-ku, Tokyo 102, Japan

Chapman & Hall Australia, Thomas Nelson Australia, 102 Dodds Street, South Melbourne, Victoria 3205, Australia

Chapman & Hall India, R. Seshadri, 32 Second Main Road, CIT East, Madras 600 035, India

First edition 1992

© 1992 David C. Glenn-Lewin, Robert K. Peet and Thomas T. Veblen

Typeset in 10/12 pt Times by Best-set Typesetter Ltd., Hong Kong
Printed in Great Britain at the University Press, Cambridge

ISBN 0 412 26900 7

A catalogue record for this book is available from the British Library

Library of Congress Cataloging-in-Publication data available

Contents

vi   Contents

Contributors

David C. Glenn-Lewin: Department of Botany, Iowa State University, Ames, IA 50011, USA

Robert K. Peet: Department of Biology, University of North Carolina, Chapel Hill, NC 27599, USA

I. Colin Prentice: Department of Plant Ecology, Lund University, S-22361 Lund, Sweden

Herman H. Shugart: Environmental Sciences Department, University of Virginia, Charlottesville, VA 22903, USA

Dean L. Urban: Department of Range Science, Colorado State University, Fort Collins, CO 80521, USA

Michael B. Usher: Scottish Natural Heritage, Edinburgh EH9 2AS, UK

Arnold G. van der Valk: Department of Botany, Iowa State University, Ames, IA 50011, USA

Eddy van der Maarel: Department of Ecological Botany, Uppsala University, S-75122 Uppsala, Sweden

Robert van Hulst: Department of Biological Sciences, Bishop's University, Lennoxville, QC, JIM 127, Canada

Thomas T. Veblen: Department of Geography, University of Colorado, Boulder, CO 80309, USA

Prologue

PREDICTION OF VEGETATION CHANGE

The broadest goals of scientific theory are the understanding and explanation of observed reality through a system of concepts, laws and empirically based generalizations. An important means of testing the fit between theory and observed phenomena is prediction (Pickett and Kolasa, 1989). Prediction is used for testing the component models and for evaluating the appropriate scope of theory in vegetation science. Prediction, either as a means for testing theory or as a goal itself, is particularly important to the field of vegetation dynamics (i.e. the study of plant succession and regeneration dynamics).

Beyond the scientific goal of relating observed patterns of vegetation change to an explanatory theoretical framework, there are numerous practical reasons for predicting vegetation change. To hunter–gatherer societies accurate prediction of fire-induced changes in vegetation that altered the availability of forage to game animals was vital for their survival. Modern forest and range production depends on our ability to predict the consequences of management activities. Similarly, the preservation of plant and animal species, biotic communities, and productive and aesthetically pleasing landscapes is unlikely without a thorough knowledge of the patterns and processes of vegetation change. In the context of growing concern over the possibility of climate change resulting from global anthropogenic atmospheric changes, the need to be able to predict vegetation change at a variety of spatial and temporal scales has never been more urgent.

HISTORICAL CONTEXT

Although humans have long recognized the phenomenon of succession, the scientific study of succession began only at the end of the nineteenth century. Many of the themes and problems recognized by plant ecologists during the first two or three decades of the twentieth century are still important research objectives in the modern field of vegetation dynamics. To understand the current advances in vegetation dynamics it is necessary to consider briefly the historical background.

The Clementsian paradigm

The concept of plant succession was developed largely in North America during the first two decades of the twentieth century. An important precursor to the development of succession as a central concept in plant ecology was the development of the theory of the 'Geographical Cycle' by the geologist and geographer William Morris Davis (Davis, 1899). Davis's theory stressed the orderly development of landforms, starting with the uplift of a new erosion surface and culminating in the 'peneplain', a rolling landscape of low relief. The progressive development of the peneplain was described by analogy to human development in which a landscape passed through stages of youth, maturity, and old age. The organismal metaphor emphasizing growth and evolution was typical of the post-Darwinian nineteenth century.

Henry C. Cowles, one of the pioneers in successional research (Cowles, 1899), studied physiography in the Geology Department at the University of Chicago under T. C. Chamberlin, a long-time collaborator of W. M. Davis (Rogers and Robertson, 1986). Subsequently, his interests shifted to plant ecology and he completed his doctorate in the Botany Department at the University of Chicago. Cowles extended the concept of the geographical cycle to the study of changes in the sand dune vegetation along the southern shore of Lake Michigan. Arthur G. Tansley (1935), an influential British botanist who himself contributed much to our understanding of plant succession, summarized Cowles's contribution as follows:

It is to Henry Chandler Cowles that we owe, not indeed the first recognition or even the first study of succession, but certainly the first thorough working out of a strikingly complete and beautiful successional series (1899), which together with later more comprehensive studies ('01, '11) brought before the minds of ecologists the reality and the universality of the process in so vivid a manner as to stimulate everywhere – at least in the English-speaking world – that interest and enthusiasm for the subject which has led and is leading to such great results.

Although Cowles and other early twentieth-century botanists (e.g. Cooper, 1926) made important contributions by pointing out the predictive power of succession as a concept, and by documenting many examples of apparent successional sequences, it was Clements (1904, 1916) who offered a comprehensive theory of plant succession. So compelling was the order and logic of Clements' theory that it dominated the field for the first half of the twentieth century. Long-disputed aspects of Clementsian successional theory revolve around the issues of predictability, convergence, and equilibrium. Just as Davis had likened the development of the peneplain to the pattern of human development, Clements viewed succession as a highly orderly and predictable process in which vegetation change represented the life history of a plant community that assumed organism-like attributes. From initially distinct, environmentally determined starting points, communities were believed to converge through succession towards a climax vegetation, the characteristics of which were controlled solely by the regional climate. According to this viewpoint, the climax was a condition of great stability in which the vegetation had reached an equilibrium with the present climate. Clementsian successional theory was an equilibrium viewpoint in its assumption that successional change necessarily progressed towards the development of a stable vegetation type in equilibrium with the regional climate. It was deterministic by postulating that the development of the climax was as orderly and as predictable as the life history of an individual organism.

Clements (1904, 1916) developed a scheme of processes that drive succession:

1. *nudation*, which is the creation of a bare area or partially bare area by the disturbance which initiates succession;
2. *migration*, arrival of organisms at the open site;
3. *ecesis*, the establishment of organisms at the site;
4. *competition*, the interaction of organisms at the site;
5. *reaction*, the modification of the site by the organisms thereby changing the relative abilities of species to establish and survive; and
6. *stabilization*, the development of a stable climax.

Stabilization is perhaps more appropriately viewed as a consequence of the first five processes rather than as a process itself (Pickett *et al.*, 1987). The other processes in Clements' scheme, however, provide a framework within which all mechanisms of successional change can be incorporated. Clements emphasized the importance of reaction in creating environmental conditions less favourable to early colonists and more favourable to late seral and climax species. He viewed reaction as the main driving force of succession, but today reaction, or facilitation (*sensu* Connell and Slatyer, 1977), is emphasized much less.

Two early critics of Clementsian successional theory were Henry A. Gleason and Arthur G. Tansley. Gleason (1917, 1926, 1927, 1939) challenged Clements' assertion that plant communities were highly integrated organic entities and stressed the unique, individualistic behaviour of plant species and the role of chance events. He viewed communities as resulting from the fortuitous overlap of distributions of species with similar environmental tolerances. Gleason emphasized the importance of stochastic processes in succession. Tansley (1935) strongly criticized Clements' assumption that all vegetation change in a particular region would converge towards the same type of climax. He argued that local factors, such as rock type and topographic position, may result in climax vegetation types that differ from that associated with the regional climate. Ultimately, Robert H. Whittaker (1953) merged the views of Gleason and Tansley to describe climax vegetation as varying continuously across a continuously varying landscape.

In search of global generalities

In the 1960s, Ramon Margalef and Eugene Odum attempted to draw together the disparate observations of successional phenomena into unifying theories of succession. Their views were widely repeated in textbooks, even if not fully accepted by many researchers. Margalef (1958, 1963, 1968) applied information theory to ecological systems as a strategy for seeking and explaining universal patterns of successional change. He argued that the linkages among trophic levels and populations represented information and that succession represents a natural trend towards the accumulation of greater information in an ecosystem. Thus, he believed succession is driven from simple ecosystems towards more complex ecosystems with more trophic levels and greater diversity of species and life-forms. Despite the tenuous link between information channels and structural complexity, Margalef's ideas were attractive to ecologists seeking a post-Clementsian universal explanation for succession.

Odum (1969) summarized a number of successional trends in community- and ecosystem-level properties (e.g. biomass, diversity) which he postulated result from the tendency of ecosystems to develop towards greater homeostasis. He argued that within the limits set by the physical environment, succession necessarily proceeds towards an ecosystem of maximum biomass and diversity. The contributions of Odum and Margalef, and the ecologists who followed their lead, were stimulating, but to a large extent their ideas lacked empirical support. Few of their proposed general trends had ever been carefully tested. In the end their most lasting contribution was to focus attention on the need for careful testing of the mechanisms and patterns they proposed. In most cases subsequent tests have failed to provide support for their hypotheses. Perhaps the type of global generalities they sought will ultimately be found, but it is now recognized that such generalities must account for the importance of site-specific initial conditions.

The Margalef–Odum synthesis of succession theory, although couched in terms of thermodynamics and ecosystem parameters, is philosophically remarkably similar to Clements' theory. Both bodies of theory include developmental (*sensu* Drury and Nisbet, 1971) schemes where successional change is viewed largely as the consequence of relationships and interactions within a community, whereas external influences such as large-scale disturbance, climate variation, and immigration of new species are either relegated to minor roles or are assumed to be constant. Such views are equilibrial in their assumption that the physical site and biotic resources remain constant for a long period during which there is progressive development towards a type of ecosystem in which certain properties are maximally expressed within the constraints set by the physical environment. The deterministic driving forces and trends towards greater homeostasis or maximal information content are analogous to the deterministic life history of Clements' super-organismic climax and represent more a philosophy of how nature should be than a theory derived from empirical data and tested by careful experimentation.

Non-equilibrium and mechanistic approaches to succession

By the early 1970s, ecologists recognized the inadequacy of both Clementsian successional theory and the Margalef–Odum synthesis as exemplified by Odum's (1969) paper on the strategy of ecosystem development. There was, however, no other readily available paradigm to replace the global explanations of Clements, Odum, and Margalef. Instead, there was an increased appreciation of the need for site-specific information on the mechanisms, or proximate causes, of vegetation change.

Since the mid-1970s two major conceptual trends have dominated research on vegetation dynamics: (1) a shift away from holistic explanations of successional phenomena towards reductionist and mechanistic approaches emphasizing proximate causes of vegetation change; and (2) a shift away from equilibrium towards non-equilibrium paradigms. The corresponding shift from broad attempts at holistic theorization to a reductionist concern with mechanisms of change does not imply a lack of concern with prediction; prediction is still central to studies of vegetation dynamics. However, predictions are derived empirically from knowledge of the mechanisms of vegetation change which apply to a particular local habitat instead of being deductively derived from a universally applicable theory (Pickett et al., 1987). Generalizations about the relative importance of different patterns and mechanisms of vegetation dynamics are the result of induction rather than deduction.

The view of succession as primarily a species replacement process driven by reaction, or plant-controlled (i.e. autogenic) environmental modification, has been rejected (e.g. Drury and Nisbet, 1973; Connell and Slatyer, 1977; Peet and Christensen, 1980). It has been superseded by several non-mutually exclusive hypotheses that may all apply in varying degrees to any one successional sequence. These hypotheses are presented in subsequent chapters, and variously represent succession as gradients in time or resource availability (e.g. Drury and Nisbet, 1971, 1973; Pickett, 1976; Tilman, 1985, 1988), the consequence of differential longevity and other population processes (e.g. Egler, 1954; Peet and Christensen, 1980), the result of differences in life history characteristics (Drury and Nisbet, 1973; Pickett, 1976; Noble and Slatyer, 1980) or as a stochastic process (e.g. Horn, 1975, 1976). Common to all of these hypotheses is a reductionist perspective emphasizing life histories and competitive interactions of the component species rather than emergent properties of communities (Peet and Christensen, 1980). Contemporary ecologists view vegetation change as the outcome of populations interacting within fluctuating environmental conditions. Thus, parameters such as birth and death rates, and plant growth rates under varying environmental conditions form the core of quantitative models of vegetation dynamics. For long-lived plants, the lack of data on demographic parameters is often a serious constraint for the calibration of these models.

A conceptual trend which has paralleled the shift away from holistic towards mechanistic explanations of succession is the shift from equilibrium towards non-equilibrium paradigms. This trend was presaged by Hugh M. Raup of Harvard Forest and Alex S. Watt of Cambridge University several decades before the 1970s' explosion of interest in ecological disturbance. Raup's work on vegetation change, begun about 1930, was proudly inductive, rejecting deductive Clementsian theory (Raup, 1957; Stout, 1981). Raup (1957) pointed out that the time period required for succession from a pioneering community to a climax community may be so long that the assumption of a constant physical environment is not reasonable. Over the many centuries required for a climax forest of long-lived trees to develop, climatic variability may affect the course of vegetation change.

Watt (1919, 1934), in his studies of beech forests in southern England, showed how the death of individual or small groups of trees and subsequent gap-phase processes explained the apparent stability at a stand or community scale while allowing for instability at a small-patch scale. In his powerful and seminal synthesis on 'pattern and process' (1947) he showed how many plant communities are best seen as a mosaic of structural phases which are always changing as one phase grades into the next. Watt's work heralded the detailed studies of treefall gap dynamics that have been conducted largely since the late 1970s (e.g. Runkle, 1985).

The emphasis in Clementsian theory on a long disturbance-free period is not consistent with what we now know about vegetation dynamics (Pickett and White, 1985; see also Chapters 1, 3, and 8). In many landscapes successional change rarely proceeds all the way to a climax endpoint due either to changing climate or to the frequency of disturbance. The contemporary view of vegetation dynamics may be termed a 'dynamic' or 'kinetic' view in which there is no assumption of either long-term site stability or of the existence of an endpoint to succession (Drury and Nisbet, 1971). The modern view of vegetation change emphasizes the importance of repeated, relatively frequent disturbance and accepts continuous change in vegetation as the norm (Pickett and White, 1985). Indeed, we have come to accept a view very much like the one articulated by Cowles in 1901, that succession is a 'variable approaching a variable rather than a constant'.

THE SCOPE OF THIS BOOK

Ecologists have been trying to develop a theoretical framework for understanding and predicting vegetation change since ecology was first recognized as a discipline. Nonetheless, only recently have these efforts begun to yield the depth of understanding necessary to predict the consequences of human and natural impacts on the landscape. We had many reasons for assembling this book, but certainly one of the foremost was to convey the excitement and urgency which we attach to this line of research. We wanted to demonstrate how this venerable field has found new focus and new direction, and how it can play a pivotal role in decision making at many scales.

The chapters in this book are not designed to be retrospective or to provide an exhaustive overview of all that we know about vegetation dynamics. Rather, they are intended to introduce the reader as quickly as possible to the issues and approaches that occupy the minds of ecologists actively advancing this field. Our intent is to be forward looking, to show what the field has accomplished and where it might go in the near future. Our strategy is to examine empirically derived generalizations about the nature and mechanisms of vegetation dynamics, the conceptual framework for organizing these generalizations, and the major approaches to modelling vegetation dynamics quantitatively.

The first half of the book examines the conceptual basis for understanding and predicting vegetation change. In Chapter 1, Glenn-Lewin and van der Maarel discuss the basic concepts and typologies of succession. They consider the variety of observed patterns of succession and provide an overview of the mechanisms postulated as proximate causes of successional change.

Van der Valk (Chapter 2) takes the perspective of the individual plant, or plant species, and considers the factors that influence establishment and persistence of plants. He examines the patterns of establishment, reproduction, persistence, and colonization for a wide variety of community types.

In Chapter 3, Peet examines the extent to which there are predictable trends in vegetation development that are independent of the particular species present. He considers successional changes in properties such as diversity and stability, biomass and production, and nutrient accumulation and cycling efficiency.

In the final chapter of this section, Veblen (Chapter 4) considers the internal dynamics that account for the maintenance of relatively mature and stable forest communities. He examines the importance of regeneration niches and fine-scale treefall gap dynamics in permitting the coexistence of species in forests that appear to be in a state of compositional equilibrium.

Predictions about the dynamics of complex systems typically require the construction of quantitative simulation models. In the second half of the book we present three chapters that illustrate quite different strategies for constructing such models. The first chapter, by van Hulst (Chapter 5) discusses a variety of simple, analytical models of successional processes, models which often are amenable to analytical solution. He discusses the inadequacies of models based solely on the dynamics of plant populations and suggests alternative approaches. This chapter stresses models that are derived from first principles and which have the advantage of potentially providing a theoretical underpinning for community dynamics, in contrast to the more phenomenological alternatives considered in Chapters 6 and 7.

In Chapter 6, Usher examines in detail the application of Markov models for both predicting short-term vegetation change, and also for statistically testing certain hypotheses about vegetation change. These are the best-known models that incorporate stochasticity in succession. They require only minimal knowledge of underlying ecological mechanisms and simply project forward, based on changes that have occurred in the recent past.

For examining change over time scales greater than the longevity of an individual adult plant, the most popular models are those based on projections of individual plant growth within small areas approximating the size of natural gaps. Urban and Shugart (Chapter 7) discuss this family of models and how they have been adapted to provide long-range projections of compositional change in ecologically and geographically diverse vegetation types. These models are based on computer simulation of multiple instances of simple and known processes for a particular stand of vegetation.

Most theory of vegetation change has been developed for spatial scales of $1-100\,000\,m^2$ and temporal scales of $1-1000$ years. Sometimes it is necessary to work at larger scales than these. In the final chapter, Prentice (Chapter 8) looks at the difficulties in discovering and predicting vegetation dynamics over time scales of $1000-100\,000$ years.

REFERENCES

Clements, F. E. (1904) The development and structure of vegetation. *Botanical Survey of Nebraska 7. Studies in the Vegetation of the State*, Lincoln, Nebraska.

Clements, F. E. (1916) *Plant Succession*, Carnegie Institute Washington Publication 242, Washington, D.C.

Connell, J. H. and Slatyer, R. O. (1977) Mechanisms of succession in natural communities and their role in community stability and organization. *Amer. Nat.*, **111**, 1119–44.

Cooper, W. S. (1926) The fundamentals of vegetational change. *Ecology*, **7**, 391–413.

Cowles, H. C. (1899) The ecological relations of vegetation on the sand dunes of Lake Michigan. *Bot. Gazette*, **27**, 95–117, 167–202, 281–308, 361–91.

Cowles, H. C. (1901) The physiographic ecology of Chicago and vicinity. *Bot. Gazette*, **31**, 73–108.

Cowles, H. C. (1911) The causes of vegetative cycles. *Bot. Gazette*, **51**, 161–83.

Davis, W. M. (1899) The geographical cycle. *Geogr. J.*, **14**, 481–504.

Drury, W. H. and Nisbet, I. C. T. (1971) Inter-relations between developmental models in geomorphology, plant ecology, and animal ecology. *Gen. Syst.*, **16**, 57–68.

Drury, W. H. and Nisbet, I. C. T. (1973) Succession. *J. Arnold Arboretum*, **54**, 331–68.

Egler, F. E. (1954) Vegetation science concepts. I. Initial floristic composition – a factor in old-field vegetation development. *Vegetatio*, **4**, 412–17.

Gleason, H. A. (1917) The structure and development of the plant association. *Bull. Torrey Bot. Club*, **44**, 463–81.

Gleason, H. A. (1926) The individualistic concept of the plant association. *Bull. Torrey Bot. Club*, **53**, 1–20.

Gleason, H. A. (1927) Further views on the succession concept. *Ecology*, **8**, 229–326.

Gleason, H. A. (1939) The individualistic concept of the plant association. *Amer. Mid. Nat.*, **21**, 92–110.

Horn, H. S. (1975) Markovian properties of forest succession, in *Ecology and Evolution of Communites* (eds M. L. Cody and J. Diamond), Belknap Press, Cambridge, Mass., pp. 196–211.

Horn, H. S. (1976) Succession, in *Theoretical Ecology* (ed. R. M. May), Blackwell, Oxford, pp. 187–204.

Margalef, R. (1958) Information theory in ecology. *Gen. Syst.*, **3**, 36–71.

Margalef, R. (1963) On certain unifying principles in ecology. *Amer. Nat.*, **97**, 357–74.

Margalef, R. (1968) *Perspectives in Ecological Theory*, Univ. of Chicago Press, Chicago.

Noble, I. R. and Slatyer, R. O. (1980) The use of vital attributes to predict successional changes in plant communities subject to recurrent disturbances. *Vegetatio*, **43**, 5–21.

Odum, E. P. (1969) The strategy of ecosystem development. *Science*, **164**, 262–70.

Peet, R. K. and Christensen, N. L. (1980) Succession: a population process. *Vegetatio*, **43**, 131–40.

Pickett, S. T. A. (1976) Succession: an evolutionary interpretation. *Amer. Nat.*, **110**, 107–19.

Pickett, S. T. A. and White, P. S. (1985) *The Ecology of Natural Disturbance and Patch Dynamics*, Academic Press, Orlando.

Pickett, S. T. A., Collins, S. L. and Armesto, J. J. (1987) Models, mechanisms and pathways of succession. *Bot. Rev.*, **53**, 335–71.

Pickett, S. T. A. and Kolasa, J. (1989) Structure of theory in vegetation science. *Vegetatio*, **83**, 7–15.

Raup, H. M. (1957) Vegetational adjustment to the instability of the site. *Proc. and Papers of the Tech. Meeting, 6th Internat. Union for the Conserv. of Natural Resources, Edinburgh*, pp. 36–48.

Rogers, G. and Robertson, J. (1986) Henry Chandler Cowles: 1869–1939. *Geographers: Biobiblio. Studies*, **10**, 29–33.

Runkle, J. R. (1985) Disturbance regimes in temperate forests, in *The Ecology of Natural Disturbance and Patch Dynamics* (eds S. T. A. Pickett and P. S. White), Academic Press, Orlando, pp. 17–34.

Stout, B. B. (1981) *Forests in the Here and Now*, Montana Forest and Conservation Experiment Station, Missoula, Montana.

Tansley, A. G. (1935) The use and abuse of vegetational concepts and terms. *Ecology*, **16**, 284–307.

Tilman, D. (1985) The resource ratio hypothesis of succession. *Amer. Nat.*, **125**, 827–52.

Tilman, D. (1988) *Plant Strategies and the Dynamics and Structure of Plant Communities*, Princeton Univ. Press, Princeton, New Jersey.

Watt, A. S. (1919) On the causes of failure of natural regeneration in British oakwoods. *J. Ecol.*, **7**, 173–203.

Watt, A. S. (1934) The vegetation of Chiltern with special reference to the beechwoods and their seral relationship. Part I: *J. Ecol.*, **22**, 230–70; Part II: *Ibid.*, **22**, 445–507.

Watt, A. S. (1947) Pattern and process in the plant community. *J. Ecol.*, **35**, 1–22.

Whittaker, R. H. (1953) A consideration of climax theory: the climax as a population and pattern. *Ecol. Monogr.*, **23**, 41–78.

1 Patterns and processes of vegetation dynamics

David C. Glenn-Lewin and
Eddy van der Maarel

1.1 INTRODUCTION

Since its initial development in the late nineteenth century, the study
of vegetation dynamics has cycled through periods of active study and
quiescence (the history of the succession concept is nicely detailed by
McIntosh (1981) and Miles (1987)). In the last two decades, advances in
our knowledge of the nature of communities, the dynamics of populations, and the structure and function of ecosystems, along with our
greatly increased capability for quantitative analysis and modelling
of vegetation phenomena, have led to another re-examination of the
conceptual basis of vegetation dynamics. The field now finds itself with a
large body of data and hypotheses, innumerable definitions and terms,
and several areas of unresolved debate. In recent years, the description
and analysis of vegetation dynamics has generated a huge literature, a
literature that 'reflects the confusion of the subject matter' (Miles, 1987).
Here, we undertake to sort out some of the major themes of vegetation
dynamics, in order to provide the conceptual basis for what follows in this
book.

We begin by reviewing some broad classes of vegetation dynamics. This
is followed by a brief review of the patterns of community change that
have been observed, and by a discussion that reflects our increasing
recognition of the importance of disturbance in ecological systems. We
then consider vegetation dynamics as population-based phenomena in
time and space, and summarize typology, pattern and disturbance in a
framework of vegetation dynamics considered over the temporal scale.
The mechanisms of vegetation dynamics are taken up, and the chapter
concludes with an overview of the data and methods used in studying vegetation dynamics. Our approach throughout is predominantly
population-based and mechanistic, multivariate and multifarous.

Our goal is theory, in both this chapter and this book. Nevertheless,
ecological management and management of natural areas is primarily
the manipulation of succession (Luken, 1990; Miles, 1987), and thus a

Plant Succession: Theory and prediction Edited by David C. Glenn-Lewin, Robert K. Peet and Thomas
T. Veblen © 1992 Chapman & Hall, London ISBN 0 412 26900 7

thorough understanding of vegetation dynamics is necessary for applications as well as for theoretical studies.

1.2   SUCCESSIONAL DICHOTOMIES AND POPULATION-BASED PROCESSES

Ecologists have formulated numerous typologies of vegetation dynamics, trying to increase understanding by classifying patterns, processes, forces or mechanisms. Three of the most enduring typologies are reviewed here because they illustrate the multifaceted nature of succession, and because they reveal the importance of underlying assumptions about mechanism and equilibrium. After considering the importance of disturbance in ecological systems, we shall recast these dichotomies in a framework based on the scale of time.

1.2.1   Progression and retrogression

Progressive succession is what is normally intended when one uses such terms as 'succession', and 'community development'. 'Progression' implies a direction, variously perceived as a series of stages (the sere) or a continuum from an initial or pioneer community to a well-developed, mature, perhaps stable community. It is succession as Clements (1916, 1928) and Cowles (1899) meant it. Odum (1969) and Whittaker (1975) codified many of the features of the classical progressive model, for instance, increasing species diversity, increasing complexity, greater biomass, and floristic stability.

Retrogression is a reversing of developmental trends, towards earlier, more simple stages, usually with fewer species, lower productivity and less biomass (Clements, 1916; Gleason, 1927; Tansley, 1935; Iversen, 1964; Woodwell, 1967, 1970; Woodwell and Whittaker, 1968; Whittaker, 1975). Iversen (1964) associated retrogression with long-term changes in the physical environment, such as soil leaching or water table elevation. Subsequently, the concept has been used in the context of severe stress. For instance, Woodwell (1967) and Woodwell and Whittaker (1968) described the simplification of an oak–pine forest under continual radiation from a gamma source as a retrogression from the forest towards a vegetation of fewer species and lower stature, the actual diversity and biomass depending upon the intensity of the radiation (as measured by distance from the gamma source).

Miles (1987) argued that vegetation dynamics (progressive or retrogressive) should be seen fundamentally as floristic change, but others (e.g. Odum, 1969; Whittaker, 1975; Austin, 1981) have allowed broader concepts, including community structure and ecosystem function. The virtue in Miles's limitation is that the more parameters allowed in the concept, the more confusing it will get, and the more exceptions will be noted. Bakker (1989) illustrated the problem: in the same vegetation, he showed progressive floristic change (*sensu* Braun-Blanquet, 1964), but retrogression in terms of species diversity and vegetational complexity.

The concepts of progressive and retrogressive succession imply attributes about beginnings and ends that may not be true (sections 1.3 and 1.5). Community dynamics are characterized by a complexity of processes, patterns and mechanisms, and repeated disturbance is important in many systems. The notions of progressive and retrogressive succession may have more heuristic usefulness than theoretical importance.

1.2.2   Primary and secondary succession

Primary succession, defined as vegetation development on newly formed or exposed substrate, proceeds on raw parent material rather than a developed or modified soil, and is usually characterized by low fertility, especially of nitrogen (Gorham *et al.*, 1979). A primary site contains no biological legacy, that is, no previous vegetation, no seed bank, and no organic matter derived from prior vegetation. Propagules of colonizing organisms must arrive by immigration. Dispersal distances will vary, but may be great. In Chapter 2, van der Valk details colonization of primary sites in terms of seed sources, dispersal, and the distribution of seeds on the landscape.

The literature contains numerous descriptions of primary succession on many different kinds of substrates. Examples include succession in river delta development (Gill, 1973; Waldemarson Jensén, 1979; Rejmánek *et al.*, 1987), elevating seashores (Cramer and Hytteborn, 1987), salt marsh deposits (Roozen and Westhoff, 1985; Clark and Patterson, 1985), volcanic deposits (Paijmans, 1973; Beard, 1976; Tagawa *et al.*, 1985; Halpern *et al.*, 1990; Rejmánek *et al.*, 1982; Whittaker *et al.*, 1989), dunes (Cowles, 1899; Pidgeon, 1940; Olson, 1958; van der Maarel *et al.*, 1985; van Dorp *et al.*, 1985), glacial forefields (Cooper, 1923a,b,c, 1931, 1939; Matthews, 1979), land slips (Wood and del Moral, 1987; Guariguata, 1990; Veblen and Ashton, 1978), granite outcrops (Rundel, 1975; Shure and Ragsdale, 1977), coal and other waste tips (Glenn-Lewin, 1979; Wagner *et al.*, 1978; Thatcher and Westman, 1975; Titlyanova and Mironycheva-Tokareva, 1990), abandoned quarries (Bradshaw *et al.*, 1982; Borgegård, 1990), peatlands (Sjörs, 1980; Drury, 1956; Casparie, 1972), and reefs (Harris *et al.*, 1984).

The concepts of xerarch and hydrarch succession are subsets of primary succession. These two terms refer to primary succession beginning on dry land, or in aquatic environments or wetlands, respectively (Cooper, 1913). In most usages, they imply a directional development toward a mature and stable vegetation, in the one case from bare, dry land to (usually) a forest, in the other from lake, pond or marsh to a wetland filled with sediment, to wet forest, to (finally) a mesic forest.

Secondary succession is the replacement of pre-existing vegetation following a disturbance that disrupts the vegetation. There is a developed soil present, and a biological legacy from the previous vegetation exists, often as a seed or seedling bank. As for primary succession, the literature contains many examples, and only a few will be used here for illustration. Much secondary succession research has been done on old fields, that is, agricultural land that has been abandoned and allowed to revegetate naturally. The work of Billings (1938), Oosting (1942) and Bard (1952) can serve as examples of the hundreds of studies of old-field succession. A great deal has been learned from old-field studies, but conversely, perhaps we should be concerned that much of our understanding of vegetation dynamics depends upon old-field succession models, which may be actually rather special. Most old-field sites, being former agricultural fields, are on relatively fertile soils, and, of course, have been subjected to human management, often for long times.

Another genre of secondary succession studies includes the thousands of reports of vegetation development following fire. This is such an important topic that a whole subdiscipline of fire ecology has arisen, ranging from physiological and population studies to ecosystem analysis and landscape ecology. Some important models of succession, for example that of Noble and Slatyer (1980), have been developed specifically to deal with secondary succession following fire.

Primary succession seems easy to recognize, leaving all else to secondary succession, which makes the concept of secondary succession very broad and consequently murky. Primary and secondary succession probably form a continuum, depending upon the characteristics and duration of the disturbance that initiates succession, and also upon the position of the site in the landscape (Walker and Chapin, 1987; see also Chapter 3). For example, river delta development and salt marsh deposits were cited above as examples of primary succession. However, such sites may be the recipients of biological information from many upstream sites before succession starts or during its course, and their substrates may incorporate organic material from upstream. In one sense, this downstream dispersal appears to be just another form of dispersal into primary sites. On the other hand, since propagules and organic matter would always be moved downstream, delta and salt marsh deposits form in the continual presence or availability of these, and the resulting process of development may therefore look more like secondary succession. Like many other dichotomies in ecology, the concepts of primary and secondary are helpful ways of organizing our observations of nature, but not every observation will fit neatly into one or the other category; intermediate situations exist.

Vegetation change can be generated by the removal of an environmental disturbance or stress, as well as the imposition of one: fire suppression may cause a change from a pyrophytic vegetation to another assemblage of plants that are fire sensitive. The historical change from oak (*Quercus*) forests to those dominated by more shade-tolerant species, especially species of *Acer*, in the north-central USA can be ascribed to the cessation of fire (Anderson and Adams, 1978). Control of river flooding by dams, which alters another kind of disturbance regime, will lead to a different vegetation. For example, alterations of the flood patterns along the Platte River in the central USA have led to dramatic increases of riparian forest such that habitat for important populations of two species of crane has been much changed (Anonymous, 1981). Cessation of fertilization, and a subsequent change to hay making along low brooksides in The Netherlands, have produced significant floristic changes along independent, and in some cases divergent, paths (van Duuren *et al.*, 1981).

1.2.3   Autogenic and allogenic succession

Autogenic succession is vegetation change due to forces of biotic inter-actions and biotic modification of the environment (Tansley, 1935). It implies 'internal' forces and mechanisms, for example competition, shade generation, and soil modification by plants. Allogenic succession is vegetation change due to environmental conditions and environmental change, or 'external' forces. Long-term vegetation response to climate change is an example. Other examples include the vegetational dynamics observed on river deltas that continuously receive sediment from up-stream, vegetation dynamics largely influenced by disturbances and those that are dependent upon immigration of plant propagules. The distinction is 'artificial', to use the characterization of Sharik *et al.* (1989; following Miles, 1987), in that both autogenic and allogenic forces likely play a role in most successions, and the relative importance of each changes with the changes in vegetation and depends upon the kind of vegetation. As examples of this last point, salt marsh development is largely allogenic (Clark and Patterson, 1985), but forests show characteristics of autogenic processes (Finegan, 1984). It might be possible, although not especially useful, to label individual processes as auto- or allogenic, but it seems inappropriate and even misleading to label a whole succession as either.

1.2.4   Pathways and directions

In his re-examination of the Lake Michigan dune succession described by Cowles (1899), Olson (1958) produced a diagram of multiple successional pathways which perfectly captured the simile of Cooper (1926) that succession is like a braided stream. In Olson's case, the particular path that succession followed depended upon the environment and the nature of the initiating force. Despite Cooper's analogy, Olson's confirmation, and other hints, it is only in the last two decades or so that the idea of multiple pathways has become incorporated into treatments of vegetation dynamics theory (Drury and Nisbet, 1973; Connell and Slatyer, 1977; Cattelino *et al.*, 1979; Miles, 1987). For example, multiple pathways in primary succession have been shown for dunes by Londo (1974, whose diagram has about the same degree of complexity as that of Olson (1958)), van der Maarel and Westhoff (1964), van der Maarel *et al.* (1985) and van Dorp *et al.* (1985, who also reported some retrogression pathways), and for glacier forefields (Matthews, 1979), salt marshes (van Noordwijk-Puyk *et al.*, 1979) and hydrarch succession (Walker, 1970). Clark and Patterson (1985) wrote that salt marsh development was not 'orderly', but rather a 'mosaic of vegetation type shifts in a non-predictable way'. Bradshaw *et al.* (1982) argued that in primary succession in quarries, nutrient differences explained the multiple pathways that they observed, whereas Borgegård (1990) concluded that, in abandoned gravel pits, soil development was a significant process, but the multiple pathways were largely due to variable colonization that reflected the differences in the surrounding vegetation.

Miles (1987) emphasized that secondary succession also shows multiple pathways, and gave a convincing illustration of the dynamics of British uplands in response to fire and grazing. There are other examples. Repeated colonizations of white pine (*Pinus strobus*) in Michigan, USA, behaved quite differently in their subsequent population dynamics (Sharik *et al.*, 1989). Disturbed stands of *Pinus banksiana* in Michigan, USA, revealed different developmental pathways (Abrams *et al.*, 1985). Australian rain forest succession is 'not unidirectional' (Webb *et al.*, 1972). Cattelino *et al.* (1979) modelled different possible pathways of forest succession as functions of species life histories and fire frequencies. At the smaller scale of patches, Herben *et al.* (1990) documented multiple pathways, including reciprocal replacements of patches in a montane grassland. Botkin (1979) argued that succession is fundamentally a stochastic process; indeed, the current version of the forest model JABOWA (Botkin, 1992) can produce quite different results from individual runs, even though the initial parameters are fixed from one run of the model to the next. This variation and stochasticity, and the dependence of succession upon local conditions and chance, led Glenn-Lewin (1980), Myster and Pickett (1988) and Pickett (1983) to characterize the 'individualistic' nature of vegetation dynamics, just as Gleason (1927) had more than a half century earlier (see Chapter 3).

In any multiple pathway scheme both convergence and divergence can be observed. With respect to a particular community property, such as composition, diversity, physiognomy, or structure, convergence means that dissimilar sites become more similar with time; divergence means that for similar sites these properties become more dissimilar. Physiognomic convergence is probably the rule, but convergence of composition, diversity or other community properties is more debatable. An appearance of convergence or divergence may depend upon the spatial scales of observations. For instance, the absence of convergence for landscape elements mentioned by Glenn-Lewin (1988), and for large, early-successional fields reported by Facelli and D'Angela (1990), contrast with the results of Inouye and Tilman (1988), who reported convergence at the scale of individual experimental plots. The appearance of convergence may also depend upon the different histories of sites (Myster and Pickett, 1990).

Conditions likely to lead to convergence include low diversity, relatively few species available from the immigration pool, or strong dominance by one or a few species. Johnson *et al.* (1981) point out that if there is an important spatial element in succession that affects propagule distribution, for instance, in forest fragments following landscape modification, sites are likely to show divergent successional patterns. (Direction and convergence are explored in more depth in Chapter 3.)

Cycles repeat not only some previous stage or stages of vegetation development, but also the sequence; that is, they are cycles of species replacements. Cycles have been described or postulated across a range of spatial scales. Probably the best studied small or patch-scale cycles are those in dwarf heathlands (Watt, 1947, 1955; Gimingham, 1988). The evidence does not generalize, however, since other heathlands do not seem to show the cycles (Prentice and Prentice, 1983; Prentice *et al.*, 1987; de Smidt, 1977; Gimingham, 1988). Other examples of cycles include *Scalesia* communities on the Galapagos (Lawesson, 1988) and cycles between *Larrea tridentata* and *Opuntia leptocaulis* in the Chihuahuan Desert (Yeaton, 1978). Regeneration 'waves', such as described for *Abies* forests by Sprugel (1976), produce cycles at a particular fixed point. Mueller-Dombois (1986, 1987) explained cycles of tree species as repeated age cohorts that become established within a limited time, grow and undergo a common senescence (Mueller-Dombois calls this 'cohort senescence'). When such a cohort occurs only once, it appears as a 'stage' in succession, for example, the *Pinus taeda* stage of forest succession in the Piedmont, southeastern USA (Peet and Christensen, 1980).

Cycles on the community scale, that is, cycles of community replacement rather than species replacement, have been described for bogs (Drury, 1956), and for some tropical forests (Webb *et al.*, 1972). In northwest Africa, *Pinus halepensis* forests after fire (Moravec, 1990) have a cyclical appearance because many of the angiosperm species in this vegetation resprout from roots. The pines, however, regenerate only from seeds, and therefore regenerate more slowly than the sprouting species. There is thus an alternation between resprouted vegetation soon after a fire, and the more slowly developing pines. Indeed, differential regeneration seems to underlie many vegetation cycles (Miles, 1985).

1.3 DISTURBANCE

We have to distinguish among (and live with) a number of different ways in which 'disturbance' has been defined. Grime (1979) defines it as 'the mechanisms which limit the plant biomass by causing its partial or total destruction'. Forman and Godron (1986) identify disturbance as 'an event that causes a significant change from the normal pattern in an ecological system'. This is not too different from the 'consensus' definition offered by van Andel and van den Bergh (1987): ' . . . a change in conditions which interferes with the normal functioning of a given biological system . . . disturbance is a cause (a change in condition) measured by its effects (change in normal functioning of a system)' (see also Rykiel, 1985). We are left, though, with the question of: What constitutes a 'normal' pattern or functioning in an ecological system?

A third approach to disturbance defines it as processes which lead to increased availability of resources to which either the survivors of disturbance or new colonists respond (Marks, 1974; Mooney and Godron, 1983; White and Pickett, 1985; Sousa, 1984; Tilman, 1985; Denslow, 1987; Runkle, 1989). Thus, accelerated growth of small individuals that are released from competition is one pattern of response to disturbance. Another is the creation of bare ground, loose soil, light gaps, or other situations which form microsites in which recruitment can occur. The choice of definition depends upon one's emphasis, whether on causes (changes in the environment), effects (responses of organisms, populations or communities), or mechanisms. In any case, disturbance is multifaceted, its role in vegetation dynamics depending upon a number of interacting characteristics. It is the subject of a large recent literature in ecology, and an understanding of it is assumed in much of what follows in this book. Disturbance and succession are inextricably linked, but we should be careful not to overextend the case by making disturbance the only foundation of vegetation science. Grubb (1988), for instance, illustrates the process gradient from recruitment coupled to disturbance, to recruitment that is independent of disturbance. He goes on to show how, although many modes of both regional and local persistence are disturbance related, others are not, and he illustrates with examples of what he calls 'shifting clouds of abundance' of plant species that are not coupled to disturbance.

Three not entirely independent dimensions of disturbance are space, time and magnitude. The spatial dimension is the extent of the disturbance, the physical dimensions of area and volume, and location, in particular in relation to environmental gradients. The temporal dimension includes frequency and predictability. These temporal factors combine in the concept of turnover and return time (especially if we can calculate the variance of return time: White and Pickett, 1985). In the case of fire, and probably other kinds of disturbance as well, the season of the disturbance is another important aspect of the temporal dimension of disturbance. The third axis of disturbance is magnitude, which is described either as the force of the event, or more commonly as the severity as reflected by the effects on the vegetation (White and Pickett, 1985). The severity of disturbance will depend upon the character of the disturbance force (e.g. major storms or individual treefalls), and the nature of the existing vegetation, especially its sensitivity to disturbance. Many different kinds of disturbance have been studied (White, 1979; Pickett and White, 1985); examples and illustrations for what follows come primarily from the literature on fire and grazing disturbances.

(a)   Extent

The dynamics of a disturbance patch are, in part, determined by its size (Oliver and Stephens, 1977; Miller, 1982; Spies and Franklin, 1989); different sized patches allow for different life histories. Furthermore, patch size will influence the interactions between a patch and its edge. For treefalls in tropical forests, Brokaw (1982) reported average gap sizes of 85 to 110 m^2, depending upon whether the minimum was chosen to be 20 or 40 m^2. Liu and Hytteborn (1991) measured gap sizes from 9 to 360 m^2 in a primeval spruce (*Picea abies*) forest, and found an almost normal distribution of relative gap sizes calculated as the ratio between gap diameter and mean height of gap-bordering trees. In sand grasslands, the spatial extent of disturbance will vary from the dimensions of blowouts and dunes, that is, tens of metres, to the few centimetres of insect burrows and the like (Loucks *et al.*, 1985).

Fires that cover a large area of vegetation in which plants are killed will initiate a succession, in which the internal composition and function will vary according to local environment, distance to seed sources, character of the seed bank, number and position of surviving plants, the inherent patchiness of fire itself (Christensen, 1987), and many of the other properties discussed in this chapter and book. Extensive fire in vegetation composed of plants that survive fire, such as perennial grasslands, is less an initiator of succession than an invigoration of plant growth and reproduction, and ecosystem function. Indeed, fire in perennial grasslands probably serves to retard succession by repeatedly eliminating invading woody plants (see, for one example among many, the forest invasion of the Patagonian steppe over the past century: Veblen and Markgraf, 1988; Veblen and Lorenz, 1988). However, in very dry, unproductive grasslands, Loucks *et al.* (1985) considered fire to be a disturbance 'if it fosters a community the following year with a higher number of disturbance-adapted species'. This is a means of defining disturbance by its biological effect, here somewhat circularly, rather than by the disturbance process itself. Some systems develop a propensity for disturbance. For example, systems such as Mediterranean shrublands and certain *Pinus* forests accumulate fuel which leads to an increased probability of fire (see also Macphail, 1980). Fires that produce a mosaic of fire patches may lead to, or maintain, higher community diversity (Davidson and Bratton, 1988).

The extent of disturbance by grazing or browsing will reflect whether herbivores are sedentary, migratory or mobile, as well as the number of herbivores. For example, Beavers (*Castor canadensis*) generally have local direct effects as herbivores, although their ability to alter water levels often has profound effects (Barnes and Dibble, 1988). Muskrats (*Ondatra zibethicus*), on the other hand, may clear the emergent and shore vegetation of entire marshes (Fritzell, 1989). Prairie dogs (*Cynomys ludoviciana*) affect the areas covered by their colony, up to hundreds of hectares (Whicker and Detling, 1988), while snow geese (*Chen caerulescens*) may graze up to several square kilometres (Smith and Odum, 1981). Large herbivores, especially migratory species, will affect very large areas. It should be pointed out that, to the degree that herbivores and plants have evolved together, it may be inappropriate to consider grazing as a disturbance.

(b) Time

Canham and Loucks (1984) estimated that the return time of catastrophic windthrow in northern Wisconsin was 1210 years. In the case of disturbance by fire in boreal forests, Johnson (1979) modelled recurrence in terms of the lag before a fire could recur, the mean recurrence time, and the variation in recurrence times. He estimated return times from about 37 years to about 102 years in subarctic forests. Return times of about 100 years were reported for coniferous forests in northern Minnesota, USA (Heinselman, 1973). Stein (1988) reviewed the temporal patterns of fires in the *Pinus ponderosa* communities of western North America, reporting fire intervals of from 4 to 47 years, depending upon the particular community. In the Cascade Range of Oregon, USA, steep low-elevation sites burned more often but with less severity than cooler and more moist high-elevation sites that had gentle slopes (Morrison and Swanson, 1990). It is also possible to estimate tropical forest turnover rates from the rates of treefall gap formation. For lowland tropical forests in Panama, Brokaw (1982) estimated a turnover time of from 113 years to 126 years, depending upon the minimum gap size that one assumed. Fire intervals in perennial grasslands are often very short, as little as 1–3 years (Loucks *et al.*, 1985). It is important, when calculating fire return times or intervals, to know whether the value is calculated on a per area basis, or refers to individual points. A fire interval quoted per area will be substantially shorter than that for a particular point on the ground.

Vegetation change associated with changes in disturbance frequency provides a strong argument for the importance of the temporal axis of disturbance. Two instances may serve to illustrate this: due probably to human influences, the regional fire interval in the Pine Barrens of New Jersey, USA, has changed from 20 to 65 years since the beginning of the twentieth century (Forman and Boerner, 1981). During this same period, tree dominance in the swamps shifted from eastern white cedar (*Chamaecyparis thyoides*) to hardwoods. In western Minnesota, USA, fire frequency has varied between 8 and 45 years over the last 750 years, apparently in response to climatic variation (Clark, 1990).

(c)   Magnitude

Disturbance magnitude can be measured as the force of the disturbance (intensity), or as the degree of effect on the biological elements under investigation (severity: Sousa, 1984; White and Pickett, 1985; Malanson, 1984). In some instances it is possible to quantify disturbance intensity, for instance by measuring fire temperatures, grazing pressure, or shore wave intensity. Often, however, this cannot, or has not, been done; especially when, as has been the case for many studies, the disturbance(s) has occurred at some time, known or unknown, in the past.

Fire intensity, and therefore the degree of destruction at a point, depends upon such factors as wind, temperature, atmospheric and fuel moisture, and fuel loads. Spurr and Barnes (1980) order forest fire intensity as: surface fires, which burn the herbaceous and shrubby vegetation but do not kill trees; crown fires, which kill trees; and ground fires, which burn down into the soil organic layers, destroying also roots, rhizomes, etc. In terms of its biological effect, fire intensity is related to the susceptibility or resistance of plants to fire (reviewed by Trabaud, 1987), indicated by, as one example among many, bark thickness (Fox and Fox, 1987).

Grazing intensity will depend upon the quality of the forage (Smith and Kadlec, 1985), the density of the grazers, and the behaviour of the grazers, especially whether they are selective feeders or not. Harper (1977) and Bobbink and Willems (1988) argue, for instance, that the high species richness of northern European chalk grasslands is due to 'selective grazing of the potentially dominant grasses'.

The effects of herbivory in perennial grasslands differ, depending upon herbivore behaviour. Those herbivores that produce well-defined patches with the greatest intensity of disturbance at the centre lead to increased community diversity. In other cases, less distinct patches will be formed, and the effect of the grazers on community structure and biogeochemical cycles will be more through feeding strategies and physical disturbance (Naiman, 1988).

Grazing increases tillering of grasses (McNaughton, 1976, 1979), while fire may increase or decrease tillering, or have no effect, depending upon the time of year in which the fire occurs (Glenn-Lewin *et al.*, 1990). Grazers also leave patches of unpalatable species (Thalen *et al.*, 1987), which increases the complexity of the community patch structure.

In North American tallgrass prairie, the frequency (Kucera and Koelling, 1964) and the timing (i.e. season) of fire are important (Glenn-Lewin *et al.*, 1990). Spring burning increases the vigour of prairie species, but tends also to favour biennial weeds. Woody species are damaged by fire; some are killed, but others resprout and maintain or increase their abundance, unless fires are repeated with a high frequency. Frequent burning favours grasses over forbs, while less frequent burning favours the forbs (Kucera and Koelling, 1964). After fire, prairie grass seeds germinate at a higher rate, while forbs germinate at a decreased rate (Rohn and Bragg, 1989). In any case, however, few seedlings survive to adulthood (Glenn-Lewin *et al.*, 1990).

Grazing and fire combined serve to illustrate the multidimensionality of disturbance in ecological systems (e.g. Fox and Fox, 1987.) Both vary greatly in extent and intensity. Disturbances which happen rarely or which influence a wide area (or both) act as a matrix for more frequent disturbances or disturbances of smaller size (Allen, 1987). For instance, large herbivores are credited, along with fire, with helping to create and maintain the savannah ecosystems of Africa (McNaughton *et al.*, 1988). In the intermountain west of the USA, overgrazing led to a replacement of native grasses by the exotic *Bromus tectorum*. The continuing dominance of *B. tectorum* is now in part maintained by frequent fires which the grass encourages (Mack, 1981). In other instances fire provides the matrix for grazing. For examples, fossorial rodents such as pocket gophers (family Geomyidae) and larger predators such as badgers (*Taxidea taxus*) cause mosaics of change over the microtopography of vegetation that is burned over larger areas (Anderson and MacMahon, 1985; Platt, 1975; Huntly and Inouye, 1988).

Grazing and fire are interacting forces in grasslands. When choice is available, ungulates tend to spend more time grazing upon burned areas than upon unburned areas (Larson and Murdock, 1989), an interaction that in many human cultures was important for improving forage or attracting game. Also, grazing may affect fire intensity, extent, and frequency by reducing fuel loads (e.g. Savage and Swetnam, 1990). Since grazing intensity may be patchy, fire intensity and extent may also become patchy (Turner and Bratton, 1987).

Whether a particular environmental force should be considered to be a disturbance or not depends upon the measure of ecological effect. If fire disrupts the biotic organization (e.g. burns off and kills trees in a woodland), then it would appear to be a disturbance. However, if fire maintains biotic organization (i.e. prevents changes that would be detrimental to present species or favourable to future species, as in mulch removal in productive grasslands), then according to some views it is not a disturbance (Forman and Godron, 1986; Lawton, 1987; van Andel and van den Bergh, 1987; Evans *et al.*, 1989).

1.4   VEGETATION DYNAMICS IN TIME AND SPACE

Vegetation dynamics have temporal and spatial aspects. Indeed, time and space are related, in that forcing functions for vegetation change over large areas tend to be the same as those causing change over long time periods, and likewise for small areas and short time spans the causes of change are similar (Chapter 8; Austin, 1981; Delcourt *et al.*, 1983; Falinski, 1988).

1.4.1   Temporal aspects of dynamics

Since the temporal aspects of vegetation dynamics are of prime interest, we shall begin with an elaboration on the time scale. This can be roughly (and arbitrarily) divided into three segments: (1) short term, or the domain of fluctuations; (2) long term, the domain of vegetation history on time scales of millennia; and (3) the middle range, or what is usually the domain of succession.

Fluctuations are impermanent vegetation changes that comprise the natural temporal variation in plant abundance or aspect in a community over short time periods (Miles, 1979; Rabotnov, 1974; van der Valk, 1981, 1985; Vale, 1982). Some good data-based examples come from grasslands (e.g. Watt, 1971; van der Bergh, 1979; Williams, 1970). Variation in an equilibrium state would be around a mean composition, but it is perhaps more likely that fluctuations will be around some kind of temporal trend in the vegetation (i.e. fluctuations superimposed on a succession). Many fluctuations in vegetation occur because of environmental changes on the same time scales, such as fluctuations in precipitation (e.g. van der Maarel, 1978, 1981; Albertson and Tomanek, 1965) or growing season temperature. Mild disturbances on a short time scale might also produce what appear to be fluctuations, as in the response of heathlands in Brittany to mild fires (Gloaguen, 1990).

In a shifting mosaic, species increase or decrease at any particular position, but the original average remains more or less constant. What appears to be a fluctuation on a small or meso-scale, might at the same time be part of a shifting mosaic such that over a large scale, the average composition changes little. This is true at the scale of individual plants (e.g. Watt, 1947) and at the scale of landscapes (Bormann and Likens, 1979; Romme, 1982; Heinselman, 1973).

Succession is compositional change, usually evident over a few decades to a few centuries. The change is in some way directional (Miles, 1979, 1987), uni- or multidirectional, although the direction(s) followed may be determined only after the fact. Long-term change occurs over many centuries or millennia, and results from equally long-term environmental changes such as climatic change or soil development, from species' migrations, or from other long-term forces (see Chapter 8).

The boundaries between these temporal patterns (and associated processes) are indefinite (Miles, 1987). The distinction between fluctuation and succession is arbitrary, in that a large enough fluctuation over a long enough period of time becomes succession (Austin, 1981; Bornkamm, 1988). Vegetation cycles over a long enough period can, nevertheless, be thought of as fluctuations. Distinguishing succession from long-term changes requires a similar arbitrariness. Species range changes are individualistic and probabilistic; vegetational composition and structure are not stable over periods of centuries to millennia, nor are community boundaries stable (West, 1964; Davis, 1976, 1981, 1986, 1987; Davis et al., 1986; Webb, 1981, 1987; Jacobson, 1979; Macphail, 1980; Delcourt et al., 1983; Huntley, 1990; see also Chapter 8). The underlying processes of plant population dynamics are the same for fluctuation, succession and long-term change.

Thus, we can ask whether fluctuations are really different from succession and, if so, how they can be recognized. The answer seems to be that there is no fundamental difference. Both are functions of temporal and spatial scales. Directionality has been cited as a characteristic of succession that distinguishes it from fluctuation, but the distinction is arbitrary. Succession involves changes in dominant or diagnostic species, and new species enter the community and become diagnostic species. In contrast, during fluctuations the dominant or diagnostic species do not change, and new species do not enter the community, or if they do they do not become diagnostic species (Miles, 1979). However, intermediate situations between these extremes cannot logically be included with only one or the other. If only some important species change, or a few new species enter, is it succession or a dramatic fluctuation?

The same problem of arbitrary distinction applies to defining succession as qualitative change (change in species composition) and fluctuations as quantitative change (change in relative abundances), unless qualitative change includes all species migration into, or extirpation from, the community, regardless of its importance or the length of time of its residence. Vegetation cycles, to the extent that they occur, resemble fluctuations in the sense that vegetation returns to something like its state at a previous time, but there is no logical 'mean' in a vegetation cycle, since there are qualitative changes.

Further, we cannot distinguish between succession and fluctuation until after the fact. That is, projecting past and present trends leads to predictions that might seem directional, but could be just part of fluctuations. If we can ever tell the difference, it is only after it has happened. Complicating this are the problems that arise from the frequency of observations, which, if widely spaced in time, may reveal what appears to be a direction, but in fact mask fluctuations (e.g. Austin, 1981).

A related question is whether communities return to some state that existed before the succession began. In other words, is succession a repeatable process of development of identical communities? Indeed, this was the thrust of Clements' (1916, 1928) proposition. The response of the central North American grassland vegetation to the Great Drought of the 1930s indicates the complexity of the answer. After substantial drought-induced changes, some communities apparently returned, or are returning, to something resembling their original composition, but after a relatively long time (Albertson and Tomanek, 1965; Weaver, 1954). Is this succession, or a 30–40 year fluctuation (cf. Austin, 1981)? After all, the returning composition is not identical with that prior to the drought, but it has certain similarities. In addition, it is clear that some sites have not yet returned to their pre-drought condition; they may require more time to do so, or they may never do so. Although allogenic forces appear to be similar, differences in autogenic forces, due to different dynamic pathways, could be an explanation as to why some have returned and some have not. Similar vegetation will not necessarily develop on similar sites, because of anomalous distribution patterns of plants, broad overlap of plant abundances in environmental space, and chance (Glenn-Lewin, 1980; McCune and Allen, 1985). In short, with a non-equilibrium theoretical framework, there probably is no need for a climax, or stable endpoint of succession (Raup, 1957; Drury and Nisbet, 1971; Connell, 1978; Pickett, 1980; Wiens, 1984; Chesson and Case, 1986; Davis, 1986; DeAngelis and Waterhouse, 1987).

1.4.2 Spatial aspects of dynamics

One can conceive of vegetation dynamics as a regional process, as development and change of vegetation in a landscape. Or, if we wish to consider only an area large enough for compositional change actually to occur, succession can occur in very small areas. Thus, again vegetation dynamics exhibits a continuum, this one of horizontal scale: fine-scale gap phase processes, regeneration, community-wide dynamics, landscape dynamics. A community can be viewed as a changing mosaic of patches of different sizes, ages, structures, and composition (Watt, 1947; Webb *et al.*, 1972; Sousa, 1984; Pickett and White, 1985; Hubbell and Foster, 1986; Martinez-Romos *et al.*, 1989). This means that spatial patterns are important for an understanding of community change and cannot be ignored as random noise (Austin, 1981; Austin and Belbin, 1981).

The significance of spatial processes in vegetation dynamics can be illustrated in several ways. In the nucleation concept of Yarranton and Morrison (1974), succession proceeds by the expansion and coalescence of initally small nuclei of a vegetation element, in this case, nuclei of

Juniperus on sand dunes. Archer *et al.* (1988) describe a similar process in a subtropical *Prosopis glandulosa* woodland. Other illustrations include: (1) differential colonization resulting from both environmental heterogeneity (e.g. Andersen and MacMahon, 1985; Halpern and Harmon, 1983; McDonnell, 1988) and the locations of seed sources (e.g. Wood and del Moral, 1987); (2) neighbourhood effects on establishment at a particular point (e.g. Lippe *et al.*, 1985; Ryser, 1990); and (3) the influence of spatial heterogeneity on disturbances such as fire and grazing (e.g. Turner and Bratton, 1987). Some of the most successful models of vegetation dynamics function by summing over small-scale or gap processes (Botkin, 1992; see also Chapter 7) to produce a profile of community succession. Szwagrzyk (1990) argued that explicit spatial prediction from gap dynamics theory is elusive. However, Busing (1991) has developed a spatial model of forest dynamics that is based on gap theory, and the heathland dynamics model of van Tongeren and Prentice (1986) and Prentice *et al.* (1987) is explicitly spatial in that points given by spatial coordinates are covered by individual species.

Spatial patterns themselves change with succession, for instance in early successional Spanish grasslands from fine-grained to coarse-grained patterns, both for floristics (de Pablo *et al.*, 1982; Sterling *et al.*, 1984; Pineda *et al.*, 1981) as microtopographical influences on colonization processes give way to the influence of geomorphology, and for biomass (Casado *et al.*, 1985). In primary succession on glacial moraines, both single species and multispecies patterns changed (Dale and Blundon, 1990, 1991), and the scales of pattern changed, but the number of scales at which patterns occurred remained the same.

1.4.3 Vegetation dynamics as a gradient in time

One of the difficulties in dealing with vegetation dynamics theory is that more than one thing may be changing. Miles (1979, 1987), for instance, considers floristic change to be the essence of succession. Contrast this with, for instance, Odum (1969) or Austin (1981), who treat succession more broadly as change in any measure of ecosystem character.

If a community is recognized primarily on the basis of a single attribute, for example, species composition, a change in this attribute is succession. While using just one criterion simplifies the problem, it also ignores the dynamics possible in other community aspects.

Grime (1979) considers succession as change in the predominant kinds of life histories of the plant species, succession being a shift in the relative importance of ruderal, competitive and stress-tolerant species. Grime illustrated his ideas with data from grasslands, and Clément and Touffet (1990) reported a heathland illustration. Here, again, the scope seems rather limited, and rests on the validity and universality of Grime's tripartite model, which has been criticized for being simplistic and too local in its application.

Most communities exhibit multiple vertical strata. Some layers may undergo significant change while others show little change. For instance, a ground fire may initiate or modify change in the herbaceous layer, while having little direct effect on the canopy layer. (Of course, a ground fire may affect the young plants of potential canopy species, and thereby have a delayed or indirect effect.) If the description of succession is restricted to the canopy (i.e. canopy change is 'required' for 'succession'), then how should ecologists deal with change that is occurring in the lower layers? Once again we see that communities show individualistic behaviour. If we accept that vegetation dynamics are population-based processes, we conclude that there is no fundamental ecological reason to exclude any change in a community, regardless of kind or vertical distribution, from our concepts and models of vegetation dynamics.

Succession can be thought of as a compositional gradient in time, either of species or of other community characters (Drury and Nisbet, 1973; Whittaker, 1975; Pickett, 1976; Peet and Christensen, 1980; Walker and Chapin, 1987). If the spatial axis of traditional gradient analysis is replaced by a temporal one, then the rise and fall of the species' abundance curves represent the dynamics of plant populations (or of community characteristics), and the shape of the curves will reveal the rates and degrees of change. With the concept of a temporal gradient, fluctuations and cycles will show up as repeating patterns: fluctuations will appear as periods of relative abundance change, while cycles will appear as periodic compositional change. Species replacement during succession will appear as aperiodic compositional change.

Changing communities exhibit individualistic behaviour, and vegetation dynamics are population-based processes (Drury and Nisbet, 1973; Glenn-Lewin, 1980; Peet and Christensen, 1980; Christensen and Peet, 1984; Miles, 1987; Liljelund et al., 1988; Myster and Pickett, 1988; see also Chapter 3). Consequently, community dynamics can be made more sensible if we directly address population-based change in the context of what is changing (e.g. structure, composition, proportion of plant 'strategies'), the character of the environment and the scales, both physical and temporal, at which changes occur. For instance, treating changes in relative abundances (due to differential plant growth) separately from species replacement processes has the advantage of separating growth-based, quantitative changes from those requiring immigration and establishment in an existing vegetation, or in gaps in an existing vegetation (van der Valk, 1981, 1985; Glenn-Lewin, 1988). Such immigration and establishment reflect the invasive ability of the plants and the resistance to invasion of the existing vegetation.

1.4.4 A framework for vegetation dynamics

The notion of a gradient in time provides us with a basis for a comprehensive framework for vegetation dynamics; Major (1974) and Miles

(1979, 1987) took just this approach in their treatments. Building on this idea, and the concept of a 'hierarchy of successional causes' (Pickett *et al.*, 1987b), we summarize the different approaches to vegetation dynamics in a temporal scheme with seven classes (adapted from van der Maarel, 1988): fluctuation, gap dynamics, patch dynamics, cyclic succession, secondary succession, primary succession and secular or long-term succession. Along this sequence the time scale increases, although we recognize that each class may overlap substantially those adjacent to it on the time scale. The spatial scale of observation increases as well, again with considerable overlap.

Fluctuation is the shortest division of the vegetation dynamics time scale, being non-directional, quantitative change resulting from population responses to short-term changes in environmental conditions, or what Pickett *et al.* (1987a) term stochastic environmental stress. As we emphasized above, fluctuations may have a cycle long enough to make it difficult to distinguish them from longer-term trends. Another complication is that drastic environmental stress may easily change the course of longer-term successional changes (e.g. Austin and Williams, 1988).

Fine-scale gap dynamics result from ontogenetically determined or externally imposed death of individual plants or local populations, and implies qualitative changes in vegetation rather than merely the quantitative changes of fluctuation. As a consequence, species have to regenerate in order to maintain themselves in the community, while at the same time new possibilities for such re-establishment are provided. The importance of this aspect of vegetation dynamics has been increasingly recognized, especially since Grubb (1977) emphasized the importance of the regeneration niche in community dynamics. The concept goes back to Sernander (1936), who emphasized that the mosaic structure of a forest was the result of repeated treefalls that created 'storm gaps'.

Patch dynamics results from the disappearance of local populations or patches of one species, or of several species together, and the establishment of other populations. As with gap dynamics, changes may be cyclical in the sense that a given patch structure may reappear over time. The time span involved may (but need not) be longer than in the case of gap dynamics, but the main, gradual difference is the size of the patch relative to the size of the community.

Cyclic succession concerns the cyclic replacement of vegetation components which usually are recognized as separate communities. The vegetation component here may not be more than a dominant population, as in the classical case of *Calluna vulgaris* heath (recently reappraised by Gimingham, 1988). Cyclic succession may be linked to stand dieback (i.e. the nearly synchronous death of a canopy dominant) either by endogenous processes such as tree senescence or by environmental stresses. This phenomenon has recently been claimed to be more universal than hitherto assumed and, moreover, spreading as a result of direct and indirect effects of air and soil pollution (Mueller-Dombois, 1986, 1987).

❋ **Secondary succession** is the recovery of a mature community from a major disturbance. Such recovery may follow disturbance by natural agents, such as fire, storm or insect attack, or after human impact such as burning and clearing. For such straightforward vegetation redevelopment, van der Maarel (1988) used the term 'regeneration succession', and noted that it differed from cyclic succession by exhibiting a greater duration of the mature stage relative to the time needed for recovery. Another form of secondary succession, widely studied, is vegetation development following a specific sequence of events: total disturbance of vegetation, followed by a different type of land use, followed in turn by abandonment of the land use. Thus, the succession starts from a semi-natural or cultural situation.

Primary succession occurs on a newly exposed substrate. Such new substrates are usually poor in nitrogen, which implies that nitrogen-fixing organisms are important in early stages, provided phosphorus is available (Gorham *et al.*, 1979; Vitousek and Walker, 1987). If the disturbance of an existing vegetation leads to a new substrate, a primary succession rather than a secondary succession will occur (e.g. van Dorp *et al.*, 1985).

Secular succession concerns long-term changes in a landscape as a result of long-term environmental changes, often climate change. The apparent changes during long-term succession include those in the geographic distributions of individual species. Through palaeoecology, much has been learned about postglacial secular changes in forest vegetation, especially in temperate areas (see section 1.6.1). Long-term soil changes accompany secular succession (e.g. Beard, 1974; Walker *et al.*, 1981), probably as an integrated process rather than as cause or effect.

1.5 MECHANISMS

If the underlying processes of vegetation dynamics are to be found in the population processes of the constituent plants, then it is fruitful to examine the population-based mechanisms that lead to community-level patterns. Research of this kind has developed rapidly in the last two decades. The mechanisms may result, on the one hand, from the properties of the plants themselves, such as colonization abilities, growth and development and life-history characteristics. On the other hand, the mechanisms may result from plant interactions, which may be studied at the plant–plant level, or as net effects, or may be mediated by third parties (Grubb, 1988).

1.5.1 Colonization

Colonization results from the interaction between the presence or immigration of propagules, the spatial patterns of the environment and existing vegetation, and the morphology and physiology of the propagules. The presence or immigration of propagules will reflect the varying contributions (including the absence) of seed banks, seedling banks and differential seed dispersal. Distances to seed sources (Cooper, 1923b; Frenzen et al., 1988; see also Chapter 2), spatial distribution of seed source (Wood and del Moral, 1987), neighbourhood influences on establishment (Lippe et al., 1985; Ryser, 1990) and the movement of seeds from productive areas into unproductive ones, thereby maintaining populations in the unproductive sites (Shmida and Whittaker, 1981; Shmida and Ellner, 1984), are all important in determining the outcome of colonization.

The presence of spatial discontinuities exerts a selective influence on the colonization success of different species. To illustrate, remnant and standing dead trees (Guevara et al., 1986; Frenzen et al., 1988), vertical structure that facilitates dispersal of seeds by birds (McDonnell and Stiles, 1983; McDonnell, 1988; Archer et al., 1988), and heterogeneity created by burrowing rodents (Anderson and MacMahon, 1985; Huntly and Inouye, 1988) all result in differential plant colonization. At a larger scale, environmental heterogeneity of landscape elements (e.g. road banks, sand bars, streams) likewise affects the spatial arrangement of successional species (Halpern and Harmon, 1983).

Finally, the colonization ability of organisms depends upon their morphological, physiological and reproductive characteristics. Although there are numerous illustrations of this (Bazzaz, 1979; see also Chapter 2), a few will suffice here. For example, Grubb (1986) reviews the successes of many woody species in colonizing harsh, nutrient-poor, primary sites. Adams et al. (1987) documented high proportions of geophytes, plants regenerating from fragments and annuals following the eruption of Mount St Helens, Washington, USA, in 1980. Some species have a narrow temporal 'window' in which to become established (Grubb, 1977; Rankin and Pickett, 1989). Chapter 2 discusses the processes of colonization and the reproductive characteristics of colonizing species in detail.

1.5.2 Initial floristics and pre-emption

Pre-emption occurs when a species, group of species, or vegetation creates an environment which precludes potential invaders. Observations of pre-emption led Egler (1954) to conclude that the initial floristic composition of a site may largely determine the subsequent vegetation. In this context, seed banks and seedling banks, or rapid and efficient dispersal into a site, are important (Chapter 2). DeSteven (1991a,b) has argued that differential seed rain is a large factor in tree succession in old fields, and the model of Liljelund et al. (1988) used differential seed production as one of the two inputs into the model. Resprouting of some woody plants is a mechanism for prolonging their presence in a habitat. Some, such as various species of *Tilia*, seem to maintain themselves in an intact vegetation by continuous resprouting, while others, such as species of *Betula* and *Acer* (Solomon and Blum, 1967), and many Australian rain forest trees (Webb *et al.*, 1972), resprout following disturbance, especially fire.

1.5.3 Competition

In the 'competitive hierarchy' model of Horn (1981), late succession plants are increasingly dominant by virtue of their competitive success with early species, but the late successional species can also invade in the earliest stages of succession. Patterns of replacement, then, are determined by the outcome of competition among the various species. However, if in a changing or non-constant environment the competitive relationships of the species are likely to change, the outcome of succession by this mechanism cannot be predicted (Pickett, 1980; Walker and Chapin, 1987; cf. Connell, 1978).

The 'competitive hierarchy' might be seen as the primary–secondary continuum, in that a complete, non-invasible initial floristics is one extreme, and initial composition with few species that are rapidly replaced is the other extreme.

Tilman (1987, 1988, and earlier), argued that the role of competition in succession must be put into operational terms, and that more than one environmental resource is important to plant competition. Thus, he hypothesized that if resource gradients varied inversely, then a particular species would be competitively most successful over some particular range of ratios of environmental resources. If the ratios of resources change over time, then the series of species reaching competitive ascendency as the resource ratios change produces the observed succession. Tilman's emphasis on the importance of competition in succession, and especially the resource-ratio hypothesis, has provided an important direction for succession research. It was Tilman, himself, and his colleagues, whose experimental results confirmed the importance of competition in succession but who refuted the resource-ratio hypothesis when their experimental observations on relative competitive success (Tilman and Wedin, 1991a) and changes in root and shoot allocations, at least on poor soils (Gleeson and Tilman, 1990), contradicted the predictions of the hypothesis.

1.5.4 Life-history factors and 'vital attributes'

If succession is a population process, an understanding of succession requires an understanding of the life histories of species (Pickett, 1976; Drury and Nisbet, 1973). For example, Gomez-Pompa and Vázquez-

Yanes (1974) and Whitmore (1990) list some of the life history properties of secondary species in tropical lowland forests, including short life cycle, high growth rate, high reproductive resource allocation, continuous and early seed production, small seed size, long seed viability and chemical protection against herbivory. Bazzaz's (1979) ecophysiological approach is similar, and various life history characteristics are either explicit or implicit in some of the chapters that follow in this book.

Modelling succession as arising out of important life history properties was undertaken by Noble and Slatyer (1980), who thus demonstrated the power of such an approach. They termed theirs the 'vital attributes' model, because it is based upon a small number of highly significant life history traits, including propagule persistence or dispersal, age at first reproduction and longevity or life span. They classified each life history trait into a few elements, for instance whether propagules migrate into a site, are stored in a seed bank, or are vegetative. The result is a set of species with specific dynamic properties that come into play at specific times or under specific conditions following a disturbance. The vital attributes model was originally intended for application to vegetation dynamics after fire (Noble and Slatyer, 1980), and it has been useful for natural area management by illustrating the likely outcomes following fires (e.g. Cattelino et al., 1979). Its success has been with relatively species-poor forests subject to fire. It becomes more cumbersome as the number of species increases, and as additional vital attributes, or more complex displays of them, become involved (cf. Austin, 1981).

1.5.5   Changes due to species interactions

Vegetation change resulting from the effects of particular species on other species, either directly or through environmental modification, has long been a central theme of succession. Connell and Slatyer (1977), in a seminal work, summarized species' interactions during succession into three processes: facilitation, inhibition and tolerance. Each of these processes is really a composite of several mechanisms that often act together to produce the observed effect.

(a)   Facilitation

Facilitation (reaction, relay floristics) describes a situation in which one or more species enable the establishment, growth or development of other species. Facilitation may be caused by environmental changes that are favourable to future species, which amounts to the reaction model of Clements (1916, 1928). Soil development is often used as an illustration. Bradshaw et al. (1982) showed that the direction of primary succession in abandoned quarries is strongly controlled by the presence or abundance of the nitrogen-fixing legume shrub, Ulex europaeus, emphasizing the

importance that a single species can assume. Nurse plants clearly facilitate
the establishment of other species. In the New Forest, UK, for example,
Ilex aquifolium appears to favour seedlings of *Fagus sylvatica* and *Quercus*
spp. by protecting them from heavy deer browsing (Morgan, 1991).
Sumac (*Rhus* spp.) apparently facilitates late successional trees by
decreasing the abundance of herbaceous plants which inhibit the woody
plants (Werner and Harbeck, 1982; Petranka and MacPherson, 1979).
In deserts, young plants of succulents often benefit from nurse shrubs
(Steenbergh and Lowe, 1977; Valiente-Banuet *et al.*, 1991), which
provide shelter from small rodents as well as from the hot sun.

(b)   Inhibition

Inhibition is the prevention of plant maturation or growth, or, especially,
the prevention of plant establishment, by existing plants. All of these
constitute negative effects. Inhibition generally results from environmental changes that are detrimental to potential future species (e.g.
inhibition of nitrogen fixation by pioneer species in old fields; Rice, 1964,
1968; Parks and Rice, 1969).

(c)   Tolerance

Tolerance describes the situation where one or more species neither
inhibit nor enable other species during succession. It represents a zero
net outcome of species' interactions. One would assume tolerance in
the absence of evidence of either facilitation or inhibition. DeSteven
(1991b) pointed out that tolerance in the sense of Connell and Slatyer
(1977) holds only if the species involved reach maturation. Simple
coexistence alone (especially for the short time of many successional
studies) is not sufficient to label an interaction as tolerance. This is one of
the confusions which may explain the difficulties some have had in testing
the applicability of Connell and Slatyer's three processes.

(d)   Combinations and evaluation

Connell and Slatyer (1977) recognized that facilitation, inhibition and
tolerance were not mutually exclusive in a community. Indeed, they
reckoned that all three processes occurred during most successions.
Subsequent observations have borne them out (Peet and Christensen,
1980; Christensen and Peet, 1981; Bornkamm, 1988; Gill and Marks,
1991; Walker *et al.*, 1986; DeSteven, 1991b).
    Connell and Slatyer's processes have been important for reshaping
our thinking and impelling new research into successional mechanisms.
They have been criticized, however, because facilitation, inhibition and
tolerance are not mechanisms at the level of species-by-species replace-

ment (Walker and Chapin, 1987; Pickett *et al.*, 1987b). Each of the three processes subsumes many elements and interactions, and the importance of each changes during succession (Walker and Chapin, 1987), and so these processes are characterizations of 'net effects' at the community level (Connell *et al.*, 1987) rather than proximate mechanisms. Furthermore, the three processes address species interaction, but were not intended to cover other successional processes such as those that result from life history differences, colonization or herbivory (Connell *et al.*, 1987).

Interpretation of the three processes also depends upon the spatial scale of observation. In a tropical dry forest succession on dunes, Campbell *et al.* (1990) described soil development on a very small scale (that of individual saplings), which seemed to be a facilitation process. At a larger scale, they reported no evidence of soil development, because of the averaging effect of large-scale sampling on small-scale differences.

1.5.6 'Third party' effects

Grubb (1986) pointed out that what he calls a 'third party,' which may be a species of plant, animal or micro-organism, may change the relative success of establishment of two species. In the absence of the third party, one species will always be more successful than the second, but in the presence of the third party, the second will be the more successful colonizer. Such a third party, for instance, can act by changing the light regime of a site, thereby altering the relative success of a pair of species, one of which is shade tolerant and the other of which is intolerant (Kohyama, 1984; Grubb, 1986), or indirectly through its litter (Facelli and Pickett, 1991) or through the effects that animals associated with early species exert on later species (Gill and Marks, 1991). Gaps caused by animal activity, when such gaps are essential for establishment by a particular species, constitute another example of third party facilitation (Grubb, 1986).

1.5.7 The mechanisms of vegetation dynamics

It is obvious from this discussion (see also from among numerous examples: Pickett *et al.*, 1987a,b; DeSteven, 1991a,b) that a multiplicity of interacting mechanisms operates during succession, including, among others, differential seed rain and propagule availability and other factors resulting from life history differences, competition, herbivory and changes in environmental factors. Pickett *et al.* (1987a) have summarized these in a hierarchical manner, by showing which processes contribute to site availability, species availability and species performance, and the environmental and ecological factors which modify the processes (Table 1.1).

Table 1.1 A hierarchical summary of the causes, processes and factors of vegetation dynamics (from Pickett *et al.*, 1987a)

General causes of succession	Contributing processes or condition	Modifying factors
Site availability	Coarse-scale disturbance	Size, severity, time, dispersion
Differential species availability	Dispersal	Landscape configuration, dispersal agents
	Propagule pool	Time since last disturbance, land use treatment
	Resource availability	Soil condition, topography, microclimate, site history
Differential species performance	Ecophysiology	Germination requirements, assimilation rates, growth rates, population differentiation
	Life history	Allocation pattern, reproductive timing, reproductive mode
	Environmental stress	Climate cycles, site history, prior occupants
	Competition	Hierarchy, presence of competitors, identity of competitors, within-community disturbance, predators, herbivores, resource base
	Allelopathy	Soil chemistry, soil structure, microbes, neighbouring species
	Herbivory, predation and disease	Climate cycles, predator cycles, plant vigour, plant defences, community composition, patchiness

1.6   THE STUDY OF VEGETATION DYNAMICS

1.6.1   Observation and inference in vegetation dynamics

Before taking up the patterns and processes detailed in the following chapters, we need to review the nature of the data and methods used in the study of vegetation dynamics. The quality of description, analysis and modelling depends directly on the quality of the data and the reliability of the methods of studying vegetation dynamics.

(a)   Direct observations

Direct observations generally mean comparisons of the vegetation on a portion of ground over time using permanent plots. Such study plots have the advantage of accurate measurements of the parameters of interest, and at any particular spatial scale and temporal period. Although it generally takes a long time to see the whole pattern, carefully designed permanent plot studies yield useful data on mechanisms, initial conditions and environmental changes during succession. Indeed, Austin (1981), among others, argued that permanent plots were essential to an understanding of vegetation dynamics. There are now a number of permanent plot studies available or underway, although most are, not surprisingly, of relatively short duration.

A few examples will serve to illustrate the range of permanent plot research. The Buell succession study, a permanent plot study of 'old-field' succession in central New Jersey, USA, was begun in 1958 (Buell et al., 1971; Pickett, 1982). Species' covers and the number of tree stems have been recorded yearly or every other year. The plots have been useful for the analysis of mechanisms, as well as understanding temporal patterns. For instance, the validity of the initial floristics hypothesis (Egler, 1954; see above) was confirmed (Buell et al., 1971; Myster and Pickett, 1988) using these plots.

Some other examples of permanent plot studies show their usefulness over a range of circumstances. Measurements of changes in abundances and individual tree dimensions have been made in the Duke Forest, North Carolina, USA, approximately every five years since 1934 (Peet and Christensen, 1980, 1987). These data have been important for detailing population dynamics and the species-by-species replacement processes of succession. J. T. de Smidt has mapped the canopy cover, herb position and cryptogram distribution in permanent plots of *Calluna vulgaris* heathlands in The Netherlands yearly for 30 years (as of 1990), providing data useful for testing and parametrizing models of heathland dynamics (Lippe et al., 1985; Prentice et al., 1987; van Tongeren and Prentice, 1986). Beeftink (1977, 1987) has plotted salt marshes for about 25 years to document how environmental change affects the course of succession, and since 1972 Bakker (1989) has maintained a large number of permanent plots in a variety of communities, including several grasslands, heathlands and salt marshes, in order to study the effects of treatments such as grazing, mowing and fertilization on vegetation change.

(b)   Chronosequences, or 'space-for-time substitution'

Probably the most common means of describing succession is to order vegetation data taken from sites in similar environments but of different ages, an approach termed a chronosequence, or 'space-for-time substitution' by Pickett (1989). The resulting vegetation sequence is assumed to be the successional sequence. Summarizing a good deal of previous research, Miles (1979) and Austin (1981) emphasized that space-for-time could be misleading because the implicit assumption of 'similiar environments' (i.e. the same soil conditions, microclimate, history (especially important), availability of propagules), may well not be valid. Indeed, subsequent work (e.g. Collins and Adams, 1983; Jackson et al., 1988; Schmidt, 1988; and especially the thorough review by Pickett, 1989) bears out the warning. Pickett (1989) concludes that while space-for-time may be useful for qualitative purposes and for hypothesis generation, it is unreliable for a deeper understanding of successional change because site history is so often important.

(c)   Inference from population structure

For vegetation in which woody species produce annual growth rings, it is sometimes possible to employ dendrochronological techniques to understand population dynamics, and, if plant remains are available, to reconstruct vegetation history. Thus, successional change has frequently been inferred from the age structures of tree populations in temperate forests (e.g. Lorimer, 1977; Glitzenstein et al., 1986; Johnson and Fryer, 1989). Dendrochronological dating of tree death and of past disturbances that leave scars (e.g. treefall scars or fire scars) or abrupt changes in tree growth patterns can be combined with the data on tree population age structures to create detailed reconstructions of the history of disturbance and successional change in a stand (Henry and Swan, 1974; Oliver and Stephens, 1977; Lorimer, 1985; see also Chapter 4). Veblen et al. (1989) exploited tree population age structures to show the importance of one kind of disturbance (blowdown of one tree species which released the other species) following another (fire, which initated the original succession).

(d)   Historical records

Historical records include such sources as journals, photo points (e.g. Hastings and Turner, 1965; Waldemarson Jensén, 1979), repeated aerial photography and repeated mapping (e.g. van Dorp et al., 1985; van der Maarel et al., 1985) and land surveyors' records (e.g. Vale, 1982). These data can be important sources of information for comparing change over long time periods, but are often limited by such factors as infrequent observations, lack of quantitative data and differences in taxonomy. Sometimes these sources provide very general, physiognomic descriptions. Other data may be quantitative, such as the land surveyors' records of much of the United States, but comparing today's vegetation with these records jumps perhaps 150 years at once.

(e)   Micro- and macrofossil deposits

Pollen and macrofossils deposited in environments in which decomposition is negligible (e.g. anaerobic sediments in bogs, fens and lakes, or in pack rat middens) constitute an important record of ecological chronology (see Chapter 8). Although the methods of analysis are limited by selective preservation, problems of species identification and a degree of uncertainty about the provenance of the pollen or fossil material, they nevertheless are very useful. They have been most widely used in temperate and boreal environments, but are now being applied in the tropics (e.g. Bush and Colinvaux, 1990) and arid environments (e.g. Wells and Jorgensen, 1964; Betancourt *et al.*, 1990). Both the spatial and temporal resolution of the method have greatly increased in recent years (Prentice, 1986b; see also Chapter 8).

Studies of pollen and macrofossil records lead us to conclude that the nature and composition of plant communities has changed on a time scale of millennia (or longer). They have also added significantly to our understanding of the nature of communities. These methods can be useful for analysis of vegetation change in the shorter term as well, for instance over decades or a few centuries (Prentice, 1986b). As one example, Mitchell (1990) used pollen techniques to show the influence of human disturbance and grazing on certain Irish woodlands over a 250-year period. He reported substantial differences in the pollen profile from two sites only 85 m apart, simultaneously illustrating the fine spatial scale of vegetation change and the ability of the technique to resolve differences on this scale. The use of pollen and macrofossil records for the study of succession is addressed in detail in Chapter 8.

Whenever feasible, more than one type of data should be used for studying vegetation dynamics. For example, short-term observations from permanent plots can be combined with longer term data obtained from historical records or tree population age structures. Similarly, Veblen and Markgraf (1988) used stand age profiles along with pollen analysis and repeated photography to show that forest is invading the Patagonian steppe, rather than the other way around as had been proposed previously.

1.6.2 Approaches to the study of vegetation dynamics

Descriptive methods predominate in the study of succession. Fundamentally, such methods tell us what has happened, or what is happening, based on sleuthing out the history of a community, or watching permanent plots. Unfortunately, the term 'descriptive' sometimes casts the shadow of a perjorative in ecology; it is in no way intended to do so here. Many descriptive succession studies are remarkable for their data intensity, analytical elegance and depth of understanding. Good description is necessary in ecology. On it rests the accuracy and sensibility of both experimental interpretation aᵢ d modelling.

Multivariate approaches, while fundamentally descriptive, are here given separate mention because their methodology stands somewhat apart from the rest. The whole panoply of multivariate methods, from the standard statistical library to those developed or adapted especially for ecology, has been applied to analyse successional data, to sort out the temporal trends from environmental patterns and randomness, and to develop successional vectors of both communities and species (van der Maarel, 1969; Austin, 1977; van der Maarel and Werger, 1978).

We cite here a few examples to illustrate the applications of multivariate methods to succession, to show how temporal trends can be separated from environmental patterns. Cramer and Hytteborn (1987) used canonical correspondence analysis to separate the long-term trends from short-term effects of disturbance in a primary succession on a rising seashore in Sweden. Swaine and Greig-Smith (1980) and Austin et al. (1981) showed how a multivariate analysis could partition a data set of repeated measures of species' abundances into environmental variation and variation due to vegetation dynamics. Jasieniuk and Johnson (1982) and Johnson (1981) used multiple regression to partition variation in peatland vegetation and taiga, respectively, into components reflecting time since fire and habitat factors. They then applied ordination techniques to the residuals, thereby illustrating vegetation dynamics following fire, which seemed to be primarily changes in species abundances (i.e. quantitative changes) and those due to habitat changes, which seemed to be compositional changes (i.e. qualitative).

The other major use of multivariate methods in succession research is to determine the directional vectors of communities or species. Goff and Zedler (1972) combined age-class profiles with ordination to exhibit species' behaviours as vectors in multivariate space, rather than as points. Helsper et al. (1983) applied detrended correspondence analysis ordination to experimental data resulting from nutrient applications to Calluna heathlands. They used this to illustrate the different community vectors that resulted from the different resource modifications. Wildi (1988) showed how trend analysis by regression could improve the interpretation of successional vectors derived from ordination. In addition to ordination, cluster techniques have been applied, for instance, to quantify transition probabilities (van Noordwijk-Puyk et al., 1979).

Experimental studies of succession are illustrated by treatments such as plant removal or addition (as transplants, or more often by sowing seeds into the experimental community), resource supplementation and controlled disturbance or changes in disturbance regimes, such as gap creation, fire treatments and grazing exclusures. Succession experiments have been employed to study species compositional changes, changes in ecosystem properties, the importance of resource competition in succession, successional mechanisms and applied management questions. Connell et al. (1987) briefly review the warnings of others, that 'indirect effects of other interacting species may confound interpretation' of the experimental results.

A few examples of succession experiments will illustrate how experimental approaches have been taken. Studies that manipulated plant densities or composition include Armesto and Pickett (1985, 1986), Hils and Vankat (1982) and Gill and Marks (1991), who used plant removal experiments to test ideas about successional mechanisms and rates. Another group of succession experiments includes those in which resource levels are manipulated. Tilman (Tilman, 1987, 1988; Tilman and Wedin 1991a,b; and others) has used a series of resource (specifically nitrogen) supplementation experiments to explore the importance of resource competition during succession in grasslands (see section 1.5). Helsper *et al.* (1983) showed striking differences in dynamics resulting from treatment of heathland plots with different nutrients and at different frequencies. Application of NPK fertilizer to an old field strongly affected ecosystem parameters and affected species composition, but did not appear to 'set' it 'back' to some previous stage (Bakelaar and Odum, 1978). Schmidt (1988) used a factorial design of soil type, nutrient supplementation and water levels to demonstrate the control of these environmental factors on the rates of vegetation change.

Thirdly, manipulation of a disturbance regime has been used in succession experiments. Hobbs and Gimingham (1984) and Trabaud (1974) used fire experiments to understand the dynamics of heathlands. The main underlying design of the Konza Prairie Long-term Ecological Reserve (Hulbert, 1973; Gibson and Hulbert, 1987) is the application of different fire frequency treatments. Numerous exclosure experiments have demonstrated the importance of grazing and browsing as forces affecting vegetation dynamics by, for instance, selective removal of species, prevention of competitive exclusion and changes in ecosystem function.

Modelling expresses processes, in our case processes of vegetation dynamics, in the language of symbolic logic and mathematics. Modelling inevitably simplifies the processes, although modern ecological models can be quite complex. Ecological modelling continues to increase in importance, due largely to the exponential increase in the power of computers, their usefulness for ecological systems, and the need for ecological forecasting and management.

Models generate testable hypotheses that have an explicit explanation in the assumptions of the model (Levins, 1966; Pielou, 1981). They can be used to look for the consequences of predictions when tests in nature would be too complicated, take too long, or could not be performed for practical or ethical reasons. Models can also be used for sensitivity analysis, that is, to search for those parts of a system that are most likely to influence the outcome.

Levins (1966) argued that models may exhibit properties of generality, precision or reality, or any paired combination of these, but never all three simultaneously. Thus, it is necessary to sacrifice at least one of these properties in order to maximize the others. This has important consequences for one's approach to modelling, since the nature and form of the model will reflect one's choice of which of the properties should be stressed.

A major problem in ecological modelling is validation of the model. We have many more plausible models than appropriate validation tests of them. The fit of a model to some data in itself is not sufficient to demonstrate model correctness; different models with different assumptions may nevertheless show similar behaviours. It takes careful testing of the data, assumptions, processes included in the model, and prediction in order to validate a model. Conversely, however, a lack of fit is sufficient to demonstrate that at least one aspect of the model is not correct.

There are numerous ways to classify models (Pielou, 1981; Usher, 1981; Jeffers, 1982; see also Chapter 6). Our classification is specifically for vegetation dynamics, and reflects how successional models have developed.

Analytical models are theoretical, explanatory expressions based on first principles that are derived from observed characteristics and behaviour of ecological systems. Such models lead to a better analytic understanding by determining the logical consequences of certain assumptions and postulates that one can make from observations of ecological systems. Analytical models tend to be unrealistically simple, but general and precise. They are usually tractable and have few parameters. They are excellent for heuristic exploration of assumptions and their consequences (Pielou, 1981). Chapter 5 explores analytical models for vegetation dynamics in depth.

Statistical models are stochastic expressions where the parameters are probabilities of events, often replacements or transitions of species to species or state to state (where the state is defined externally, such as by community classification). They are useful for probabilistic predictions of vegetation dynamics, for tests for randomness of successional events, and for making precise estimates of the times associated with successional events when the probabilities are associated with a time step.

The form of statistical models may be general, but they typically are not very realistic. The parameters are specific, but because the models are stochastic, their precision depends upon the quality of parameter measurements. Statistical models, especially Markov models for succession, are explored in depth in Chapter 6.

Lottery models, another form of statistical model, have found some use in ecology. For instance, Liljelund et al. (1988) used lottery models to show how differences in life-history traits, specifically seed production and plant longevity, can account for succession from an initial floristics composition of a site. Chapter 5 provides more detail about lottery models.

Simulation models are attempts to duplicate the true behaviour of a process or phenomenon. They are real and precise, but not general. They typically have numerous parameters that have functional, ecological significance (e.g. light and moisture, topographic setting, substrate conditions, growth, demographic characterstics, competition, reproduction) and consequently they are usually site and species specific. Simulation models have improved to the point that they can be used for prediction and sensitivity analysis (Prentice, 1986a). Chapter 7 examines simulation models in vegetation dynamics.

All of these are models of succession in time. Less well developed, but under active investigation, are models of vegetation in space (e.g. regression and geostatistical spatial modelling, and co-krijging; Ver Hoef 1991). We can look forward to combining succession and spatial models into ecological space–time models. Successful space–time models will need to make explicit statements of scale, such as succession as a function of patches within a community, or communities in landscapes.

With our understanding of the patterns and mechanisms of succession, and an appreciation of the role of disturbance in vegetation, we would now like to know the relative contribution of environmental control and disturbance factors to vegetational variation. One powerful approach to dissecting this problem may be through spatial modelling. In an analogous sense, we would like to know the relative contribution of deterministic factors, based on plant life histories, site history or disturbance (and its extent, frequency, etc.) and stochastic factors, also including disturbance and, in addition, unknown or unmeasured factors and purely random effects. Working out the relative contributions of these over a longer time remains a challenge to vegetation scientists. That succession has important spatial components complicates the challenge even more. However, the recent rapid development of an arsenal of tools, both for measurement and analysis, gives us confidence that the study of vegetation dynamics will move ahead rapidly in the near future.

SUMMARY

1. Dichotomies that have been recognized in the discipline of vegetation dynamics, such as progression and retrogression, primary and secondary succession, and autogenic and allogenic succession, may be heuristically useful, but intermediate conditions occur, and these dichotomies are unidimensional constructs for a multidimensional problem. They also may fail to account for differences of scale in space or time.
2. Vegetation dynamics are based on the population processes of the constituent plants. It is in the population processes that we shall find the mechanisms of vegetation dynamics, and the explanations for the patterns that we observe. Typologies based on other criteria may be useful descriptions, but they will lack explanatory power.

3. Successional histories typically exhibit multiple pathways, and may also show community patterns such as convergence, divergence or cycles. Which of these occur or predominate will depend not only upon the initial conditions and mechanisms of succession, but also on community circumstances such as species diversity, landscape complexity and community isolation.

4. Disturbance is extremely important in ecological systems. Disturbance initiates succession and may interrupt or redirect succession if disturbance(s) occurs regularly. Disturbance implies that non-equilibrium dynamics characterize changing vegetation systems.

5. Disturbance has dimensions of space, time and magnitude, and the specific dimensions of a disturbance event will have important effects on the initiation and outcome of vegetation dynamics.

6. The course of vegetation dynamics depends upon the spatial scale of the vegetation and of any disturbance. Conversely, spatial patterns of vegetation change with succession.

7. Succession is a vegetation gradient in time (and space), and different dynamic patterns can be ordered over the scale of time (and to some degree space): fluctuations, fine-scale gap dynamics, patch dynamics, cyclic succession, secondary succession, primary succession, and secular (or long-term) succession.

8. Careful analytical studies are still needed to continue our understanding of vegetation dynamics, but thorough mechanistic studies will likely be the most important means for progress in the field.

9. It is clear that initial conditions are important to the course of most vegetation dynamics. We know less about the mechanisms of change once an initial set of conditions is given, however. It is here that colonization processes, life-history characteristics, species interactions and third-party effects will interact to produce the patterns that we see.

10. Sources of data for the study of vegetation dynamics include: direct observations (permanent plots), chronosequences, inference from population structure, historical records, and micro- and macrofossil deposits.

11. The means of analysing the data of vegetation dynamics include: descriptive methods, multivariate analyses, experimental studies, and modelling. Models that incorporate both spatial variation and the gradient of changes over time will ultimately be needed for reliable prediction of vegetation dynamics.

ACKNOWLEDGEMENTS

We thank R. K. Peet, I. C. Prentice, T. Rosburg, A. van der Valk, T. T. Veblen, J. M. Ver Hoef and M. J. A. Werger for their help, comments and ideas. Support for research reported in this chapter or for the preparation of the chapter was provided by the US National Park Service, Whitehall Foundation, North Atlantic Treaty Organization (NATO), the Netherlands Organization for Scientific Research (NWO) and the Swedish Natural Science Research Council.

REFERENCES

Abrams, M. P., Sprugel, D. G. and Dickmann, D. I. (1985) Multiple successional pathways on recently disturbed jack pine sites in Michigan. *For. Ecol. Mgmt*, **10**, 31–48.

Adams, A. B., Dale, V. H., Smith, E. P. and Kruckeberg A. R. (1987) Plant survival, growth form and regeneration following the 18 May 1980 eruption of Mount St. Helens, Washington. *Northwest Sci.*, **61**, 160–70.

Albertson, F. W. and Tomanek, G. W. (1965) Vegetation changes during a 30-year period in grassland communities near Hays, Kansas. *Ecology*, **46**, 714–20.

Allen, T. F. H. (1987) Hierarchical complexity in ecology: A non-Euclidean conception of the data space. *Vegetatio*, **69**, 17–25.

Andersen, D. C. and MacMahon, J. A. (1985) Plant succession following the Mount St. Helens volcanic eruption: Facilitation by a burrowing rodent, *Thomomys talpoides. Amer. Mid. Nat.*, **114**, 62–9.

Anderson, R. C. and Adams, D. E. (1978) Species replacement patterns in central Illinois white oak forests, in *Proc. Central Hardwood Forest Conf. II.* (ed. P. E. Pope), Purdue Univ., West Lafayette, pp. 284–301.

Anonymous (1981) *The Platte River Ecology Study*, Special Research Report, US Dept Interior, Fish and Wildlife Service, Jamestown, N. D.

Archer, S., Scifres, C., Bassham, C. R. and Maggio, R. (1988) Autogenic succession in a subtropical savanna: Conversion of grassland to thorn woodland. *Ecol. Monogr.*, **58**, 111–27.

Armesto, J. J. and Pickett, S. T. A. (1985) Experimental studies of disturbance in oldfield plant communities: Impact on species richness and abundance. *Ecology*, **66**, 230–40.

Armesto, J. J. and Pickett, S. T. A. (1986) Removal experiments to test mechanisms of plant succession in old fields. *Vegetatio*, **66**, 85–93.

Austin, M. P. (1977) Use of ordination and other multivariate descriptive methods to study succession. *Vegetatio*, **35**, 165–75.

Austin, M. P. (1981) Permanent quadrats: An interface for theory and practice. *Vegetatio*, **46**, 1–10.

Austin, M. P. and Belbin, L. (1981) An analysis of succession along an environmental gradient using data from a lawn. *Vegetatio*, **46**, 19–30.

Austin, M. P. and Williams, O. B. (1988) Influence of climate and community composition on the population demography of pasture species in semi-arid Australia. *Vegetatio*, **77**, 43–9.

Austin, M. P., Williams, O. B. and Belbin, L. (1981) Grassland dynamics under sheep grazing in an Australian Mediterranean type climate. *Vegetatio*, **47**, 201–11.

Bakelaar, R. G. and Odum, E. P. (1978) Community and population level responses to fertilization in an old-field ecosystem. *Ecology*, **59**, 660–5.

Bakker, J. P. (1989) *Nature Management by Grazing and Cutting*, Geobotany, **14**, Kluwer Acad. Publ., Dordrecht.

Bard, G. E. (1952) Secondary succession on the Piedmont of New Jersey. *Ecol. Monogr.*, **22**, 195–215.

Barnes, W. J. and Dibble, E. (1988) The effects of beaver in riverbank forest succession. *Canad. J. Bot.*, **66**, 40–44.

Bazzaz, F. A. (1979) The physiological ecology of plant succession. *Ann. Rev. Ecol. System.*, **10**, 351–371.

Beard, J. S. (1974) Vegetational changes on aging landforms in the tropics and subtropics, in *Vegetation Dynamics* (ed. R. Knapp), Junk, The Hague, pp. 219–24.

Beard, J. S. (1976) The progress of plant succession on the Soufriere of St. Vincent: Observations in 1972. *Vegetatio*, **31**, 69–77.

Beeftink, W. G. (1977) The coastal salt marshes of western and northern Europe: An ecological and phytosociological approach, in *Wet Coastal Ecosystems* (ed. V. J. Chapman), Elsevier, Amsterdam, pp. 109–55.

Beeftink, W. G. (1987) Vegetation responses to changes in tidal inundation of salt marshes, in *Disturbance in Grasslands. Causes, Effects and Processes* (eds J. van Andel, J. P. Bakker and R. W. Snaydon), Geobotany, **10**, Junk, Dordrecht, pp. 97–117.

Betancourt, J. L., van Devender, T. R. and Martin, P. S. (eds) (1990) *Packrat Middens: The Last 40,000 Years of Biotic Change*, Univ. of Arizona Press, Tucson.

Billings, W. D. (1938) The structure and development of old field shortleaf pine stands and certain associated physical properties of the soil. *Ecol. Monogr.*, **8**, 437–99.

Bobbink, R. and Willems, J. H. (1988) Effects of management and nutrient availability on vegetation structure of chalk grassland, in *Diversity and Pattern in Plant Communities* (eds H. J. During, M. J. A. Werger and J. H. Willems), SPB Publ., The Hague, pp. 183–93.

Borgegård, S. -O. (1990) Vegetation development in abandoned gravel pits: Effects of surrounding vegetation, substrate and regionality. *J. Veg. Sci.*, **1**, 675–82.

Bormann, F. H. and Likens, G. E. (1979) *Pattern and Process in a Forested Ecosystem*, Springer-Verlag, New York.

Bornkamm, R. (1988) Mechanisms of succession on fallow land. *Vegetatio*, **77**, 95–101.

Botkin, D. B. (1979) A grandfather clock down the staircase: Stability and disturbance in natural ecosystems, in *Forests: Fresh Perspectives from Ecosystem Analysis* (ed. R. H. Waring), Corvallis, Oregon State University Press, pp. 1–10.

Botkin, D. B. (1992) The Ecology of Forests: Theory and Evidence, Oxford Univ. Press, New York.

Bradshaw, A. D., Marrs, R. H. and Roberts, R. D. (1982) Succession, in *Ecology of Quarries. The Importance of Natural Vegetation* (ed. B. N. K. Davis), Inst. Terrestrial Ecol. Symp. No. 11, pp. 47–52.

Braun-Blanquet, J. (1964) *Pflanzensoziologie*, 3rd edn, Springer-Verlag, Wien.

Brokaw, N. V. L. (1982) The definition of treefall gap and its effect on measures of forest dynamics. *Biotropica*, **14**, 158–60.

Buell, M. F., Buell, H. F., Small, J. A. and Siccama, T. G. (1971) Invasion of trees in secondary succession on the New Jersey Piedmont. *Bull. Torrey Bot. Club*, **98**, 67–74.

Bush, M. B. and Colinvaux, P. A. (1990) A pollen record of a complete glacial cycle from lowland Panama. *J. Veg. Sci.*, **1**, 105–18.

Busing, R. T. (1991) A spatial model of forest dynamics. *Vegetatio*, **92**, 167–79.

Campbell, B. M., Lynam, T. and Hatton, J. C. (1990) Small-scale patterning in the recruitment of forest species during succession in tropical dry forest, Mozambique. *Vegetatio*, **87**, 51–7.

Canham, C. D. and Loucks, O. L. (1984) Catastrophic windthrow in the presettlement forests of Wisconsin. *Ecology*, **65**, 803–9.

Casado, M. A., de Miguel, J. M., Sterling, A., Peco, B., Galiano, E. F. and Pineda, F. D. (1985) Production and spatial structure of Mediterranean pastures in different stages of ecological succession. *Vegetatio*, **64**, 75–86.

Casparie, W. A. (1972) Bog development in southeastern Drenthe (The Netherlands). *Vegetatio*, **25**, 1–271.

Cattelino, P. J., Noble, I. R., Slatyer, R. O. and Kessell, S. R. (1979) Predicting the multiple pathways of plant succession. *Environ. Mgmt*, **3**, 41–50.

Chesson, P. L. and Case, T. J. (1986) Overview: Non-equilibrium community theories: Chance, variability, history and coexistence, in *Community Ecology* (eds J. Diamond and T. J. Case), Harper and Row, New York, pp. 229–39.

Christensen, N. L. (1987) The biogeochemical consequences of fire and their effects on the vegetation of the coastal plain of the southeastern United States, in *The role of fire in ecological systems* (ed. L. Trabaud), SPB Acad. Publ., The Hague, pp. 1–21.

Christensen, N. L. and Peet, R. K. (1981) Secondary forest succession on the North Carolina Piedmont, in *Forest Succession. Concepts and Applications* (eds D. C. West, H. H. Shugart and D. B. Botkin), Springer, New York, pp. 230–45.

Christensen, N. L. and Peet, R. K. (1984) Convergence during secondary forest succession. *J. Ecol.* **72**, 25–36.

Clark, J. S. (1990) Fire and climate change during the last 750 years in northwestern Minnesota. *Ecol. Monogr.*, **60**, 135–59.

Clark, J. S. and Patterson, W. A. III (1985) The development of a tidal marsh: Upland and oceanic influences. *Ecol. Monogr.*, **55**, 189–217.

Clément, B. and Touffet, J. (1990) Plant strategies and secondary succession on Brittany heathlands after severe fire. *J. Veg. Sci.*, **1**, 195–202.

Clements, F. E. (1916) *Plant Succession: An Analysis of the Development of Vegetation*, Carnegie Inst. Washington Publ. 242.

Clements, F. E. (1928) *Plant Succession and Indicators*, Wilson, New York.

Collins, S. L. and Adams, D. E. (1983) Succession in grasslands: Thirty-two years of change in a central Oklahoma tallgrass prairie. *Vegetatio*, **51**, 181–90.

Connell, J. H. (1978) Diversity in tropical rain forests and coral reefs. *Science*, **199**, 1302–10.

Connell, J. H., Noble, I. R. and Slatyer, R. O. (1987) On the mechanisms producing successional change. *Oikos*, **50**, 136–7.

Connell, J. H. and Slatyer, R. O. (1977) Mechanisms of succession in natural communities and their roles in community stability and organization. *Amer. Nat.*, **111**, 1119–44.

Cooper, W. S. (1913) The climax forest of Isle Royale, Lake Superior, and its development. *Bot. Gazette*, **55**, 1–44, 189–235.

Cooper, W. S. (1923a) The recent ecological history of Glacier Bay, Alaska. I. The interglacial forests of Glacier Bay. *Ecology*, **4**, 93–128.

Cooper, W. S. (1923b) The recent ecological history of Glacier Bay, Alaska. II. The present vegetation cycle. *Ecology*, **4**, 223–46.

Cooper, W. S. (1923c) The recent ecological history of Glacier Bay, Alaska. III. Permanent quadrats at Glacier Bay: An initial report upon a long-period study. *Ecology*, **4**, 355–65.

Cooper, W. S. (1926) The fundamentals of vegetational change. *Ecology*, **7**, 391–413.

Cooper, W. S. (1931) A third expedition to Glacier Bay, Alaska. *Ecology*, **12**, 61–5.

Cooper, W. S. (1939) A fourth expedition to Glacier Bay, Alaska. *Ecology*, **20**, 130–59.

Cowles, H. C. (1899) The ecological relations of the vegetation on the sand dunes of Lake Michigan. *Bot. Gazette*, **27**, 95–117, 167–202, 281–308, 361–91.

Cramer, W. and Hytteborn, H. (1987) The separation of fluctuation and long-term change in vegetation dynamics of a rising seashore. *Vegetatio*, **69**, 157–67.

Dale, M. R. T. and Blundon, D. J. (1990) Quadrat variance analysis and pattern development during primary succession. *J. Veg. Sci.*, **1**, 153–64.

Dale, M. R. T. and Blundon, D. J. (1991) Quadrat covariance analysis and the scales of interspecific association during primary succession. *J. Veg. Sci.*, **2**, 103–12.

Davidson, K. L. and Bratton, S. P. (1988) Vegetation response and regrowth after fire on Cumberland Island National Seashore, Georgia. *Castanea*, **53**, 47–63.

Davis, M. B. (1976) Pleistocene biogeography of temperate deciduous forests. *Geosci. Man*, **13**, 13–26.

Davis, M. B. (1981) Quaternary history and the stability of forest communities, in *Forest succession: Concepts and application* (eds D. C. West, H. H. Shugart and D. B. Botkin), Springer-Verlag, New York, pp. 132–53.

Davis, M. B. (1986) Climatic instability, time lags and community disequilibrium, in *Community Ecology* (eds J. Diamond and T. J. Case) Harper and Row, New York, pp. 269–84.

Davis, M. B. (1987) Invasions of forest communities during the Holocene: Beech and hemlock in the Great Lakes region, in *Colonization, Succession and Stability* (eds A. J. Gray, M. J. Crawley and P. J. Edwards), Blackwell, Oxford, pp. 373–94.

Davis, M. B., Woods, K. D., Webb, S. L. and Futyma, R. P. (1986) Dispersal versus climate: Expansion of *Fagus* and *Tsuga* into the upper Great Lakes region. *Vegetatio*, **67**, 93–103.

DeAngelis, D. L. and Waterhouse, J. C. (1987) Equilibrium and non-equilibrium concepts in ecological models. *Ecol. Monogr.*, **57**, 1–21.

Delcourt, H. R., Delcourt, P. A. and Webb, T., III (1983) Dynamic plant ecology: The spectrum of vegetational change in space and time. *Quat. Sci. Rev.*, **1**, 153–75.

Denslow, J. S. (1987) Tropical rainforest gaps and tree species diversity. *Ann. Rev. Ecol. System.*, **18**, 431–51.

de Pablo, C. L., Peco, B., Galiano, E. F., Nicolas, J. P. and Pineda, F. D. (1982) Space-time variability in Mediterranean pastures analyzed with diversity parameters. *Vegetatio*, **50**, 113–25.

de Smidt, J. T. (1977) Interaction of *Calluna vulgaris* and the heather beetle (*Lochmaea suturalis*), in *Vegetation und Fauna* (ed. R. Tüxen), Cramer, Vaduz, pp. 179–86.

DeSteven, D. (1991a) Experiments on mechanisms of tree establishment in old-field succession: Seedling emergence. *Ecology*, **72**, 1066–75.

DeSteven, D. (1991b) Experiments on mechanisms of tree establishment in old-field succession: Seedling survival and growth. *Ecology*, **72**, 1076–88.

Drury, W. H. (1956) Bog flats and physiographic processes in the upper Kuskokwim River region, Alaska. *Contrib. Gray Herb.*, **178**, 1–130.

Drury, W. H. and Nisbet, I. C. T. (1971) Interrelations between developmental models in geomorphology, plant ecology, and animal ecology. *Gen. Syst.*, **16**, 57–68.

Drury, W. H. and Nisbet, I. C. T. (1973) Succession. *J. Arnold Arboretum*, **54**, 331–68.

Egler, F. E. (1954) Vegetation science concepts. I. Initial floristic composition, a factor in old-field development. *Vegetatio*, **4**, 412–17.

Evans, E. W., Briggs, J. M., Finck, E. J., Gibson, D. J., James, S. W., Kaufman, D. W. and Seastedt, T. R. (1989) Is fire a disturbance in grasslands?, in *Proc. 11th North American Prairie Conf.* (eds T. B. Bragg and J. Stubbendieck), Univ. of Nebraska, Lincoln, pp. 159–61.

Facelli, J. M. and D'Angela, E. (1990) Directionality, convergence, and rate of change during early succession in the Inland Pampa, Argentina. *J. Veg. Sci.*, **1**, 255–60.

Facelli, J. M. and Pickett, S. T. A. (1991) Plant litter: Light interception and effects on an old-field plant community. *Ecology*, **72**, 1024–31.

Falinski, J. B. (1988) Succession, regeneration and fluctuation in the Bialowieza Forest (NE Poland). *Vegetatio*, **77**, 115–28.

Finegan, B. (1984) Forest succession. *Nature*, **312**, 109–14.

Forman, R. T. T. and Boerner, R. E. (1981) Fire frequency and the pine barrens of New Jersey. *Bull. Torrey Bot. Club*, **108**, 34–50.

Forman, R. T. T. and Godron, M. (1986) *Landscape Ecology*, Wiley, New York.

Fox, M. D. and Fox, B. J. (1987) The role of fire in the scleromorphic forests and shrublands of eastern Australia, in *The Role of Fire in Ecological Systems* (ed. L. Trabaud), SPB Academic Publ., The Hague, pp. 23–48.

Fritzell, E. K. (1989) Mammals in prairie wetlands, in *Northern Prairie Wetlands* (ed. A. G. van der Valk), Iowa State Univ. Press, Ames, pp. 268–301.

Frenzen, P. M., Krasny, M. E. and Rigney, L. P. (1988) Thirty-three years of plant succession on the Kautz Creek mudflow, Mount Rainier National Park, Washington. *Canad. J. Bot.*, **66**, 130–7.

Gibson, D. J. and Hulbert, L. C. (1987) Effects of fire, topography and year-to-year climatic variation on species composition in tallgrass praire. *Vegetatio*, **72**, 178–85.

Gill, D. (1973) Floristics of a plant succession sequence in the Mackenzie Delta, Northwest Territories. *Polarforschung*, **43**, 55–65.

Gill, D. S. and Marks, P. L. (1991) Tree and shrub seedling colonization of old fields in central New York. *Ecol. Monogr.*, **61**, 183–205.

Gimingham, C. H. (1988) A reappraisal of cyclical processes in *Calluna* heath. *Vegetatio*, **77**, 61–4.

Gleason, H. A. (1927) Further views of the succession concept. *Ecology*, **8**, 299–326.

Gleeson, S. K. and Tilman, D. (1990) Allocation and the transient dynamics of succession on poor soils. *Ecology*, **71**, 1144–55.

Glenn-Lewin, D. C. (1979) Natural revegetation of acid coal spoils in southeast Iowa, in *Ecology and Coal Resources Development* Vol. 2 (ed. M. K. Wali), Pergamon, New York, pp. 568–75.

Glenn-Lewin, D. C. (1980) The individualistic nature of plant community development. *Vegetatio*, **43**, 141–6.

Glenn-Lewin, D. C. (1988) On the notions of direction and convergence during succession, in *The Biogeography of the Island Region of Western Lake Erie* (ed. J. F. Downhower), The Ohio State Univ. Press, Columbus, pp. 221–30.

Glenn-Lewin, D. C., Johnson, L. A., Jurik, T. W., Akey, A., Leoschke, M. and Rosburg, T. (1990) Fire in central North American grasslands: Vegetative reproduction, seed germination, and seedling establishment, in *Fire in North American Tallgrass Prairies* (eds S. L. Collins and L. L. Wallace), Univ. of Oklahoma Press, Norman, pp. 28–45.

Glitzenstein, J. S., Harcombe, P. A. and Streng, D. R. (1986) Disturbance, succession, and maintenance of species diversity in an east Texas forest. *Ecol. Monogr.*, **56**, 243–58.

Gloaguen, J. C. (1990) Post-burn succession on Brittany heathlands. *J. Veg. Sci.*, **1**, 147–52.

Goff, F. G. and Zedler, P. H. (1972) Derivation of species succession vectors. *Amer. Mid. Nat.*, **87**, 397–412.

Gomez-Pompa, A. and Vázquez-Yanes, C. (1974) Studies on the secondary succession of tropical lowlands: The life cycle of secondary species, in *Unifying Concepts in Ecology* (eds W. H. van Dobben and R. H. Lowe-McConnell), Pudoc, Wageningen, pp. 336–42.

Gorham, E., Vitousek, P. M. and Reiners, W. A. (1979) The regulation of chemical budgets over the course of terrestrial ecosystem succession. *Ann. Rev. Ecol. System.*, **10**, 53–84.

Grime, J. P. (1979) *Plant Strategies and Vegetation Processes*, Wiley, Chichester.

Grubb, P. J. (1977) The maintenance of species-richness in plant communities: The importance of the regeneration niche. *Biol. Rev. Camb. Phil. Soc.*, **52**, 107–45.

Grubb, P. J. (1986) The ecology of establishment, in *Ecology and Design in Landscape* (eds A. D. Bradshaw, D. A. Goode and E. H. P. Thorp), Blackwell, Oxford, pp. 83–97.

Grubb, P. J. (1988) The uncoupling of disturbance and recruitment, two kinds of seed banks, and persistence of plant populations at the regional and local scales. *Ann. Zool. Fennici*, **25**, 23–36.

Guariguata, M. R. (1990) Landslide disturbance and forest regeneration in the Upper Luquilla Mountains of Puerto Rico. *J. Ecol.*, **78**, 814–32.

Guevara, S., Purata, S. E. and van der Maarel, E. (1986) The role of remnant forest trees in tropical secondary succession. *Vegetatio*, **66**, 77–84.

Halpern, C. B., Frenzen, P. M., Means, J. E. and Franklin, J. F. (1990) Plant succession in areas of scorched and blown-down forest after the 1980 eruption of Mount St. Helens, Washington. *J. Veg. Sci.*, **1**, 181–94.

Halpern, C. B. and Harmon, M. E. (1983) Early plant succession on the Muddy River mudflow, Mount St. Helens, Washington. *Amer. Mid. Nat.*, **110**, 97–106.

Harper, J. L. (1977) *Population Biology of Plants*, Academic Press, London.

Harris, L. G., Ebeling, A. W., Laur, D. R. and Rowley, R. J. (1984) Community recovery after storm damage: A case of facilitation in primary succession. *Sci.*, **224**, 1336–8.

Hastings, J. R. and Turner, R. M. (1965) *The Changing Mile*, Univ. of Arizona Press, Tucson.

Heinselman, M. L. (1973) Fire in the virgin forest of the Boundary Waters Canoe Area, Minnesota. *Quat. Res.*, **3**, 329–82.

Helsper, H. P. G., Glenn-Lewin, D. C. and Werger, M. J. A. (1983) Early regeneration of *Calluna* heathland under various fertilization regimes. *Oecologia*, **58**, 208–14.

Henry, J. D. and Swan, J. M. A. (1974) Reconstructing forest history from live and dead plant material – an approach to the study of forest succession in southwest New Hampshire. *Ecology*, **55**, 771–83.

Herben, T., Krahulec, F., Kovárová, M. and Hadincová, V. (1990) Fine scale dynamics in a mountain grassland, in *Spatial Processes in Plant Communities* (eds F. Krahulec, A. D. Q. Agnew, S. Agnew and J. H. Willems), Academia Prague, and SPB Acad. Publ., The Hague, pp. 173–84.

Hils, M. H. and Vankat, J. L. (1982) Species removals from a first year old-field plant community. *Ecology*, **63**, 705–11.

Hobbs, R. J. and Gimingham, C. H. (1984) Studies on fire in Scottish heathland communities. II. Post-fire vegetation development. *J. Ecol.*, **72**, 585–610.

Horn, H. H. (1981) Succession, in *Theoretical Ecology: Principles and Applications* (ed. R. M. May), Blackwell, Oxford, pp. 253–71.

Hubbell, S. P. and Foster, R. B. (1986) Biology, chance, and history and the structure of tropical rain forest tree communities, in *Community Ecology* (eds J. Diamond and T. J. Case), Harper and Row, New York, pp. 314–29.

Hulbert, L. C. (1973) Management of Konza Prairie to approximate pre-whiteman fire influences, in *Third Midwest Prairie Conf. Proc.* (ed. L. C. Hubert), Kansas State Univ., Manhattan, pp. 14–9.

Huntley, B. (1990) European post-glacial forests: Compositional changes in response to climate change. *J. Veg. Sci.* **1**, 507–18.

Huntly, N. and Inouye, R. (1988) Pocket gophers in ecosystems: Patterns and mechanisms. *BioScience*, **38**, 786–93.

Inouye, R. S. and Tilman, D. (1988) Convergence and divergence of old-field plant communities along experimental nitrogen gradients. *Ecology*, **69**, 995–1004.

Iversen, J. (1964) Retrogressive vegetational succession in the post-glacial. *J. Ecol.*, **52** (Suppl.), 59–70.

Jackson, S. T., Futyma, R. P. and Wilcox, D. A. (1988) A paleoecological test of a classical hydrosere in the Lake Michigan dunes. *Ecology*, **69**, 928–36.

Jacobson, G. L., Jr (1979) The paleoecology of white pine (*Pinus strobus*) in Minnesota. *J. Ecol.*, **67**, 697–726.

Jasieniuk, M. A. and Johnson, E. A. (1982) Peatland vegetation organization and dynamics in the western subartic, Northwest Territories, Canada. *Canad. J. Bot.*, **60**, 2581–93.

Jeffers, J. N. R. (1982) *Modelling*, Chapman and Hall, London.

Johnson, E. A. (1979) Fire recurrence in the subarctic and its implications for vegetation composition. *Canad. J. Bot.*, **57**, 1374–9.

Johnson, E. A. (1981) Vegetation organization and dynamics of lichen woodland communities in the western subarctic, Northwest Territories, Canada. *Canad. J. Bot.*, **60**, 2581–93.

Johnson, E. A. and Fryer, G. I. (1989) Population dynamics in lodgepole pine – Engelmann spruce forests. *Ecology*, **70**, 1335–45.

Johnson, W. C., Sharpe, D. M., DeAngelis, D. L., Fields, D. E. and Olson, R. J. (1981) Modeling seed dispersal and forest island dynamics, in *Forest Island Dynamics in Man-Dominated Landscapes* (eds R. L. Burgess and D. M. Sharpe), Springer-Verlag, New York, pp. 215–39.

Kohyama, T. (1984) Regeneration and coexistence of two *Abies* species dominating subalpine forests in central Japan. *Oecologia*, **62**, 156–61.

Kucera, C. L. and Koelling, M. (1964) The influence of fire on composition of central Missouri prairie. *Amer. Mid. Nat.*, **72**, 142–7.

Larson, L. and Murdock, G. K. (1989) Small bison herd utilization of tallgrass prairie, in *Proc. 11th North American Prairie Conf.* (eds T. B. Bragg and J. Stubbendieck), Univ. of Nebraska, Lincoln, pp. 243–5.

Lawesson, J. E. (1988) The stand-level dieback and regeneration of forests in the Galapogos Islands. *Vegetatio*, **77**, 87–93.

Lawton, J. H. (1987) Are there assembly rules for successional communities?, in *Colonization, Succession and Stability* (eds A. J. Gray, M. J. Crawley and P. J. Edwards), Blackwell, Oxford, pp. 225–44.

Levins, R. (1966) The strategy of model building in population biology. *Amer. Scient.*, **54**, 421–31.

Liljelund, L.-E., Ågren, G. I. and Fagerström, T. (1988) Succession in stationary environments generated by interspecific differences in life-history parameters. *Ann. Zool. Fennici*, **25**, 17–22.

Lippe, E., de Smidt, J. T. and Glenn-Lewin, D. C. (1985) Markov models and succession: A test from a heathland in The Netherlands. *J. Ecol.*, **73**, 775–91.

Liu, Q. and Hytteborn, H. (1991) Gap structure, disturbance and regeneration in a primeval *Picea abies* forest. *J. Veg. Sci.*, **2**, 391–402.

Londo, G. (1974) Successive mapping of dune slack vegetation. *Vegetatio*, **29**, 51–61.

Lorimer, C. G. (1977) The presettlement forest and natural disturbance cycle of northeastern Maine. *Ecology*, **58**, 139–48.

Lorimer, C. G. (1985) Methodological considerations in the analysis of forest disturbance history. *Canad. J. For. Res.*, **15**, 200–13.

Loucks, O. L., Plumb-Mentjes, M. L. and Rogers, D. (1985) Gap processes and large-scale disturbances in sand prairies, in *The Ecology of Natural Disturbance and Patch Dynamics* (eds S. T. A. Pickett and P. White), Academic Press, Orlando, pp. 71–83.

Luken, J. O. (1990) *Directing Ecological Succession*, Chapman and Hall, London.

Mack, R. N. (1981) Invasion of *Bromus tectorum* L. into western North America: An ecological chronicle. *Agro-Ecosystems*, **7**, 145–65.

Macphail, M. K. (1980) Regeneration processes in Tasmanian forests. *Search*, **11**, 184–90.

Major, J. (1974) Kinds and rates of changes in vegetation and chronofunctions, in

Vegetation Dynamics (ed. R. Knapp), Junk, The Hague, pp. 7–18.

Malanson, G. P. (1984) Intensity as a third factor of disturbance regime and its effect on species diversity. *Oikos*, **43**, 411–13.

Marks, P. L. (1974) The role of pin cherry (*Prunus pensylvanica* L.) in the maintenance of stability in northern hardwood ecosystems. *Ecol. Monogr.*, **44**, 73–89.

Martinez-Romos, M., Alvarez-Buylla, E. and Sarukhan, J. (1989) Tree demography and gap dynamics in a tropical rain forest. *Ecology*, **70**, 555–8.

Matthews, J. A. (1979) A study of the variability of some successional and climax plant assemblage-types using multiple discriminant analysis. *J. Ecol.*, **67**, 255–71.

McCune, B. and Allen, T. F. H. (1985) Will similar forests develop on similar sites? *Canad. J. Bot.*, **63**, 367–76.

McDonnell, M. J. (1988) Landscapes, birds, and plants: Dispersal patterns and vegetation change, in *The Biogeography of the Island Region of Western Lake Erie* (ed. J. F. Downhower), The Ohio State Univ. Press, Columbus, pp. 214–220.

McDonnell, M. J. and Stiles, E. W. (1983) The structural complexity of old-field vegetation and the recruitment of bird-dispersed plant species. *Oecologia*, **56**, 109–16.

McIntosh, R. P. (1981) Succession and ecological theory, in *Forest Succession: Concepts and Application* (eds D. C. West, H. H. Shugart and D. B. Botkin), Springer-Verlag, New York, pp. 10–23.

McNaughton, S. J. (1976) Serengeti migratory wildebeest: facilitation of energy flow by grazing. *Science*, **191**, 691–703.

McNaughton, S. J. (1979) Grassland–herbivore dynamics, in *Serengeti: Dynamics of an Ecosystem* (eds A. R. E. Sinclair and M. Norton-Griffiths), Univ. of Chicago Press, Chicago, pp. 46–81.

McNaughton, S. J., Ruess, R. W. and Seagle S. W. (1988) Large mammals and process dynamics in African ecosystems. *BioScience*, **38**, 794–800.

Miles, J. (1979) *Vegetation Dynamics*, Chapman and Hall, London.

Miles, J. (1985) The pedogenic effects of different species and vegetation types and the implications of succession. *J. Soil Sci.*, **36**, 571–84.

Miles, J. (1987) Vegetation succession: Past and present perceptions, in *Colonization, Succession and Stability* (eds A. J. Gray, M. J. Crawley and P. J. Edwards), Blackwell, Oxford, pp. 1–29.

Miller, T. E. (1982) Community diversity and interactions between the size and frequency of disturbance. *Amer. Nat.*, **120**, 533–6.

Mitchell, F. J. G. (1990) The impact of grazing and human disturbance on the dynamics of woodland in S. W. Ireland. *J. Veg. Sci.*, **1**, 245–254.

Mooney, H. A. and Godron, M. (eds) (1983) *Disturbance and Ecosystems*, Springer-Verlag, Berlin.

Moravec, J. (1990) Regeneration of N. W. African *Pinus halepensis* forests following fire. *Vegetatio*, **87**, 29–36.

Morgan, R. K. (1991) The role of protective understorey in the regeneration system of a heavily browsed woodland. *Vegetatio*, **92**, 119–32.

Morrison, P. H. and Swanson, F. J. (1990) *Fire History and Pattern in a Cascade Range Landscape*, US Forest Service General Technical Report PNW-GTR-254.

Mueller-Dombois, D. (1986) Perspectives for an etiology of stand-level dieback.

Ann. Rev. Ecol. System., **17**, 221–43.

Mueller-Dombois, D. (1987) Natural dieback in forest. *BioScience*, **37**, 575–83.

Myster, R. W. and Pickett, S. T. A. (1988) Individualistic patterns of annuals and biennials in early successional oldfields. *Vegetatio*, **78**, 53–60.

Myster, R. W. and Pickett, S. T. A. (1990) Initial conditions, history and successional pathways in ten contrasting old fields. *Amer. Mid. Nat.*, **124**, 231–8.

Naiman, R. J. (1988) Animal influences on ecosystem dynamics. *BioScience*, **38**, 750–2.

Noble, I. R. and Slatyer, R. O. (1980) The use of vital attributes to predict successional changes in plant communities subject to recurrent disturbances. *Vegetatio*, **43**, 5–21.

Odum, E. P. (1969) The strategy of ecosystem development. *Science*, **164**, 262–70.

Oliver, C. D. and Stephens, E. P. (1977) Reconstruction of a mixed-species forest in central New England. *Ecology*, **58**, 562–72.

Olson, J. S. (1958) Rates of succession and soil changes on southern Lake Michigan sand dunes. *Bot. Gazette*, **119**, 125–70.

Oosting, H. J. (1942) An ecological analysis of the plant communities of Piedmont, North Carolina. *Amer. Mid. Nat.*, **28**, 1–126.

Paijmans, K. (1973) Plant succession on Pago and Witori Volcanoes, New Britain. *Pacific Sci.*, **27**, 260–8.

Parks, J. M. and Rice, E. L. (1969) Effects of certain plants of old-field succession on the growth of blue-green algae. *Bull. Torrey Bot. Club*, **96**, 345–60.

Peet, R. K. and Christensen, N. L. (1980) Succession: A population process. *Vegetatio*, **43**, 131–40.

Peet, R. K. and Christensen, N. L. (1987) Competition and tree death. *Bio-Science*, **37**, 586–94.

Petranka, J. W. and MacPherson, J. K. (1979) The role of *Rhus copallina* in the dynamics of the forest-prairie ecotone in north-central Oklahoma. *Ecology*, **60**, 956–65.

Pickett, S. T. A. (1976) Succession: An evolutionary interpretation. *Amer. Nat.*, **110**, 107–19.

Pickett, S. T. A. (1980) Non-equilibrium coexistence of plants. *Bull. Torrey Bot. Club*, **107**, 238–48.

Pickett, S. T. A. (1982) Population patterns through twenty years of old field succession. *Vegetatio*, **49**, 45–59.

Pickett, S. T. A. (1983) The absence of an *Andropogon* stage in oldfield succession at the Hutcheson Memorial Forest. *Bull. Torrey Bot. Club*, **110**, 533–5.

Pickett, S. T. A. (1989) Space-for-time substitution as an alternative to long-term studies, in *Long-term Studies in Ecology* (ed. G. E. Likens), Springer-Verlag, New York, pp. 110–135.

Pickett, S. T. A. and White, P. S. (eds) (1985) *The Ecology of Natural Disturbance and Patch Dynamics*, Academic Press, Orlando.

Pickett, S. T. A., Collins, S. L. and Armesto, J. J. (1987a) Models, mechanisms and pathways of succession. *Bot. Rev.*, **53**, 335–71.

Pickett, S. T. A., Collins, S. L. and Armesto, J. J. (1987b) A hierarchical consideration of causes and mechanisms of succession. *Vegetatio*, **69**, 109–14.

Pidgeon, I. M. (1940) The ecology of the central coastal area of New South Wales. III. Types of primary succession. *Proc. Linn. Soc. NSW*, **65**, 221–49.

Pielou, E. C. (1981) The usefulness of ecological models: A stock-taking. *Quart. Rev. Biol.*, **56**, 17–31.

Pineda, F. D., Nicolas, J. P. Pou, A. and Galiano E. F. (1981) Ecological succession in oligotrophic pastures in central Spain. *Vegetatio*, **44**, 165–76.

Platt, W. J. (1975) The colonization and formation of equilibrium plant species associations on badger disturbances in a tall-grass prairie. *Ecol. Monogr.*, **45**, 285–305.

Prentice, H. C. and Prentice, I. C. (1983) Plant communities of Holt Heath. *Proc. Dorset Nat. Hist. Archeol. Soc.*, **105**, 127–36.

Prentice, I. C. (1986a) Some concepts and objectives of forest dynamics research, in *Forest Dynamics Research in Western and Central Europe* (ed. J. Fanta), Pudoc, Wageningen, pp. 32–41.

Prentice, I. C. (1986b) Vegetation responses to past climatic variation. *Vegetatio*, **67**, 131–41.

Prentice, I. C., van Tongeren, O. and de Smidt, J. T. (1987) Simulation of heathland vegetation dynamics. *J. Ecol.*, **75**, 203–19.

Rabotnov, T. A. (1974) Differences between fluctuations and succession. Examples in grassland phytocenoses of the U.S.S.R., in *Vegetation Dynamics*, (ed. R. Knapp), Junk, The Hague, pp. 19–24.

Rankin, W. T. and Pickett, S. T. A. (1989) Time of establishment of red maple (*Acer rubrum*) in early oldfield succession. *Bull. Torrey Bot. Club*, **116**, 182–6.

Raup, H. M. (1957) Vegetational adjustment to the instability of the site, in *Proc. Paper of the 6th Tech. Meeting, Internat. Union for the Conserv. of Nature and Natural Resources* (Edinburgh), pp. 36–48.

Rejmánek, M., Haagerova, R. and Haager, J. (1982) Progress of plant succession on the Parícutin Volcano: 25 years after activity ceased. *Amer. Mid. Nat.*, **108**, 194–8.

Rejmánek, M., Sasser, C. E. and Gosselink, J. G. (1987) Modeling of vegetation dynamics in the Mississippi River deltaic plain. *Vegetatio*, **69**, 133–40.

Rice, E. L. (1964) Inhibition of nitrogen-fixing and nitrifying bacteria by seed plants. *Ecology*, **45**, 824–37.

Rice, E. L. (1968) Inhibition of nodulation of inoculated legumes by pioneer plant species from abandoned fields. *Bull. Torrey Bot. Club*, **95**, 346–58.

Rohn, S. R. and Bragg, T. R. (1989) Effect of burning on germination of tallgrass prairie plant species, in *Proc. 11th North American Prairie Conf.*, (eds T. B. Bragg and J. Stubbendieck), Univ. of Nebraska, Lincoln, pp. 169–171.

Romme, W. H. (1982) Fire and landscape diversity in subalpine forests of Yellowstone National Park. *Ecol. Monogr.*, **52**, 199–221.

Roozen, A. J. M. and Westhoff, V. (1985) A study on long-term salt marsh succession using permanent plots. *Vegetatio*, **61**, 23–32.

Rundel, P. W. (1975) Primary succession on granite outcrops in the montane southern Sierra Nevada. *Madroño*, **23**, 209–20.

Runkle, J. R. (1989) Synchrony of regeneration, gaps and latitudinal differences in tree species diversity. *Ecology*, **70**, 546–47.

Rykiel, E. J., Jr (1985) Towards a definition of ecological disturbance. *Austral. J. Ecol.*, **10**, 361–5.

Ryser, P. (1990) *Influence of Gaps and Neighbouring Plants on Seedling Estab-lishment in Limestone Grassland. Experimental Field Studies in Northern Switzerland.* Veröffentlichungen des Geobot. Inst. der ETH, Stiftung Rübel, Zürich, Vol. 104.

Savage, M. and Swetnam, T. W. (1990) Early 19th-Century fire decline following sheep pasturing in a Navajo Ponderosa Pine forest. *Ecology*, **71**, 2374–8.

Schmidt, W. (1988) An experimental study of old-field succession in relation to different environmental factors. *Vegetatio*, **77**, 103–14.

Sernander, R. (1936) Granskär och Fiby urskog (Swedish, with English summary: The primitive forests of Granskär and Fiby: A study of the part played by storm-gaps and dwarf trees in the regeneration of the Swedish spruce forest). *Acta Phytogeographica Suecica* **8**.

Sharik, T. L., Ford, R. H. and Davis, M. L. (1989) Repeatability of invasion of eastern White Pine on dry sites in northern Lower Michigan. *Amer. Mid. Nat.*, **122**, 133–41.

Shmida, A. and Ellner, A. (1984) Coexistence of plant species with similar niches. *Vegetatio*, **58**, 29–55.

Shmida, A. and Whittaker, R. H. (1981) Pattern and biological microsite effects in two shrub communities, southern California. *Ecology*, **62**, 234–51.

Shure, D. J. and Ragsdale, H. L. (1977) Patterns of primary succession on granite outcrop surfaces. *Ecology*, **58**, 993–1006.

Sjörs, H. (1980) An arrangement of changes along gradients, with examples from successions in boreal peatland. *Vegetatio*, **43**, 1–4.

Smith, L. M. and Kadlec, J. A. (1985) Fire and herbivory in a Great Salt Lake marsh. *Ecology*, **66**, 259–65.

Smith, T. J. and Odum, W. E. (1981) The effects of grazing by snow geese on coastal salt marshes. *Ecology*, **61**, 98–106.

Solomon, D. S. and Blum, B. M. (1967) Stump sprouting of four northern hardwoods. *US Forest Service Research Paper NE-59*.

Sousa, W. P. (1984) The role of disturbance in natural communities. *Ann. Rev. Ecol. System.*, **15**, 353–91.

Spies, T. A. and Franklin, J. F. (1989) Gap characteristics and vegetation response in coniferous forests of the Pacific Northwest. *Ecology*, **70**, 543–5.

Sprugel, D. G. (1976) Dynamic structure of wave-generated *Abies balsamea* forests in the northeastern United States. *J. Ecol.*, **64**, 889–911.

Spurr, S. H. and Barnes, B. V. (1980) *Forest Ecology*, 3rd edn, Wiley, New York.

Steenbergh, W. H. and Lowe, C. H. (1977) Ecology of the saguaro. II. Repro-duction, germination, establishment, growth and survival of the young plant. *US National Park Service Sci. Monograph* **8**.

Stein, S. J. (1988) Fire history of the Paunsaugunt Plateau in southern Utah. *Great Basin Nat.*, **48**, 58–63.

Sterling, A., Peco, B., Casado, M. A., Galiano, E. F. and Pineda, F. D. (1984) Influence of microtopography on floristic variation in the ecological succession in grassland. *Oikos*, **42**, 334–42.

Swaine, M. D. and Greig-Smith, P. (1980) An application of principal com-ponents analysis to vegetation change in permanent plots. *J. Ecol.*, **68**, 33–41.

Szwagrzyk, J. (1990) Natural regeneration of forest related to the spatial structure of trees: A study of two forest communities in Western Carpathians, southern

Poland. *Vegetatio*, **89**, 11–22.

Tagawa, H., Suzuki, E. and Partomihardjoo, T. (1985) Vegetation and succession on the Krakatau islands, Indonesia. *Vegetatio*, **60**, 131–45.

Tansley, A. G. (1935) The use and abuse of vegetational concepts and terms. *Ecology*, **16**, 284–307.

Thalen, D. C. P., Poorter, H., Lotz, L. A. P. and Oosterveld, P. (1987) Modelling the structural changes in vegetation under different grazing regimes, in *Disturbance in Grasslands* (eds J. van Andel, J. P. Bakker and R. W. Snaydon), Geobotany 10, Junk, Dordrecht, pp. 167–183.

Thatcher, A. C. and Westman, W. E. (1975) Succession following mining on high dunes of coastal southeast Queensland. *Proc. Ecol. Soc. Austral.*, **9**, 17–33.

Tilman, D. (1985) The resource ratio hypothesis of succession. *Amer. Nat.*, **125**, 827–52.

Tilman, D. (1987) Secondary succession and the pattern of plant dominance along experimental nitrogen gradients. *Ecol. Monogr.*, **57**, 189–214.

Tilman, D. (1988) *Plant Strategies and the Dynamics and Structure of Plant Communities*, Princeton Univ. Press, Princeton, N. J.

Tilman, D. and Wedin, D. (1991a) Dynamics of nitrogen competition between successional grasses. *Ecology*, **72**, 1038–49.

Tilman, D. and Wedin, D. (1991b) Plant traits and resource reduction for five grasses growing on a nitrogen gradient. *Ecology*, **72**, 683–98.

Titlyanova, A. A. and Mironycheva-Tokareva, N. P. (1990) Vegetation succession and biological turnover on coal-mining spoils. *J. Veg. Sci.*, **1**, 643–52.

Trabaud, L. (1974) Experimental studies of the effects of prescribed burning on a *Quercus coccifera* L. garrigue. *Proc. Ann. Tall Timbers Fire Ecol. Conf.*, **13**, 97–129.

Trabaud, L. (1987) Fire and survival traits of plants, in *The Role of Fire in Ecological Systems*, (ed. L. Trabaud), SPB Acad. Publ., The Hague, pp. 65–89.

Turner, M. G. and Bratton, S. P. (1987) Fire, grazing, and the landscape heterogeneity of a Georgia barrier island, in *Landscape Heterogeneity and Disturbance*, (ed. M. G. Turner), Springer-Verlag, New York, pp. 85–101.

Usher, M. B. (1981) Modelling ecological succession, with particular reference to Markovian models. *Vegetatio*, **46**, 11–18.

Vale, T. R. (1982) Plants and people. Vegetation change in North America. *Resource Publs. in Geography*, Assoc. of American Geographers, Washington, D. C.

Valiente-Banuet, A., Bolongaro-Crevenna, A., Briones, O., Ezcurra, E., Rosas, M., Nuñez, H., Barnard, G. and Vazques, E. (1991) Spatial relationships between cacti and nurse shrubs in a semi-arid environment in central Mexico. *J. Veg. Sci.*, **2**, 15–20.

van Andel, J. and van den Bergh, J. P. (1987) Disturbance of grasslands. Outline of the theme, in *Disturbance in Grasslands. Causes, Effects and Processes* (eds J. van Andel, J. P. Bakker and R. W. Snaydon), Geobotany **10**, Junk, Dordrecht, pp. 3–13.

van den Bergh, J. P. (1979) Changes in the composition of mixed populations of grassland species, in *The Study of Vegetation* (ed. M. J. A. Werger), Junk, The Hague, pp. 57–80.

van der Maarel, E. (1969) On the use of ordination methods in phytosociology. *Vegetatio*, **19**, 21–46.

van der Maarel, E. (1978) Experimental succession research in a coastal dune grassland, a preliminary report. *Vegetatio*, **38**, 21–8.

van der Maarel, E. (1981) Fluctuations in a coastal dune grassland due to fluctuations in rainfall: Experimental evidence. *Vegetatio*, **47**, 259–65.

van der Maarel, E. (1988) Vegetation dynamics: Patterns in time and space. *Vegetatio*, **77**, 7–19.

van der Maarel, E., Boot, R., van Dorp, D. and Rijntjes, J. (1985) Vegetation succession on the dunes near Oostvoorne, The Netherlands: A comparison of the vegetation in 1959 and 1980. *Vegetatio*, **58**, 137–87.

van der Maarel, E. and Werger, M. J. A. (1978) On the treatment of succession data. *Phytocenosis*, **7**, 257–77.

van der Maarel, E. and Westhoff, V. (1964) The vegetation of the dunes near Oostvoorne (The Netherlands). *Wentia*, **12**, 1–61.

van der Valk, A. G. (1981) Succession in wetlands: A Gleasonian approach. *Ecology*, **62**, 688–96.

van der Valk, A. G. (1985) Vegetation dynamics of prairie glacial marshes, in *The Population Structure of Vegetation* (ed. J. White), Junk, Dordrecht, pp. 293–312.

van Dorp, D., Boot, R. and van der Maarel, E. (1985) Vegetation succession on the dunes near Oostvoorne, the Netherlands, since 1934, interpreted from air photographs and vegetation maps. *Vegetatio*, **58**, 123–36.

van Duuren, L., Bakker, J. P. and Fresco, L. F. M. (1981) From intensively agricultural practices to hay-making without fertilization. Effects on moist grassland communities. *Vegetatio*, **47**, 241–58.

van Noordwijk-Puyk, K., Beeftink, W. G. and Hogeweg, P. (1979) Vegetation development on salt-marsh flats after disappearance of the tidal factor. *Vegetatio*, **39**, 1–13.

van Tongeren, O. and Prentice, I. C. (1986) A spatial simulation model for vegetation dynamics. *Vegetatio*, **65**, 163–73.

Veblen, T. T. and Ashton, D. H. (1978) Catastrophic influences on the vegetation of the Valdivian Andes. *Vegetatio*, **36**, 149–67.

Veblen, T. T. and Lorenz, D. C. (1988) Recent vegetation changes along the forest/steppe ecotone of northern Patagonia. *Ann. Assoc. Amer. Geographers*, **78**, 93–111.

Veblen, T. T. and Markgraf, V. (1988) Steppe expansion in Patagonia. *Quat. Res.*, **30**, 331–8.

Veblen, T. T., Hadley, K. S., Reid, M. S. and Rebertus, A. J. (1989) Blowdown and stand development in a Colorado subalpine forest. *Canad. J. For. Res*, **19**, 1218–25.

Ver Hoef, J. M. (1991) Statistical analysis of spatial pattern in ecological data. PhD dissertation, Iowa State University, Ames.

Vitousek, P. M. and Walker, L. R. (1987) Colonization, succession and resource availability: Ecosystem-level interactions, in *Colonization, Succession and Stability* (eds A. J. Gray, M. J. Crawley and P. J. Edwards), Blackwell, Oxford, pp. 207–223.

Wagner, W. L., Martin, W. C. and Aldon, E. F. (1978) Natural succession on strip-mined lands in northwestern New Mexico. *Reclamation Rev.*, **1**, 67–73.

Waldemarson Jensén, E. (1979) Successions in relationship to lagoon development in the Laitaure delta, North Sweden. *Acta Phytogeographica Suecica*, **66**.

Walker, D. (1970) Direction and rate in some British post-glacial hydroseres, in *Studies in the Vegetational History of the British Isles* (eds D. Walker and R. G. West), Cambridge Univ. Press, Cambridge, pp. 117–139.

Walker, J., Thompson, C. H., Fergus, I. F. and Tunstall, B. R. (1981) Plant succession and soil development in coastal sand dunes of subtropical eastern Australia, in *Forest Succession: Concepts and Applications* (eds D. C. West, H. H. Shugart and D. B. Botkin), Springer, Berlin, pp. 107–131.

Walker, L. R. and Chapin, F. S. III. (1987) Interactions among processes controlling successional change. *Oikos*, **50**, 131–5.

Walker, L. R., Zasada, J. C. and Chapin, F. S. III. (1986) The role of life history processes in primary succession on an Alaskan floodplain. *Ecology*, **67**, 1243–53.

Watt, A.S. (1947) Pattern and process in the plant community. *J. Ecol.*, **35**, 1–22.

Watt, A. S. (1955) Bracken versus heather, a study in plant sociology. *J. Ecol.*, **43**, 490–506.

Watt, A.S. (1971) Factors controlling the floristic composition of some plant communities in Breckland, in *The Scientific Management of Animal and Plant Communities for Conservation* (eds E. Duffey and A. S. Watt), Blackwell, Oxford, pp. 137–52.

Weaver, J. E. (1954) *North American Prairie*, Johnson Publ. Co., Lincoln.

Webb, T., III (1981) The past 11000 years of vegetational change in eastern North America. *BioScience*, **31**, 501–6.

Webb, T., III (1987) The appearances and disappearances of major vegetation assemblages: Long-term vegetation dynamics in eastern North America. *Vegetatio*, **69**, 177–87.

Webb, L. J., Tracey, J. G. and Williams, W. T. (1972) Regeneration and pattern in the subtropical rain forest. *J. Ecol.*, **60**, 675–95.

Wells, P. V. and Jorgensen, C. D. (1964) Pleistocene wood rat middens and climatic change in Mohave Desert: A record of juniper woodlands. *Science*, **143**, 1171–4.

Werner, P. A. and Harbeck, A. C. (1982) The pattern of tree seedling establishment relative to staghorn sumac cover in Michigan old fields. *Amer. Mid. Nat.*, **108**, 124–32.

West, R. G. (1964) Inter-relations of ecology and Quaternary paleobotany. *J. Ecol.*, **52** (Suppl.), 47–57.

Whicker, A. D. and Detling, J. K. (1988) Ecological consequences of prairie dog disturbances. *BioScience*, **38**, 778–85.

White, P. S. (1979) Pattern, process and natural disturbance in vegetation. *Bot. Rev.*, **45**, 229–99.

White, P. S. and Pickett, S. T. A. (1985) Natural disturbance and patch dynamics: An introduction, in *The Ecology of Natural Disturbance and Patch Dynamics* (eds S. T. A. Pickett and P. S. White), Academic Press, Orlando, pp. 3–13.

Whitmore, T. C. (1990) *An Introduction to Tropical Rain Forests*, Oxford Univ. Press, New York.

Whittaker, R. H. (1975) *Communities and Ecosystems*, 2nd edn, Macmillan, New York.

Whittaker, R. J., Bush, M. B. and Richards, K. (1989) Plant recolonization and vegetation succession on the Krakatau Islands, Indonesia. *Ecol. Monogr.*, **59**, 59–123.

Wiens, J. A. (1984) On understanding a non-equilibrium world: Myth and reality in community patterns and processes, in *Ecological Communities: Conceptual Issues and the Evidence*, (eds D. R. Strong, D. Simberloff, L. G. Abele and A. B. Thistle), Princeton Univ. Press, Princeton, pp. 439–57.

Wildi, O. (1988) Linear trend in multi-species time series. *Vegetatio*, **77**, 51–6.

Williams, O. B. (1970) Population dynamics of two perennial grasses in Australian semiarid grassland. *J. Ecol.*, **58**, 869–75.

Wood, D. M. and del Moral, R. (1987) Mechanisms of early primary succession in subalpine habitats on Mount St. Helens. *Ecology*, **68**, 780–90.

Woodwell, G. M. (1967) Radiation and the patterns of nature. *Science*, **156**, 461–70.

Woodwell, G. M. (1970) Effects of pollution on the structure and physiology of ecosystems. *Science*, **168**, 429–33.

Woodwell, G. M. and Whittaker, R. H. (1968) Effects of chronic gamma irradiation on plant communities. *Quart. Rev. Biol.*, **43**, 42–55.

Yarranton, G. A. and Morrison, R. G. (1974) Spatial dynamics of a primary succession: Nucleation. *J. Ecol.*, **62**, 417–28.

Yeaton, R. I. (1978) A cyclical relationship between *Larrea tridentata* and *Opuntia leptocaulis* in the northern Chihuahuan desert. *J. Ecol.*, **66**, 651–6.

2 Establishment, colonization and persistence

Arnold G. van der Valk

2.1 INTRODUCTION

God is in the details.

(Mies van der Rohe)

What kinds of plant communities occur in an area? How does the composition of a plant community change over time? These were the kinds of questions posed by pioneering plant ecologists. As explanations began to accumulate, it became increasingly clear that the distribution and dynamics of vegetation were complex and intricate phenomena. What causes these diverse spatial and temporal patterns? Answering this question is one of the primary preoccupations of plant ecologists today. One approach is to examine the proximate causes of the distribution of populations of species in space and time. This approach assumes that all macro or community-level features represent a collection or summation of micro- or population-level phenomena. In other words, community-level changes (succession, fluctuation, maturation) are ultimately the result of populations becoming established and spreading, the growth of individuals in those populations, and populations becoming extirpated (van der Valk, 1985). The first of these three processes, the establishment and spread of populations, is the subject of this chapter. Establishment of species from seed or from vegetative propagules is examined first. Then, effects of establishment of species on colonization of an area and the persistence of species in an area are considered. The major theme of this chapter is that complex macro patterns result in a large part from a limited set of micro phenomena associated with species establishment (i.e. seed dispersal, seed germination and clonal growth patterns).

2.2 ESTABLISHMENT

A site may be colonized by a species through recruitment from seeds or vegetative propagules. These are fundamentally different processes.

Plant Succession: Theory and prediction Edited by David C. Glenn-Lewin, Robert K. Peet and Thomas T. Veblen © 1992 Chapman & Hall, London ISBN 0 412 26900 7

Because seed dispersal is only the first step needed to establish a population from seed (i.e. seed dispersal does not guarantee seed germination), seed dispersal alone only makes a species a member of the potential flora of the site, not its actual flora (Major and Pyott, 1966). Thus, establishment from seed is a two-step process: dispersal and germination. Vegetative propagation, on the other hand, is a one-step process. It establishes a population from the moment a vegetative propagule arrives at a site as part of the actual flora.

2.2.1   Seed dispersal

The effectiveness of seed dispersal (i.e. the probability of successful recruitment from seed during some period) is dependent on five different factors: seed production, vector of seed transport, timing of seed release or removal, distance of dispersal, and fate of dispersed seeds. Unfortunately, these factors seem never to have been examined together in detail in one study.

The number of seeds produced by a plant during a given growing season or year depends on such factors as the age or size of the plant, environmental conditions during the previous and current growing season, availability of pollinators, predispersal predation, and energetic trade-offs between vegetative propagation and seed production (Janzen, 1971; Harper, 1977; Silvertown, 1980a; Stephenson, 1981; Howe and Smallwood, 1982; Willson, 1983; Price and Jenkins, 1986; Zammit and Westoby, 1988; Howe and Westley, 1988; Louda, 1989). Because these factors vary temporally and spatially both within and among species, the number of seeds of a given species available for dispersal in a given year can vary enormously. In turn, the number of seeds of a species present at any time can influence both the probability of a seed being dispersed (particularly if it is animal dispersed: Gorchov, 1988), and the probability of it being lost to predators. Some perennials are facultative seed producers, and there may be no seed available for dispersal in some years. For example, flowering in some wetland sedges is triggered by flooding during the previous growing season (Bernard, 1975).

Predispersal and postdispersal seed predation is one of the most important factors regulating the number of seeds available for recruitment of new populations (Janzen, 1971; Price and Jenkins, 1986; Louda, 1989). The number of seeds escaping predation often increases significantly during years with high seed production ('mast years') or availability (Watt, 1923; Gardner, 1977; Smith *et al.*, 1989). For example, in England more seeds of the ash, *Fraxinus excelsior*, escaped predation in a year with a good seed crop than in years with poorer seed crops, even though the rate of seed predation was highest during the year with the good seed crop (Gardner, 1977; Table 2.1). The probability of seeds escaping

Table 2.1 Annual seed production (seed m^{-2}), seeds dispersed (seeds m^{-2}), seed predation (%), seed germination (%) and seedling recruitment (seedlings m^{-2}) of *Fraxinus excelsior* in woodlands in Derbyshire, England (adapted from Gardner, 1977)

	1966	1967	1968	1969
Seed produced (m^{-2})	133	0.33	26.9	1193
Seed dispersed (m^{-2})	101	0.08	16.2	998
Seed predation (%)	70	25	49	75
Seed germination (%)	4.5	0	5.3	2.5
Seedlings (m^{-2})	6.1	0	1.6	31.6

predation depends not only on fluctuations in annual seed production, but also on fluctuations in predator density (Sork, 1987; Schupp, 1990).

Many species seem to have seeds that have more or less specialized adaptations for dispersal. Among the more common dispersal syndromes are anemochory (dispersal by wind), hydrochory (dispersal by water), epizoochory (dispersal on the outside of animals), endozoochory (dispersal as a result of ingestion by animals), barochory (gravity dispersal) and autochory (the plant itself has some mechanism for seed dispersal). Many species of plants have seeds that are polychorous; that is, they have more than one potential method of dispersal. Some species produce heteromorphic seeds, and different seed types on the same plant may vary in their dispersibility (Trapp, 1988). The various dispersal syndromes, their many variants and their putative adaptive significance are described in detail (Ridley, 1930; Howe and Smallwood, 1982; van der Pijl, 1982; Willson, 1983; Murray, 1986a; Howe, 1986; Howe and Westley, 1988; Stiles, 1989).

Seed dispersal for a given species is normally restricted to some more or less fixed period after the seeds ripen (Howe and Smallwood, 1982). These periods of dispersal may or may not correspond to periods when establishment is favourable. Some species disperse their seeds over very brief periods, particularly those species with wind-dispersed seeds (e.g. species of *Salix* and *Populus*: McCleod and McPherson, 1973; Walker *et al.*, 1986). Other species may shed seeds over very long periods; the duration of seed dispersal varied from one week to six months in a single old field in New York (Morris *et al.*, 1986). Along the Tanana River in Alaska (Fig. 2.1), the starting date and the duration of dispersal by the dominant tree species were found to vary somewhat from year to year (Walker *et al.*, 1986).

Attempts to measure the distance that seeds are dispersed from the parent plant have focused on wind dispersal (e.g. Platt and Weis, 1977; Rabinowitz and Rapp, 1981; Reader and Buck, 1986) and animal dispersal (e.g. Smith, 1975; Darley-Hill and Johnson, 1981; Herrera and

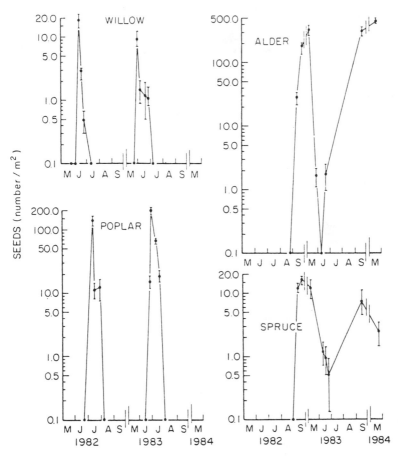

Figure 2.1 Seasonal patterns of seed dispersal of willow (*Salix alaxensis*), alder (*Alnus tenuifolia*), poplar (*Populus balsamifera*) and spruce (*Picea glauca*) from stands dominated by these species along the Tanana River in Alaska (Walker *et al.*, 1986). Note the log scale for seed density.

Jordano, 1981; McDonnell and Stiles, 1983). Experimental or observational studies of dispersal are concerned usually with primary dispersal, the movement of the seed from the parent plant to the ground. These studies reveal that the bulk of the seeds of most species are deposited only a short distance from the parent plant (Levin and Kerster, 1974). Some typical frequency distribution curves of dispersal distances from the parent plant of a group of herbaceous species from a North American prairie are shown in Fig 2.2. Seeds that are dispersed by birds, however, may be carried quite far, with few seeds deposited near parent plants (Murray, 1988). The structure of vegetation (e.g. gaps in forests) and behaviour of birds (e.g. territoriality, caching) also influence bird seed dispersal patterns (Hoppes, 1988; Johnson and Webb, 1989).

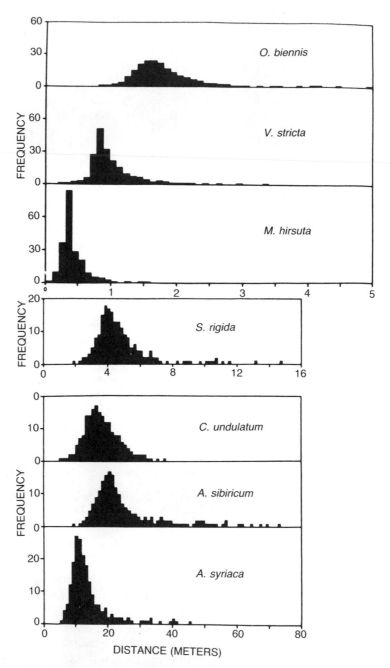

Figure 2.2 Frequency distributions of distances travelled by propagules of seven prairie species in an Iowa prairie during periods of high wind velocities (10–$15\,\mathrm{km\,h}^{-1}$). The upper three species (*Oenothera biennis*, *Verbena stricta* and *Mirabilis hirsuta*) had non-plumed seeds, the middle species (*Solidago rigida*) had seeds with small plumes, and the three lower species (*Cirsium undulatum*, *Apocynum sibiricum*, *Asclepias syriaca*) had seeds with large plumes (Platt and Weis, 1977). Note the change in the distance scale for the three groups.

Neighbouring plants often reduce the effective distance that many seeds will be dispersed. In *Erodium* species, which explosively discharge awned fruits, fruits were thrown about 1 m by plants without neighbours, but only 0.3 m by plants with neighbours (Stamp, 1989).

Even in the simplest cases of wind dispersal, the actual area over which seeds are dispersed (i.e. the seed shadow) will vary widely depending on wind speed and direction, number of seeds produced, vegetation structure and density, position of the parent plant in the landscape (e.g. top versus bottom of the hill), whether it was raining when the seeds were released, and so on (Ryvarden, 1971; Sheldon and Burrows, 1973; Burrows, 1975, 1986; Green, 1980; Verkaar *et al.*, 1983; Reichman, 1984; Reader and Buck, 1986; Augspurger and Franson, 1987; McEvoy and Cox, 1987; Okubo and Levin, 1989; Greene and Johnson, 1989). Nevertheless, studies of potential dispersal distances can be important for understanding the distribution of species within a particular vegetation type (Platt and Weis, 1977; Augspurger, 1984; Reader and Buck, 1986). For example, *Hieracium floribundum*, a wind-dispersed composite, is found in higher densities near the edges of depressions in pastures because more of its wind-dispersed seeds reach the edges than the centres of these depressions, and because openings in the vegetation, which are required for its seeds to germinate, are more frequent at the edges of depressions (Reader and Buck, 1986).

Another way to examine dispersal in the field is to start at the sites receiving seeds rather than at the seed source. In this kind of study, seed traps are used to determine what seeds are reaching the site(s) (i.e. the local seed rain: Ryvarden, 1971; Rabinowitz and Rapp, 1980; Reichman, 1984; Morris *et al.*, 1986; Reader and Buck, 1986; Walker *et al.*, 1986; Peart, 1989a). Seed rain studies have shown that seasonal and annual seed inputs into an area vary greatly, both qualitatively and quantitatively. In a study of seed rain in an old field in New York, Morris *et al.* (1986) found significant differences between years in the number of seeds in their traps for four out of the ten species examined (Table 2.2). In the most extreme case, the annual seed rain of *Trifolium agrarium* was 5000 seeds m^{-2} one year and only 20 seeds m^{-2} the next year.

The seed rain varies not only temporally, but also spatially. Ryvarden (1971) collected seeds in seed traps over three growing seasons at the base of a retreating glacier in Norway. During these three years, seeds of 57 taxa were found in his 141 traps. The seeds of 17 of these taxa (30%) were found in only one trap, and only four taxa (7%) were found in more than one-third of the traps. Of these four, seeds of only one taxon, a species of wind-dispersed *Salix*, were found in nearly all the traps (96%). Similar variability in the spatial patterns of seed dispersal have been found in a North American prairie (Rabinowitz and Rapp, 1980).

Palaeoecologists have calculated annual rates of spread of tree species during the Holocene in North America and Europe, using ponds, lakes

Table 2.2 The mean number of seeds (m^{-2}) caught in seed traps in an old field in upstate New York (Morris *et al.*, 1986). An asterisk indicates a significant difference at the 5% confidence level (based on paired *t*-tests) in the mean number of seeds of a species caught in the traps during the two sampling periods

Taxa	1982–83	1983–84
Solidago spp.	21 200	19 900
Hieracium spp.	9 500	4 700*
Trifolium agrarium	5 000	20*
Phleum pratense	3 100	2 300
Poa spp.	1 100	770
Daucus carota	600	2 600*
Anthoxanthum odoratum	280	690
Aster spp.	360	800
Taraxacum officinale	150	350*
Chrysanthemum leucanthemum	50	190
Total[a]	41 700	32 000

[a] Includes taxa not listed individually.

Table 2.3 Estimated rates of spread (m yr^{-1}) of trees in the British Isles and continental Europe (data from Birks, 1989)

	British Isles	Europe
Alnus	50–600	500–2000
Betula	250	>2000
Corylus	500	1500
Fagus	100–200	200–300
Fraxinus	50–200	200–500
Pinus	<100–700	1500
Quercus	50–500	150–500
Tilia	50–500	300–500
Ulmus	100–550	500–1000

and wetlands as continental or regional networks of pollen traps (Davis, 1981; Huntley and Birks, 1983; Birks, 1989). By examining carbon-dated pollen diagrams from these aquatic systems and determining when a species first arrived in an area, isochrone lines or migration fronts of species can be plotted on maps and used to calculate annual rates of spread. Most trees were able to migrate between 100 and 500 m yr^{-1} in Europe, but rates of spread varied both temporally and geographically (Table 2.3). In North America, average rates of spread of tree genera in the eastern United States and Canada during the Holocene were typically between 100 and 400 m yr^{-1} (Davis, 1981).

The determination of actual dispersal patterns or seed shadows usually is complicated by secondary dispersal; that is, the movement of the seed after primary dispersal. For example, a wind-dispersed seed that lands on the water may be carried further by water currents than it was by the wind and may end up many kilometres from where it was deposited originally, particularly if the seed falls into a river or stream (Ryvarden, 1971; Reichman, 1984; Symonides, 1985; Walker *et al.*, 1986; Schneider and Sharitz, 1988). Secondary dispersal of *Betula lenta* seeds across snow by wind during the winter resulted in their distribution over an area that was 3.3 times the area covered by primary dispersal, which was also by wind (Matlack, 1989). The most spectacular examples of secondary dispersal involve viable seeds and fruits carried thousands of miles by ocean currents. Seeds and fruits are routinely found on beaches all over the world, even on isolated oceanic islands (Murray, 1986b; Hacker, 1990; Smith, 1990; Smith *et al.*, 1990). Gunn *et al.* (1976) have written a key to tropical seeds and fruits found commonly along seashores. Because of secondary dispersal, it is usually difficult to predict accurately where seeds from a parent plant will finally be deposited.

A detailed examination of the distribution of seeds in the surface soil of the Sonoran Desert near Tucson, Arizona, by Reichman (1984) illustrates the spatial variability in seed density that can occur, even at the scale of a few square metres (Fig. 2.3). Densities of seeds at this site varied 78-fold. The highest densities occurred in depressions in the soil surface because seeds tended to collect where there was a wind shadow. Secondary dispersal of seeds by wind, animals and water was largely responsible for the spatial variation in seed densities found at this site. The sizes and shapes of seeds also influenced their secondary dispersal: small or round seeds tended to move further and to accumulate in denser clumps than seeds that were large or long.

Burial of seeds by litter or soil after dispersal often improves the seeds' chances of survival and germination (Stamp, 1989; Smith *et al.*, 1989), but only if they are not buried too deeply (van der Valk, 1974; van der Valk *et al.*, 1983; Galinato and van der Valk, 1986). Seeds that do not get covered are more likely to be eaten (Watt, 1923) and may lose viability sooner (Enright and Lamont, 1989). Small seeds typically germinate best on or near the surface, and the maximum depth from which recruitment from seed can occur is directly proportional to the mass of the seed (van der Valk, 1974).

2.2.2 Seed germination

Only part of the colonization process has been completed when seeds reach a site. They must still germinate. Seed germination is a complex physiological process that, as a result of natural selection, has been modified in a variety of ways to ensure that seeds germinate when there is

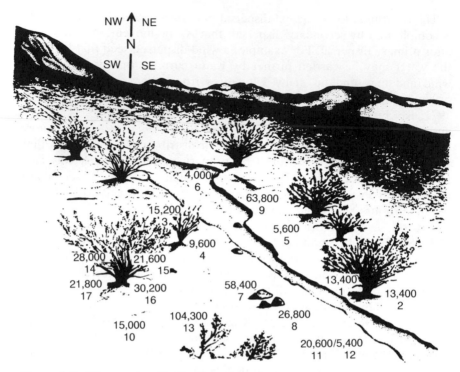

Figure 2.3 The patchy distribution of seeds in a small area in the Sonoran Desert, Arizona, USA. The density of seeds (seeds m^{-2}) in 17 microhabitats is shown (Reichman, 1984).

a high probability of seedling survival. It is the ecological significance of these physiological adaptations that is of concern here. For a discussion of seed germination syndromes and their adaptive significance, see Angevine and Chabot (1979) or Mayer and Poljakoff-Mayber (1982). Specific soil moisture levels, light conditions, temperature fluctuations, oxygen concentrations, or other environmental conditions often are required before the seeds of a particular species will germinate. Chemicals released by plants or plant litter can also inhibit or prevent seed germination (Rice, 1984).

The seed germination requirements of a species can limit its distribution in the landscape. For example, *Salix nigra* seeds are capable of germinating as soon as they are shed, but are short-lived because they lose moisture rapidly (McLeod and McPherson, 1973). Only those *S. nigra* seeds that land immediately on moist, open surfaces, such as on recently exposed river sand bars and lake shores, germinate. Seedlings and adults of *Salix nigra* are capable of growing in drier habitats, but seedlings rarely get established in such sites.

Another important feature of recruitment of species from seed is that the percentage of seeds germinating varies over time. Seed germination of *Salix alaxensis* and *Populus balsamifera* along the Tanana River in Alaska is restricted to sites with high soil moisture levels. *Salix* seed germination dropped from 36% in the second to 16% in the third week of June 1982, and was only 5% in June 1983. *Populus* germination, on the other hand, increased from 3% in 1982 to 24% in 1983 (Walker *et al.*, 1986; see also Baker, 1990). This variation is due to different environmental conditions at a site during the study and differences in seed germination requirements between species.

(a) Seed dormancy

Seeds, which after being shed still do not germinate when exposed to temperature and moisture conditions suitable for germination, are said to be **innately dormant**. Such seeds must be exposed to specific environmental conditions, often in sequence, before dormancy is broken. Seeds may be innately dormant because they contain chemical compounds that inhibit germination (physiological dormancy), because they have impervious seed coats (physical dormancy), because their embryos are undeveloped (morphological dormancy) or because of some combination of the above (Baskin and Baskin, 1989). Because all seeds do not mature simultaneously on a plant, they can be exposed to different conditions as they develop. Consequently, seeds from the same plant may require exposure to different environmental conditions to break their innate dormancy (Cavers and Harper, 1966; Stamp, 1989).

Most seeds in the temperate zones of the earth exhibit **physiological dormancy** and require exposure to light, fluctuating temperatures, wet and cold conditions (stratification) or other specific environmental conditions to break dormancy (Harper, 1977; Mayer and Poljakoff-Mayber, 1982; Baskin and Baskin, 1989). After physiological dormancy, **physical dormancy** is the most common type of dormancy. For example, the seed of *Acacia sieberia*, a savannah tree species in east Africa, has a thick, impermeable seed coat (testa) that has to be softened or cracked before it can germinate. Consequently, passage of seeds through the digestive tracts of browsers, or exposure to fire greatly increases germination (Sabiiti and Wein, 1987). Passage of many kinds of seeds through the digestive tracts of birds and mammals (Krefting and Roe, 1949; Izhaki and Safriel, 1990) often breaks their dormancy, although to varying degrees (Dinerstein and Wemmer, 1988).

Seeds that are not dormant when they are shed and are unable to germinate because of unfavourable environmental conditions may become dormant (Baskin and Baskin, 1989; Hazebroek and Metzger, 1990). This is **induced** or **secondary dormancy**. Seeds that were dormant when shed may encounter environmental conditions that break their dormancy, but do not allow them to germinate. These formerly dormant seeds can also acquire secondary dormancy. Seeds that develop secondary dormancy usually will germinate only under a more limited range of environmental conditions. Thus, as long-lived seeds in the soil undergo more dormant–non-dormant cycles, their chances of germinating may gradually decline to zero (Baskin and Baskin, 1989).

The ecological advantage of dormancy is that it reduces the probability that seeds will germinate during a period of the year with unsuitable environmental conditions (Densmore and Zasada, 1983). In other words, dormancy enables species to synchronize their seed germination and seedling growth with seasons of the year when environmental conditions are likely to be favourable. For example, at Barro Colorado Island, Panama, different dormancy syndromes among species that dispersed their seeds at different times of the year synchronized their germination with the rainy season (Garwood, 1983). There were three dormancy syndromes: (1) 'delayed-rainy season' seeds, which disperse during one rainy season but do not germinate until the beginning of the next (18% of the species); (2) 'intermediate-dry' seeds, which disperse during the dry season but do not germinate until the rainy season (42%); and (3) 'rapid-rainy' seeds, which disperse during the rainy season and germinate later on during the same rainy season (40%).

(b)   Shared versus unique germination syndromes

Seeds with similar germination syndromes (shared syndromes), all else being equal, should be expected to be in the same locations in the landscape, while species with divergent syndromes should be found in different locations. When the seeds of species with different germination requirements arrive at a site, environmental conditions at the site act as a sieve, selecting those species whose germination requirements are met (Harper, 1977; van der Valk, 1981). If each species has a different germination syndrome, this alone could determine the distribution of the species in a landscape. To determine the degree to which seed germination syndromes influence the distribution of species, we can examine whether species found growing on different portions of an environmental gradient have different germination syndromes, and whether species growing along the same section of the gradient have similar syndromes.

Keddy and Ellis (1985) studied the recruitment of 11 species that normally grow at different water depths in southern Canadian wetlands (Fig. 2.4). For 6 of the 11 species, there was a correlation between optimal water depths for seed germination and water depths at which these species grew in the field. The remaining five species did not differ in their seed germination responses along the water depth gradient. A similar study of ten lakeshore species along a soil moisture gradient by Keddy and Constabel (1986) showed that most of the species germinated

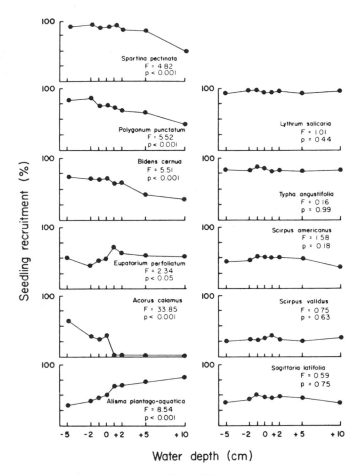

Figure 2.4 Seedling recruitment (%) at different water depths for 11 wetland species in southern Canada. Water depths ranged from −5 cm to +10 cm above the water surface. Species towards the top of the figure grow in shallow water or moist soils, while species at the bottom are found in deeper water. Seed germination of species in the left column varied significantly over the water depth gradient (Keddy and Ellis, 1985).

best under the same conditions; they had shared seed germination syndromes. These species, however, were usually distinctly zoned along lakeshores. The distribution of species along environmental gradients can be accounted for only partly by seed germination syndromes (Rozema, 1975; Welling *et al.*, 1988b), since species with the same seed germination syndromes can be found along different portions of an environmental gradient. If seeds are similarly dispersed and seed germination occurs uniformly along the gradient, post-establishment events must determine zonation patterns in these cases.

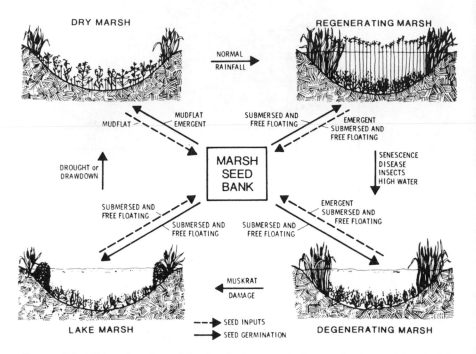

Figure 2.5 The role of seed banks during vegetation cycles in prairie glacial marshes in central North America. Recruitment of different groups of species from the seed bank is restricted to certain phases of the cycle (van der Valk and Davis, 1978).

All the species found at a site do not usually have the same seed germination syndrome (Went, 1949; van der Valk, 1974, 1981, 1985; Garwood, 1983; Schenkeveld and Verkaar, 1984; Baskin and Baskin, 1985; Walker *et al.*, 1986). Seasonal and annual variations in rainfall and solar energy inputs cause changes in environmental conditions such as soil moisture, temperature and salinity. Changes in plant cover can occur due to grazing and a variety of disturbances such as animals digging burrows, fire, wind storms, flooding, etc. These environmental changes and disturbance events allow species with different seed germination syndromes to become established on an area at different times and in different microsites. In prairie wetland areas that are free of vegetation (Fig. 2.5), for example, fluctuations in water levels allow mud-flat annuals and emergent perennials to become established during drawdowns, while during periods of standing water only seeds of submersed species can germinate (van der Valk and Davis, 1978). Differences in seed germination syndromes may be important not only in determining *where* species become established in an area, but also *when* they can become established (van der Valk, 1974; Grime *et al.*, 1981; Garwood, 1983; Baskin and Baskin, 1985).

(c)   Seed and seedling banks

Seeds that do not germinate immediately after dispersal enter and remain in the seed bank (the pool of viable seeds found in the soil at a given time) until they finally germinate, lose viability or are eaten. Many seeds in seed banks are eaten before they germinate (Crawley, 1983; Stiles, 1989). For example, mice and voles have been observed to eat 60% of the seeds of *Cirsium vulgare* within one year of their dispersal (Klinkhamer *et al.*, 1988). Dormant seeds may remain viable in the soil for only a few days, weeks, months, years or even decades after they have been dispersed (Thompson and Grime, 1979; Granström, 1987). Thompson and Grime (1979) distinguish between transient seed banks (none of the seeds remains viable for more than one year) and persistent seed banks (at least some of the seeds remain viable for more than one year). Persistent seed banks can further be divided into subtypes that are defined by the season of the year when seeds germinate and the proportion of seeds that remain in the seed bank for more than one year.

The composition of seed banks is usually spatially heterogeneous (Kellman, 1978; Reichman, 1984; Price and Reichman, 1987; Bigwood and Inouye, 1988); that is, the seeds of most species have patchy distributions. This presumably reflects the microtopographic heterogeneity of most sites, the vagaries of both primary and secondary dispersal (Ryvarden, 1971; Reichman, 1984), seed predation patterns (Reichman, 1979) and past disturbances that created a patchy distribution of the vegetation.

Species with long-lived seeds can survive in the persistent seed bank, even when the species is no longer part of the vegetation (actual flora). Thus, long-lived seeds enable species to disperse over time. Whenever conditions suitable for seed germination occur, such species can again become re-established in the vegetation. Mud-flat annuals in prairie pothole marshes can become established only when the marshes are free of standing water. This happens as a result of periodic droughts every 10 to 20 years (Fig. 2.5). Because these annuals have long-lived seeds, they are able to survive in the persistent seed banks of the marshes between droughts when mature plants of these species are absent from the wetland because they cannot tolerate flooding (van der Valk and Davis, 1978). Seeds of emergent species are also found in the persistent seed banks of these wetlands during periods when they are absent from the vegetation due to high water levels or overgrazing by muskrats (van der Valk and Davis, 1978).

Recruitment of species from the seed bank is governed by environmental conditions (soil moisture, soil temperatures, light conditions, etc.). In an experimental study in which soil moisture was varied (van der Valk and Pederson, 1989; van der Valk *et al.*, 1992), recruitment from a composite seed bank from the Delta Marsh in Manitoba, Canada, was

Table 2.4 Relative density (%) of seedlings of annual (A), wet meadow (M) and emergent (E) species recruited from a seed bank from the Delta Marsh under five different soil moisture regimes. The soil moisture regimes were shallowly flooded (FL), saturated soil (SA), watered every day (W1), watered every other day (W2) and watered every third day (W3)

Species	Soil moisture regime				
	FL	SA	W1	W2	W3
Aster laurentius (A)	12	19	33	45	20
Atriplex patula (A)	0	2	2	2	6
Chenopodium rubrum (A)	0	29	27	34	66
Annual total	12	50	62	81	92
Carex atherodes (M)	0	1	2	2	4
Hordeum jubatum (M)	0	†	1	2	3
Wet Meadow total	0	1	3	4	7
Scirpus maritimus (E)	1	1	1	2	†
Typha glauca (E)	77	44	29	10	0
Scirpus lacustris (E)	10	5	4	4	†
Emergent total	88	50	34	16	†

† Relative density less than 0.5%

dominated by annual species in the drier soil–moisture treatments (watered only every second or third day) and by emergent species in the wetter treatments (shallowly flooded and saturated soil) (Table 2.4). In fact, differential recruitment from seed banks during drawdowns in the Delta Marsh in different years has resulted in different dominant species becoming established (Welling *et al.*, 1988b). Differences in the *a priori* predicted composition of the post-drawdown vegetation in this marsh and the actual vegetation were also due to differences between environmental conditions under which the seed bank composition was determined and those in the field during the drawdown (van der Valk *et al.*, 1989). Environmental conditions during periods of recruitment can determine to a large extent the initial composition of vegetation recruited from seed banks. In other words, knowing the composition of the seed bank of an area is not sufficient to predict the composition of vegetation recruited from that seed bank.

Both field and experimental studies have demonstrated that the presence of fallen litter can significantly reduce recruitment from seed banks (van der Valk, 1986b; Bergelson, 1990; Middleton *et al.*, 1991). For example, a single layer of cattail litter reduced seedling density over 99% and the number of species recruited by 80% during a drawdown in the Delta Marsh, Canada (van der Valk, 1986b). Not only can the presence of litter directly influence recruitment from seed banks, but chemicals released by decomposing litter may also inhibit seed germination (Rice, 1984).

There are species, particularly of *Banksia* and *Eucalyptus* in Australia and *Pinus* in North America, that store seeds in their canopies in serotinous cones or similar fruits that do not open until they are exposed to high temperatures during a fire (Zammit and Westoby, 1988). Thus, canopy storage of seeds functions in many ways like storage of seeds in the soil. Further, storage need not be in the form of seeds. In many forest tree species seeds germinate immediately, but their seedlings fail to grow significantly as long as they remain in the shade of an intact forest canopy. These suppressed seedlings may survive for years, often by utilizing reserves of carbohydrates or other storage materials in the endosperm of the seed. Such seedling populations are often referred to as seedling banks, and they function in much the same way as seed banks (Whitmore, 1984; Silvertown, 1987).

2.2.3   Vegetative propagation

The majority of species in many types of vegetation (e.g grasslands, tundra, wetlands, shrublands and heathlands) are perennials with vegetative propagation. Even though individual shoots or stems are fairly short-lived, more or less continuous replacement of shoots by vegetative propagation makes a clone of such a species nearly immortal and capable of moving over the landscape, in some cases for long distances (see Table 2.6). Species capable of vegetative propagation do so using a variety of mechanisms: layering, plantlets, rhizomes, runners, stolons, bulbils and turions. All of these mechanisms allow some portion of an established individual to give rise to a new, self-replicating version of itself.

For a clonal species, all the individuals that are derived from a single seed are collectively referred to as a genet. Each vegetatively produced individual of the genet that is actually or potentially independent is referred to as a ramet. Although it is possible to determine which ramets belong to a particular genet (Silander, 1985), this is rarely feasible or necessary in community studies. What matters at this level of resolution is the behaviour of the total population of individuals or ramets of each species in an area.

The effectiveness of vegetative propagation for colonizing new sites depends upon how far new ramets can become established from their parents and whether the ramets remain attached to them. Two basic clonal growth forms, 'phalanx' and 'guerrilla', have been identified in vegetatively propagating species (Lovett Doust, 1981), most clonal species falling somewhere in between these two extremes. The phalanx growth form describes situations in which the ramets remain closely packed together and physically attached to each other. Tussock-forming grasses and sedges are good examples of this growth form. These produce

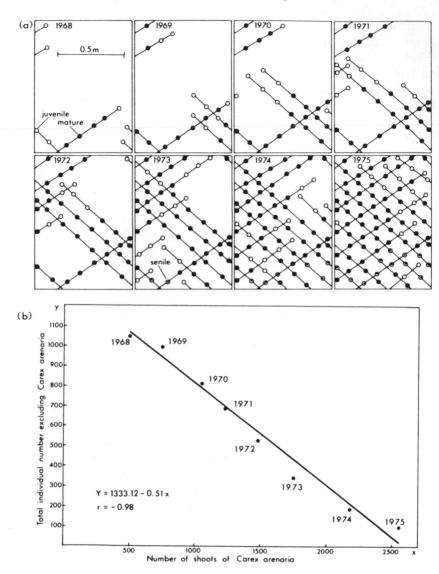

Figure 2.6 (a) Spread of *Carex arenaria* by vegetative propagation over eight years into a permanent quadrat on an inland sand dune in Poland. (b) The total number of individuals of other species in this permanent quadrat during the same period (Symonides, 1985).

new ramets from very short rhizomes or from basal buds. The guerrilla growth form has ramets that become established some distance from the parent plant and become detached from it (Figs. 2.6 and 2.7). Plants with this growth form (e.g. strawberries (*Fragaria* spp.) or bracken fern (*Pteridium aquilinum*)) usually produce new ramets from stolons or

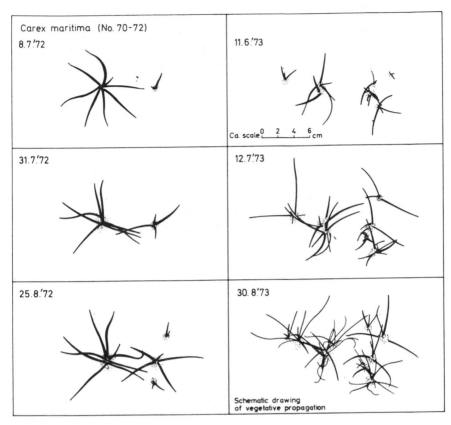

Figure 2.7   The spread of *Carex maritima* by vegetative propagation on the new volcanic island of Surtsey during 1972 and 1973 (Fridriksson, 1975).

runners. While phalanx species hold onto a particular piece of ground until they are eliminated for some reason, guerrilla species are always on the move. The differences in these two growth forms can have a major impact on the structure of grasslands, wetlands and other types of vegetation dominated by clonal species.

Simulation models of clonal growth have been and continue to be developed for better prediction of rates and direction of the clonal spread. These models to date have been highly species and location specific, so they have little or no general applicability (Cain, 1990). Nevertheless, models of clonal growth do have great potential for improving our understanding of its ecological consequences.

2.3   COLONIZATION

The term 'safe site' was suggested by Harper *et al.* (1961) to describe a microsite in an area that had environmental conditions suitable for the germination of seeds and survival of seedlings of a particular species. The more microtopographic heterogeneity there is in an area, the more different kinds of safe sites present, and the larger the number of species that should be able to become established. Unfortunately, determination of the safe site characteristics of a given species is difficult. Field studies of seedling establishment have shown considerable place-to-place variation in a given year (Huenneke and Sharitz, 1990), and that characteristics of microsites necessary for successful establishment vary from year to year (Fowler, 1988). Some of this interannual and spatial variability in recruitment is because microsite suitability is a continuum, and because all seeds of a given species need not have the same safe site requirements.

Colonization may occur only when existing vegetation or other barriers that exclude a species are removed; that is, when a safe site is created. Two types of barriers have been recognized (Johnstone, 1986a): vegetation barriers where establishment is prevented by the presence of plants; and environmental barriers, where establishment is prevented by some factor other than the presence of plants. Also, barriers can be species specific or universal.

The colonization of a site over time depends on the type of barrier that must be removed to create a safe site and the selectivity of that barrier. Four types of potential 'invasion windows', or opportunities for establishment, can be recognized on the basis of barrier selectivity and type:

1. No window (universal, environmental barrier) – probability of establishment is zero; e.g. plants cannot colonize the surface of a shifting sand dune because they cannot become established on such an unstable and inhospitable surface.
2. Stable window (selective, environmental barrier) – establishment may occur when some barrier other than the presence of plants on the site is removed; e.g. the removal of grazers allows the establishment of species that are susceptible to grazing or allows seed dispersal to a site.
3. Temporary window (universal, vegetation barrier) – establishment is possible after some disturbance creates a plant-free area; e.g. a gap in a forest caused by windthrow is rapidly colonized by whatever species are first established.
4. Future window (selective, vegetation barrier) – the dispersal of seeds to a site is not prevented by the presence of plants, but the germination of these seeds is dependent on some future disturbance to create a plant-free opening; e.g. the opening of a gap due to the death of a plant that allows a new species that has been present only in the seed or seedling bank to become established in the canopy.

This scheme provides a useful framework for determining the probability of future invasions, and, hence, of successional changes caused completely, or in part, by colonization. Stable invasion windows are most important for the colonization of primary sites (sites previously free of vegetation), while stable, temporary and future windows of invasi
be important for the colonization at secondary sites (sites tha
previously been vegetated). The revegetation of secondary sites, how-
ever, is often only partly a process of colonization. The colonization
of primary sites will be examined next, and then the colonization of
secondary sites.

2.3.1 Primary sites

Differences among species in the distances that their seeds are dispersed have a significant effect on patterns of colonization of primary sites (those undergoing primary succession) and on the distribution of species on a newly created or exposed surface (Cooper, 1923, 1931, 1939; Docters van Leeuwen, 1936; Feekes, 1936; Bakker and van der Sweep, 1957; Fridriksson, 1975; van der Pijl, 1982; Walker *et al.*, 1986; Wood and del Moral, 1987; Whittaker *et al.*, 1989). For instance, species that colonized road verges in a newly empoldered (drained) area in The Netherlands had different dispersal syndromes and became established at different times (Nip-van der Voort *et al.*, 1979). An examination of roads of different ages in the polders showed that the first species to become established along these roads were those whose seeds were spread by water; they were already present in the polder when the water was removed (Fig. 2.8). The next groups of species to become established had

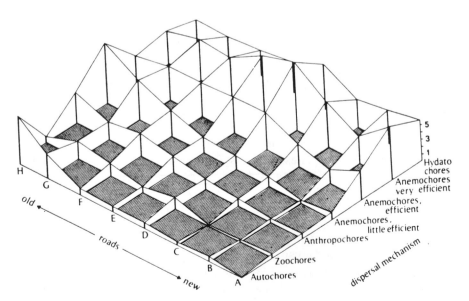

Figure 2.8 Number of species with different kinds of dispersal mechanisms found along roads of different ages in a polder in The Netherlands (Nip-van der Voort *et al.*, 1979).

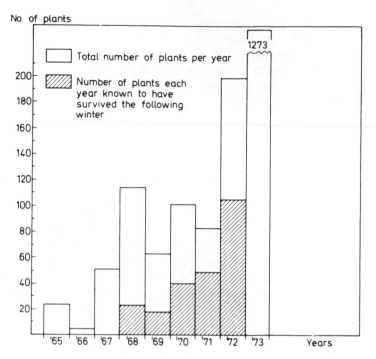

Figure 2.9 The total number of vascular plant of all species growing on the new volcanic island Surtsey from 1965 till 1973 (Fridriksson, 1975).

wind-dispersed seeds. These were followed by species brought in by human activities and by animals. The last species to arrive were those with autochorous seeds. The initial composition of the vegetation of these newly exposed roadsides was a function of dispersal capabilities and little else. Studies of the initial colonization of other primary sites have consistently shown that species with different dispersal syndromes reach the site at different times. Some dispersal syndromes are not adequate to reach some isolated sites, such as remote oceanic islands, and consequently the floras of these isolated areas lack species with certain syndromes (Ridley, 1930).

Studies of the colonization of new islands, such as the volcanic island Surtsey (Fridriksson, 1975), and of islands whose previous vegetation was obliterated by volcanic activity, such as Krakatoa (Docters van Leeuwen, 1936; Whittaker *et al.*, 1989), have shown that seeds can reach these islands by several different modes of dispersal almost simultaneously. Surtsey, for example, was colonized within a few years by species whose seeds were carried to the island by ocean currents, by wind, and by birds (Fig. 2.9). Once species become established on a new island, their population sizes can grow rapidly due to the recruitment of new individ-

uals from seeds produced by the initial colonizers, by vegetative propagation, or both. An examination of the material that was deposited upon
the shores of Surtsey from 1964 to 1972 showed that not only seeds were
being carried to the island by ocean currents, but also vegetative parts of
many species. Although most of the species that became established on
Surtsey during its first decade seem to have been recruited from seed,
it seems likely that vegetative propagation also played a role in the
colonization of the island.

The longest and most complete record of colonization of unvegetated
islands is that of the Krakatoa Islands, which were created by the eruption of the Krakatoa volcano in 1883 (Whittaker et al., 1989). During the
first 25 years, these islands were colonized primarily by sea-dispersed
species (c. 50% of their total flora). Subsequently, new colonizers had
primarily animal- and wind-dispersed seeds. Thus, by 1983 there were
about 99 seed plant species with bird- or bat-dispersed seeds and about 80
more with wind-dispersed seeds, but only about 60 species with sea-
dispersed seeds on the islands. After 100 years, various vegetation types
have developed on the islands. Coastal communities were the first to
become established and have changed very little, except that they have
become less common because of shoreline erosion. Forest covers most of
the islands, but they are species-poor and vary in composition from island
to island. Time of arrival of various species on each island seems to be
responsible for differences in the composition of forest vegetation from
island to island, but there are not enough pertinent data to show this
unequivocally. There is relatively little differentiation of forest types,
except where there has been volcanic activity that destroyed some of the
vegetation, and with increasing elevation on one island (Rakata).

Many other studies of primary succession on volcanoes have examined
early colonization and primary succession in detail (e.g. del Moral,
1983; Wood and del Moral, 1987, 1988; Tsuyuzaki, 1989). Studies of the
primary succession on the pumice plains on the north side of Mount St
Helens, USA (Wood and del Moral, 1988) indicate that after only one
year a few individuals of two species had become established as a result of
long-distance dispersal. By 1986, six years after the eruption, over 30
species were found, but overall species density was still only one or two
plant species per $100 \, m^2$. These species were a mixture of natives and
introduced ruderals normally found in disturbed areas at lower elevations. In other words, this was a new assemblage of species for this
region. Most of the species were perennial herbs with wind-dispersed
seeds, but there were also some woody species, including *Salix* (willow),
and a few individuals of *Pseudotsuga menziesii* (Douglas fir), *Tsuga
heterophylla* (Western hemlock) and *Abies concolor* (Noble fir).

To summarize, colonization of primary sites is an extremely variable
phenomenon which depends greatly on seed dispersal from adjacent
areas, local environmental conditions, and year-to-year variation in

climatic conditions. Consequently, colonization patterns of primary sites are always highly site specific.

2.3.2 Secondary sites

Secondary sites differ from primary sites in that the sites were previously vegetated. The previous vegetation may have altered environmental conditions on the site, particularly the chemical and physical properties of the soil, in a variety of ways, but its chief impact on future vegetation is normally the legacy of a seed bank (Fig. 2.5), and sometimes already established plants. Seed banks play a role in revegetation of secondary sites previously covered by most kinds of vegetation, including arctic and alpine tundra (Leck, 1980; Fox, 1983; McGraw and Vavrek, 1989), deserts (Henderson *et al.*, 1988; Kemp, 1989), temperate forests (Livingston and Allessio, 1968; Kellman, 1970; Marks, 1974; Nakagoshi, 1985; Pickett and McDonnell, 1989), boreal forests (Archibold, 1989), tropical forests (Guevara and Gomez-Pompa, 1972; Cheke *et al.*, 1979; Whitmore, 1984; Saulei and Swaine, 1988; Garwood, 1989), grasslands (Major and Pyott, 1966; Rabinowitz, 1981; Shenkeveld and Verkaar, 1984; Symonides, 1985; Rice, 1989), chaparral (Keeley, 1977; Keeley and Zedler, 1978; Parker and Kelly, 1989), heathlands (Stieperaere and Timmerman, 1983) and wetlands (van der Valk and Davis, 1978; Leck and Graveline, 1979; Leck *et al.*, 1989; Wienhold and van der Valk, 1989). When compared with primary sites, secondary sites are usually revegetated very quickly because of rapid recruitment from their seed or seedling banks, and/or clonal growth from extant plants on their peripheries.

Species that become established from seed banks on secondary sites, however, are typically only a subset of those that composed the previous vegetation, and species may colonize that were not previously components of the vegetation (Livingston and Allessio, 1968; Harper, 1977; Thompson, 1978; Thompson and Grime, 1979; Roberts, 1981). Often, species recruited from the seed bank are annuals or herbaceous perennials, but trees and shrubs may also occur at sites previously covered by forests (Marks, 1974; Whitmore, 1984).

When the re-establishment of a vegetation type is dependent on the establishment of species whose seeds are not present in the seed bank, its re-establishment is far from certain on secondary sites (Egler, 1954; Holt, 1972). This is due to the fact that, for many species, seeds that arrive after the canopy has closed have greatly reduced chances of establishment (i.e. there is only a temporary invasion window). Thus, successful establishment is favoured by early arrival. Egler (1954) pointed out the significance of this for understanding and predicting vegetation dynamics in abandoned farm fields. The recognition that in the flora of a given region there are species whose seeds can germinate in established vegetation (i.e. whose colonization is through stable invasion windows) and other species whose seeds cannot (i.e. whose colonization is through temporary invasion windows) was an important conceptual breakthrough for theoretical and applied ecologists (Noble and Slatyer, 1980; van der Valk, 1981; Johnstone, 1986a). In this way, inexplicable abnormalities, such as stable shrub communities in areas that should be dominated by trees, could be recognized for what they were, a vegetation type that was caused by local dispersal and recruitment patterns. For example, if after a fire or other major disturbance, shrubs become established at a site first because their seeds arrived first, or because their seeds were already present in the seed bank, it may be difficult or even impossible for tree species that arrive later to become established from seed under the shrub canopy (Niering and Egler, 1955; Niering and Goodwin, 1974; Glenn-Lewin, 1980). This scenario could result in the establishment of a stable shrub community surrounded by forest vegetation.

Niering and Goodwin (1974) have utilized the phenomenon of site pre-emption to produce stable shrub communities where they are desired, such as along power line rights-of-way. Tree species are not able to become established from seed under these shrub canopies, but they can become established along the edges (Table 2.5). Experimental studies of the timing of establishment (e.g. Harper, 1961) have shown that species can coexist in an area even though, on the basis of competition experiments, they would not be expected to coexist. This may occur if the weaker competitor can become established before the stronger competitor. Thus, the order of arrival of species on a site can influence both the composition and physiognomy of the vegetation.

Conversely, many species, particularly forest species, can become established within existing vegetation (Smith, 1975; Bierzychudek, 1982;

Table 2.5 Recruitment of tree species from seed over a five-year period within and at the edge of two shrub communities in New England, USA. Seedling density is expressed as number of seedlings per 0.1 ha (adapted from Niering and Goodwin, 1974)

Tree taxa	Shrub type			
	Gaylussacia baccata		Vaccinium vacillans	
	Within	Edge	Within	Edge
Acer rubrum	0	43	0	0
Betula populifolia	0	0	0	0
Juniperus virginiana	0	27	0	0
Prunus serotina	0	243	0	133
Quercus velutina	0	43	0	434
Total	0	356	0	567

Garwood, 1983; McDonnell and Stiles, 1983; Walker *et al.*, 1986). Understorey herbs, epiphytes and many bird-dispersed shrubs, in fact, do not become established until some type of vegetation is present at a site. During old-field succession, the bird dispersal of seeds into a site is a function of the physical structure of the vegetation. Seed inputs via birds in old fields is initially very low, because the birds do not use these areas. As the plants get older and taller, bird use increases because plants provide perching sites for the birds, and, as a consequence, there is a significant increase in the number of bird-dispersed species that reach the field (McDonnell and Stiles, 1983). Locations of the seeds deposited by birds can also influence recruitment. In New England, *Juniperus* seeds dropped near stones by birds had the best chance of germinating, and the resulting seedlings of surviving (Livingston, 1972).

After 1600 to 2900 years of succession on sand dunes along the shore of Lake Huron, Ontario, Canada, plant species new to the vegetation have become established under *Juniperus virginiana* trees (Yarranton and Morrison, 1974). Under these trees, environmental conditions, particularly soil conditions, have been modified sufficiently that the establishment of these new species is possible. This initiates a whole new phase in the development of the vegetation on these dunes. The process was termed 'nucleation' by Yarranton and Morris. A similar phenomenon occurs in the Sonoran Desert, where 'nurse plants' are required for the recruitment of saguaro cactus (*Carnegia gigantea*) from seed (Niering *et al.*, 1963; Steenberg and Lowe, 1977, 1983).

Although the recruitment of species from seeds reaching a site or already present at a site is generally believed to be the major mechanism of colonization for both primary and secondary sites, this is not always the case, particularly in aquatic and wetland systems. Aquatic plants, nearly all of which can propagate vegetatively, can spread clonally long distances. This is because vegetative propagules, either whole ramets or specialized vegetative structures, are readily carried by water currents within lakes and by streams among lakes (Holm *et al.*, 1969; Fridriksson, 1975; Mitchell, 1985; van der Valk, 1986a; Johnstone, 1986b). As a rather spectacular example, female plants of the dioecious submersed aquatic, *Elodea canadensis*, a native of North America, have spread vegetatively all over Europe and Asia since its escape from cultivation in England and western Europe in the later nineteenth century (Sculthorpe, 1967), clearly without any sexual reproduction. Likewise, the European submersed aquatic, *Potamogeton crispus*, has spread vegetatively over much of North America.

The rate of vegetative spread depends on the propagation mechanism(s) and the nature of the terrain. In wetlands, clones of *Phragmites australis* and other rhizomatous species may advance as much as 50 m per year (Feekes, 1936). In terrestrial vegetation, the distance a clone advances annually is often very small, usually less than 0.5 m per year (Table 2.6).

Table 2.6 Age, size (diameter in metres) and mean annual spread (metres per year) of clones of vegetatively propagating species in terrestrial habitats (adapted from Cook, 1985)

Species	Age (yr)	Size (m)	Annual spread
Picea mariana	300+	14	<0.05
Calamagrostis epigeeios	400+	50	<0.13
Convallaria majalis	670+	83	<0.12
Lycopodium complanatum	850	250	0.29
Festuca rubra	1 000+	220	<0.22
Festuca ovina	1 000+	8	<0.01
Holcus mollis	1 000+	880	<0.88
Pteridium aquilinum	1 400	489	<0.35
Larrea tridentata	11 000+	8	<0.001
Gaylussacia brachycerium	13 000	1980	0.15

Nevertheless, vegetative propagation may be the only way in which some species can colonize existing vegetation. Glenn-Lewin (1980) studied the establishment of native prairie species in an area of former prairie that had been heavily grazed. In areas of the former pasture immediately adjacent to native prairie, a number of prairie species had become established. The native prairie was encroaching on the pasture, evidently solely as a result of the spread of native species by vegetative propagation. In 30 years the native prairie species had advanced about 10–20 m into the pasture, a spread rate of between 0.3 and 0.6 m per year.

The colonization of a secondary site by vegetatively propagating species can have a major impact on the structure and composition of the vegetation. Symonides (1985) documented the spread of the rhizomatous sedge, *Carex arenaria*, into areas already vegetated on sand dunes in Poland. As the sedge colonized the area (Fig. 2.6), most other species were eliminated locally.

Typically, perennials with vegetative propagation initially become established from seed (Fig. 2.7). The subsequent spread of such a species, however, is largely determined by its vegetative propagation. In North American prairie wetlands (Fig. 2.5), emergent species become established from seed during drawdowns. During this period, shoot densities of these emergents are often very low. After the marsh refloods, emergents (*Typha glauca* and *Sparganium eurycarpum*) spread vegetatively over the marsh for the next several years until all of the available area is colonized (van der Valk, 1985). Similar patterns of colonization and subsequent vegetative spread of species were reported by Cooper (1923, 1931, 1939) in his permanent quadrats located on new moraines at Glacier Bay, Alaska.

The relative importance of seed dispersal, seed germination, seedling and adult survival, and clonal growth in determining the position of species along a new wetland coenocline (compositional gradient) were investigated experimentally by van der Valk and Welling (1988) and Welling *et al.* (1988a,b). In their study, the existing coenocline in a prairie wetland was destroyed by flooding it to 1 m above normal for two years. A new wetland coenocline was then initiated by drawing down water levels to 0.5 m below normal. This enabled seeds in the seed bank to germinate on exposed substrates. Because of secondary dispersal of seeds by water currents, seeds of all emergent species had similar, but not identical, distributions along the coenocline. Differential seed germination along the coenocline resulted in two species, *Scirpus lacustris* and *Phragmites australis*, becoming established primarily in the lower part and one, *Scolochloa festucacea*, in the upper part. Differential survival of plants after reflooding caused the elimination of *Scolochloa* and *Phragmites*, in the lower part of the range where they had occurred as seedlings. All three species continued to adjust their distributions along the coenocline by clonal growth for five years after reflooding. Each species along the new coenocline reached its final position as a result of a unique combination of seed dispersal, seed germination and differential survival of seedlings and adults at different water depths. In other words, the reason why a population of a given species came to dominate a certain portion of the new coenocline varied from species to species. This result indicates that factors controlling the position of species along coenoclines cannot be determined unequivocally from an analysis of the distributions of established populations.

2.4 PERSISTENCE

After a population is established, the ability of the population to replace individuals as they are lost determines whether or not it will persist. For species without a means of vegetative propagation, the only mechanism for persistence is recruitment of new individuals from seed. For species whose seeds can germinate under a plant canopy, such as forest herbs in deciduous forests (Bierzychudek, 1982), this requires only a suitable site free of vegetation where the seeds can germinate. For species whose seeds cannot germinate under a plant canopy, recruitment of new populations is restricted to gaps or openings in the vegetation. Such gaps can occur because of the normal death of mature plants, or because of some type of disturbance that removes part of the plant canopy.

2.4.1 Gap phase replacement

Conditions within established vegetation prevent the germination of seeds of most species. This has been shown by studies in which seed has been sown into different types of plant communities (Cavers and Harper, 1967; Peart and Foin, 1985; Bakker *et al.*, 1985; Peart, 1989b,c). Experimental

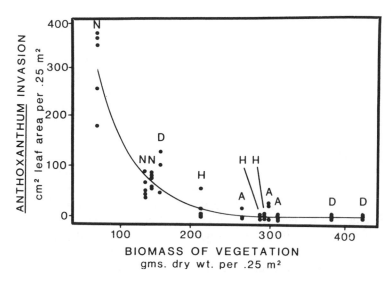

Figure 2.10 The colonization success of *Anthoxanthum odoratum* seedlings recruited from seed sown into vegetated areas with differing biomass (Peart and Foin, 1985). Seedling success is measured by the total leaf area of the seedlings in an area of 0.25 m². Patch types: N – annual, D – *Deschampsia holciformis*, A – *Anthoxanthum odoratum* and H – *Holcus lanatus*.

studies by Peart and Foin (1985) and Peart (1989b), in which seeds of various species were sown into four different vegetation types, demon-strated that the establishment of species from seed within vegetation is primarily a function of the biomass of the vegetation and is largely independent of its composition (Fig. 2.10). Only when the biomass was reduced experimentally, usually through canopy disruption, did any significant recruitment occur.

   Why is the environment within existing vegetation not conducive to the germination of most seeds? Experimental studies of seed germination have shown that under a plant canopy the germination of some species is inhibited because of the spectral composition of the light reaching the seed (King, 1975; Silvertown, 1980b; Vázquez-Yanes and Smith, 1982). Under plant canopies far-red light predominates and acts to inhibit the germination of seeds of shade-intolerant species. This light requirement for seed germination ensures that seeds germinate only when a gap in the canopy occurs and there is enough light for seedlings to grow. Many other factors may also inhibit the germination of seed within established vegeta-tion, including the thickness of the litter layer (Goldberg and Werner, 1983; Walker *et al.*, 1986; van der Valk, 1986b; Middleton *et al.*, 1991), a reduction in the magnitude of day–night temperature fluctuations (Meredith, 1985), and unsuitable soil moisture levels (Thompson and Grime, 1983; Meredith, 1985; van der Valk and Welling, 1988). In general, the poor recruitment from seed under plant canopies has been little studied, considering its importance (Harper, 1977).

Because of inhibition by plant canopies, establishment from seed within existing vegetation in most vegetation types is an episodic or stochastic event related to disturbance. Recruitment is restricted to sites where the vegetation is totally, or at least partially, destroyed as a result of factors like fire, frost heaving, treefall and animal activity. This is known as gap phase replacement. It occurs not only in forests, but also in grasslands, tundras, wetlands and shrublands (Chapter 4; Niering and Goodwin, 1974; van der Valk and Davis, 1978; van der Valk, 1981; Gartner et al., 1986; Pickett and White, 1985; Chambers et al., 1990). The revegetation of newly formed gaps is highly variable, since it is a function of such site-specific, unpredictable factors as past and present seed dispersal, seed germination and clonal growth. Complicating the picture, the majority of seeds that reach a site are either eaten by animals (ants, rodents, birds, etc.) or eventually lose viability (Janzen, 1971; Louda and Zedler, 1985), and the numbers of species that become established in gaps, and their abundances, are often functions of gap size (Davis and Cantlon, 1969; Webb et al., 1972; Goldberg and Werner, 1983; Whitmore, 1984).

Vegetative propagation is often involved in the recolonization of gaps. Rhizomes, tubers, bulbs and other below-ground plant parts may survive during the formation of the gap and quickly grow new shoots. Species may also invade gaps by vegetative propagation from the adjacent vegetation (Cremer and Mount, 1965; Webb et al., 1972; Platt, 1975; Hibbs, 1983; Brokaw, 1987; McConnaughay and Bazzaz, 1987). Ferns that recolonized gaps in a Eucalyptus regnans forest in Tasmania, Australia, did so from rhizomes that had survived after a fire. These ferns covered 50% of the ground within four years. In grasslands and wetlands, most gaps seem eventually to be recolonized by vegetative spread of species from around the gap, although this may take several years; in the meantime, the gap may be colonized by annuals and perennials (Platt, 1975; Platt and Weis, 1977; Peart, 1989c; Middleton et al., 1991).

2.4.2 Vegetative maintenance

Some species can, in theory, maintain themselves indefinitely by vegetative propagation alone. However, there is only limited, and generally only inferential, information on the longevity of clones. The available data indicate that individual clones (genets) of some species can survive for hundreds or even thousands of years (see Table 2.6). The stability of the environment may determine ultimately how long a clone will survive; a change in environmental conditions can lead to very rapid reduction in the size of a clone and even to its elimination (Harper, 1977).

Vegetative propagation is an important mechanism for the persistence of species in sites subject to periodic disturbances, particularly fire.

Many species are able to resprout from below-ground organs after fire has killed their above-ground parts (Noble and Slatyer, 1980). Most perennial grasses, sedges, many shrubs and some trees have this ability. Not surprisingly, such species dominate the vegetation in areas of the world where fires occur frequently, such as prairies in central North America and shrublands in California and Australia.

Recruitment of new individuals by vegetative propagation is usually more successful than recruitment from seed, primarily because vegetative offspring are initially larger and can receive resources from their parents until they are well established (Cook, 1985; Eriksson, 1986). Nevertheless, many populations of species with vegetative propagation also show significant recruitment of new individuals from seed (Bierzychudek, 1982; Cook, 1985).

Terminology for describing the local population status of clonal species has been developed by Russian ecologists, particularly T. A. Rabotnov and his co-workers (Rabotnov, 1978, 1985; Vorontzova and Zaugolnova, 1985). The status of a population is characterized by the number of plants (ramets) belonging to different life- or age-states (see Fig. 2.6): typical life-states include seeds (se), seedlings (pl), juveniles (j), immatures (im), mature vegetative plants (v), sexually reproducing plants (g_1, g_2, g_3), subsenile plants (ss) and post-sexually reproductive (senile) plants (s).

By examining the frequencies of different age-states in an area, three different developmental stages of a clone can be recognized: (1) the invasive stage, during which the juvenile life-state predominates; (2) the normal stage, during which all age-states are present; and (3) the regressive stage, during which the generative and senile states predominate. For clonal plant species, particularly of the guerrilla type, these developmental stages may be reversible, such that a population that is in a regressive stage may, because of a change in environmental conditions, return to a normal stage or even conceivably to an invasive stage after a major upheaval. For example, a population of *Agrostis tenuis* in a meadow in Russia was in an invasive stage (the largest percentage of plants was in the juvenile state) in 1967 (Fig. 2.12). In the three subsequent years, the population was in a normal stage with reproducing plants being the most common age-state. By 1971, the population was in a regressive state (senile plants predominated in the age-state spectrum). Another period of recruitment from seed occurred in 1973, and the *A. tenuis* population was again in an invasive state (Kurchenko, 1985). This ability of species with vegetative propagation to adjust their growth to environmental conditions can result in changes in dominant species in the meadow from year to year.

SUMMARY

Macro-patterns in the distribution and dynamics of vegetation are, in part, the result of species establishment patterns. Establishment of a

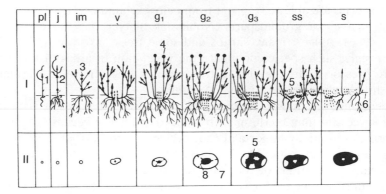

Figure 2.11 Diagnostic features of the age-states of tussock-forming grasses. The different age-states (pl, j, im, etc.) are defined in the text. I illustrates tussock ontogeny; II shows horizontal tussock projection. Note: 1, vegetative first-year shoot; 2, vegetative perennial shoot; 3, dead shoot; 4, reproductive shoot; 5, part of a dead tussock; 6, soil level; 7, boundary of a clone (genet); 8, boundary of vegetative individuals (ramets) (Vorontzova and Zaugolnova, 1985).

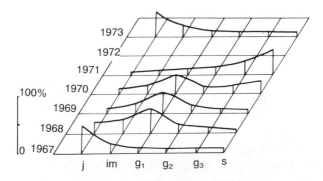

Figure 2.12 Age-spectra of a population of *Agrostis tenuis* in a grassland in Russia from 1967 to 1973 (Kurchenko, 1985). No data were collected in 1972. Note the change in the age-spectra between 1971 (regressive) and 1973 (invasive). The different age-states are defined in the text (see p. 89).

species in an area is a function of its seed-dispersal capacity, its seed-germination requirements and/or its clonal growth.

Establishment

1. The seed rain at a site varies seasonally and annually, and both qualitatively and quantitatively, because of seasonal and annual variations in seed production, seed predation and seed dispersal.
2. No matter what the primary dispersal vector, most seeds are deposited within a short distance of the parent plant.

3. Secondary dispersal of seeds by wind, water or animals can carry some seeds much further from parent plants than they are typically carried by primary vectors. Secondary dispersal also typically results in a more patchy distribution of seeds.
4. Seed dormancy ensures that seed germination occurs primarily when chances of seedling survival are high.
5. A safe site is an area that has an environment suitable for the germination of seeds of a particular species. The characteristics of safe sites, however, often are difficult to establish.
6. The temporal and spatial distribution of safe sites determines when and where a species can become established from seed.
7. In most types of vegetation, there are species whose seeds can germinate within existing vegetation and other species whose seeds require special conditions or disturbance.
8. The pool of dormant seeds in the soil is called the seed bank. Some communities, particularly in tropical and boreal forest, also have seedling banks.
9. The majority of plant species in most types of vegetation are perennials with some means of vegetative propagation.
10. Species with vegetative propagation normally colonize new sites by seed. The abundance of these species at a site, however, is largely determined by vegetative propagation.

Colonization

11. Primary sites (those with newly created, unvegetated substrates) are usually colonized as a result of seed dispersal and germination, but can also be colonized as a result of vegetative propagation.
12. Primary succession is often a highly variable and unpredictable phenomenon because of the vagaries of seed dispersal.
13. Secondary sites (large or small areas that were previously vegetated) are often revegetated by species recruited from the seed bank or as a result of the clonal growth of individuals that survived below ground or along the periphery of the site.
14. Each species becomes established along a certain portion of a coenocline because of a unique combination of factors, including dispersal of its seed along the gradient, differential seed germination along the gradient, and differential seedling and adult survival along the gradient.

Persistence

15. Because seeds of many species will not germinate under a plant canopy, these species are recruited into the vegetation in openings

or gaps created by various kinds of disturbances. Gaps are also colonized by vegetative propagation, particularly in grasslands and wetlands.

16. The age-state spectrum (numbers of seedlings, juveniles, sexually reproducing adults, etc.) of a clonal species determines whether it is expanding, stable, or declining in an area.
17. The absolute abundance of ramets and the age-state spectrum of a clonal species can vary from year to year because of differences in environmental conditions, litter cover, grazing pressure, etc.

ACKNOWLEDGEMENTS

Support for the preparation of this chapter was provided by the Science and Humanities Research Institute of Iowa State University, Ducks Unlimited Canada, the Delta Waterfowl and Wetland Research Station, the Department of Plant Ecology, University of Utrecht and a grant from the Nederlandse Organisatie voor Zuiver-Wetenschappelijk Onderzoek (ZWO) (B 84-249). This chapter was begun while I was on a faculty-improvement leave in the Department of Plant Ecology, University of Utrecht, The Netherlands, and I would like to thank Professor Peter van der Aart for allowing me to spend a year in his department. Jos Verhoeven deserves special thanks for his assistance, good humour and patience while I was at Utrecht. Tom Jurik was kind enough to enough to read over this chapter and make a number of suggestions for improvement. This is paper no. 66 of the Marsh Ecology Research Program, a joint project of Ducks Unlimited Canada and the Delta Waterfowl and Wetlands Research Station.

REFERENCES

Angevine, M. W. and Chabot, B. F. (1979) Seed germination syndromes in higher plants, in *Topics in Plant Population Biology* (eds O. T. Solbrig, S. Jain, G. B. Johnson and P. H. Raven), Columbia University Press, New York, pp. 188–206.

Archibold, O. W. (1989) Seed banks and vegetation processes in coniferous forests, in *Ecology of Soil Seed Banks* (eds M. A. Leck, V. Thomas Parker and R. L. Simpson), Academic Press, New York, pp. 107–22.

Augspurger, C. K. (1984) Seedling survival among tropical tree species: Interactions of dispersal distance, light-gaps, and pathogens. *Ecology*, **65**, 1705–12.

Augspurger, C. K. and Franson, S. E. (1987) Wind dispersal of artificial fruits varying in mass, area, and morphology. *Ecology*, **68**, 27–42.

Baker, W. L. (1990) Climatic and hydrologic effects on the regeneration of *Populus angustifolia* James along the Animas River, Colorado. *J. Biogeog.*, **17**, 59–73.

Bakker, D. and van der Sweep, W. (1957) Plant-migration studies near the former island of Urk in the Netherlands. *Acta Bot. Neerland.*, **6**, 60–73.

Bakker, J. P., Dijkstra, M. and Russchen, P. T. (1985) Dispersal, germination and early establishment of halophytes and glycophytes on a grazed salt-marsh gradient. *New Phytol.*, **101**, 291–308.

Baskin, J. M. and Baskin, C. C. (1985) Life cycle ecology of annual plant species of cedar glades of southeastern United States, in *The Population Structure of Vegetation* (ed. J. White), Junk, Dordrecht, pp. 371–98.

Baskin, J. M. and Baskin, C. C. (1989) Physiology of dormancy and germination in relation to seed bank ecology, in *Ecology of Soil Seed Banks* (eds M. A. Leck, V. T. Parker and R. L. Simpson), Academic Press, New York, pp. 53–66.

Bergelson, J. (1990) Life after death: site pre-emption by the remains of *Poa annua*. *Ecology*, **71**, 2157–65.

Bernard, J. M. (1975) The life history of shoots of *Carex lacustris*. *Canad. J. Bot.*, **53**, 256–60.

Bierzychudek, P. (1982) Life histories and demography of shade-tolerant temperate forest herbs: A review. *New Phytol.*, **90**, 757–76.

Bigwood, D. W. and Inouye, D. W. (1988) Spatial pattern analysis of seed banks: an improved method and optimized sampling. *Ecology*, **69**, 497–507.

Birks, H. J. B. (1989) Holocene isochrone maps and patterns of tree-spreading in the British Isles. *J. Biogeog.*, **16**, 503–40.

Brokaw, N. V. L. (1987) Gap-phase regeneration of three pioneer species in a tropical forest. *J. Ecol.*, **75**, 9–19.

Burrows, F. M. (1975) Wind-borne seed and fruit movement. *New Phytol.*, **75**, 405–18.

Burrows, F. M. (1986) The aerial motion of seeds, fruits, spores and pollen, in *Seed Dispersal* (ed. D. R. Murray), Academic Press, Sydney, pp. 1–47.

Cain, M. L. (1990) Models of clonal growth in *Solidago altissima*. *J. Ecol.*, **78**, 27–46.

Cavers, P. B. and Harper, J. L. (1966) Germination polymorphisms in *Rumex crispus* and *Rumex obtusifolius*. *J. Ecol.*, **54**, 357–82.

Cavers, P. B. and Harper, J. L. (1967) Studies in the dynamics of plant populations. I. The fate of seed and transplants introduced into various habitats. *J. Ecol.*, **55**, 59–71.

Chambers, J. C., MacMahon, J. A. and Brown, R. W. (1990) Alpine seedling establishment: the influence of disturbance type. *Ecology*, **71**, 1323–41.

Cheke, A. S., Nanakorn, W. and Yankoses, C. (1979) Dormancy and dispersal of seeds of secondary forest species under the canopy of a primary tropical rain forest in northern Thailand. *Biotropica*, **11**, 88–95.

Cook, R. E. (1985) Growth and development in clonal plant populations, in *Population Biology and Evolution of Clonal Organisms* (eds J. B. C. Jackson, L. W. Buss and R. E. Cook), Yale University Press, New Haven, Connecticut, pp. 259–96.

Cooper, W. S. (1923) The recent ecological history of Glacier Bay, Alaska: III. Permanent quadrats at Glacier Bay: An initial report upon a long-period study. *Ecology*, **4**, 355–65.

Cooper, W. S. (1931) A third expedition to Glacier Bay, Alaska. *Ecology*, **12**, 61–95.

Cooper, W. S. (1939) A fourth expedition to Glacier Bay, Alaska. *Ecology*, **20**, 130–55.

Crawley, M. J. (1983) *Herbivory: The Dynamics of Animal–Plant Interactions*, Blackwell, Oxford.

Cremer, K. W. and Mount, A. B. (1965) Early stages of plant succession following the complete felling and burning of *Eucalyptus regnans* forests in the Florentine Valley, Tasmania. *Austral. J. Bot.*, **13**, 303–22.

Darley-Hill, S. and Johnson, W. C. (1981) Acorn dispersal by the Blue Jay (*Cyanocitta cristata*). *Oecologia (Berlin)*, **50**, 231–2.

Davis, M. B. (1981) Quaternary history and the stability of forest communities, in *Forest Succession: Concepts and Application* (eds D. C. West, H. H. Shugart and D. B. Botkin), Springer-Verlag, New York, pp. 132–53

Davis, R. M. and Cantlon, J. E. (1969) Effect of size area open to colonization on species composition in early old-field succession. *Bull. Torrey Bot. Club*, **96**, 660–73.

del Moral, R. (1983) Initial recovery of subalpine vegetation on Mount St Helens, Washington. *Amer. Mid. Nat.*, **109**, 72–80.

Densmore, R. and Zasada J. (1983) Seed dispersal and dormancy patterns in northern willows: ecological and evolutionary significance. *Ecology*, **61**, 3207–16.

Dinerstein, E. and Wemmer, C. M. (1988) Fruits Rhinoceros eat: dispersal of *Trewia nudiflora* (Euphorbiaceae) in lowland Nepal. *Ecology*, **69**, 1768–74.

Docters van Leeuwen, W. M. (1936) Krakatua, 1883–1933. *Ann. Jardin Bot. Buitenzorg*, **56–57**, 1–506.

Egler, F. E. (1954) Vegetation science concepts. I. Initial floristic composition, a factor in old-field vegetation development. *Vegetatio*, **4**, 412–17.

Enright, N. J. and Lamont, B. B. (1989) Seed banks, fire season, safe sites and seedling recruitment in five co-occurring *Banksia* species. *J. Ecol.*, **77**, 1111–22.

Eriksson, O. (1986) Survivorship, reproduction and dynamics of ramets of *Potentilla anserina* on a Baltic seashore meadow. *Vegetatio*, **67**, 17–25.

Feekes, W. (1936) De ontwikkeling van de natuurlijke vegetatie in de Wieringermeerpolder. *Ned. Kruidk. Arch.*, **46**, 1–295.

Fowler, N. L. (1988) What is safe site?: neighbor, litter, germination date, and patch effects. *Ecology*, **69**, 947–61.

Fox, J. F. (1983) Germinable seed banks of interior Alaska tundra. *Arctic Alpine Res.*, **15**, 405–11.

Fridriksson, S. (1975) *Surtsey,* Butterworths, London.

Galinato, M. I. and van der Valk, A. G. (1986) Seed germination traits of annuals and emergents recruited during drawdowns in the Delta Marsh, Manitoba, Canada. *Aquatic Bot.*, **26**, 89–102.

Gardner, G. (1977) The reproductive capacity of *Fraxinus excelsior* on the Derbyshire limestone. *J. Ecol.*, **65**, 107–18.

Gartner, B. L., Chapin, F. S. III. and Shaver, G. R. (1986) Reproduction of *Eriophorum vaginatum* by seed in Alaskan tussock tundra. *J. Ecol.*, **74**, 1–18.

Garwood, N. C. (1983) Seed germination in a seasonal tropical forest in Panama: A community study. *Ecol. Monogr.*, **53**, 159–81.

Garwood, N. C. (1989) Tropical soil seed banks: A review, in *Ecology of Soil Seed Banks* (eds M. A. Leck, V. T. Parker and R. L. Simpson), Academic Press, San Diego, pp. 149–209.

Glenn-Lewin, D. C. (1980) The individualistic nature of plant community development. *Vegetatio*, **43**, 141–6.

Goldberg, D. E. and Werner, P. A. (1983) The effects of size of openings in vegetation and litter cover on seedling establishment of goldenrods (*Solidago* spp.). *Oecologia (Berlin)*, **60**, 149–55.

Gorchov, D. L. (1988) Does asynchronous fruit ripening avoid satiation of seed dispersers?: a field test. *Ecology*, **69**, 1545–51.

Granström, A. (1987) Seed viability of fourteen species during five years storage in a forest soil. *J. Ecol.*, **75**, 321–31.

Green, D. S. (1980) The terminal velocity and dispersal of spinning samaras. *Amer. J. Bot.*, **67**, 1218–24.

Greene, D. F. and Johnson, E. A. (1989) A model of wind dispersal of winged or plumed seeds. *Ecology*, **70**, 339–47.

Grime, J. P., Mason, G., Curtis, A. V., Rodman, J., Band, S. R., Mowforth, M. A. G., Neal, A. M. and Shaw, S. (1981) Comparative study of germination characteristics in a local flora. *J. Ecol.*, **69**, 1017–59.

Guevara, S. and Gomez-Pompa, A. (1972) Seeds from surface soils in a tropical region of Veracruz, Mexico. *J. Arnold Arboretum*, **53**, 312–35.

Gunn, C. R., Dennis, J. V. and Paradine, P. J. (1976) *World Guide to Tropical Drift Seeds and Fruits*, Quadrangle/New York Times Book Company, New York.

Hacker, J. B. (1990) Drift seeds and fruit on Raine Island, northern Great Barrier Reef, Australia. *J. Biogeog.*, **17**, 19–24.

Harper, J. L. (1961) Approaches to the study of plant competition. *Symp. Soc. Exp. Biol.*, **15**, 1–39.

Harper, J. L. (1977) *Population Biology of Plants*, Academic Press, New York.

Harper, J. L., Clatworthy, J. N., McNaughton, I. H. and Sagar, G. R. (1961) The evolution of closely related species living in the same area. *Evolution*, **15**, 209–27.

Hazebroek, J. P. and Metzger, J. D. (1990) Environmental control of seed germination on *Thlaspi arvense* (Cruciferae). *Amer. J. Bot.*, **77**, 945–53.

Henderson, C. B., Petersen, K. E. and Redak, R. A. (1988) Spatial and temporal patterns in the seed bank and vegetation of a desert grassland community. *J. Ecol.*, **76**, 717–28.

Herrera, C. M. and Jordano, P. (1981) *Prunus mahaleb* and birds: the high efficiency seed dispersal system of a temperate tree. *Ecol. Monogr.*, **51.**, 203–18.

Hibbs, D. E. (1983) Forty years of forest succession in central New England. *Ecology*, **64**, 1394–1401.

Holm, L. G., Weldon, L. W. and Blackburn, R. D. (1969) Aquatic weeds. *Science*, **166**, 699–709.

Holt, B. R. (1972) Effect of arrival time on recruitment, mortality, and reproduction in successional plant populations. *Ecology*, **53**, 668–73.

Hoppes, W. G. (1988) Seedfall patterns of several species of bird-dispersed plants in an Illinois woodland. *Ecology*, **69**, 320–9.

Howe, H. F. (1986) Seed dispersal by fruit-eating birds and mammals, in *Seed Dispersal* (ed. D. R. Murray), Academic Press, Sydney, pp. 123–89.

Howe, H. F. and Smallwood, J. (1982) Ecology of seed dispersal. *Ann. Rev. Ecol. System.*, **13**, 201–28.

Howe, H. F. and Westley, L. C. (1988) *Ecological Relationships of Plants and Animals*, Oxford University Press, New York.

Huenneke, L. F. and Sharitz, R. R. (1990) Substrate heterogeneity and regeneration of a swamp tree, *Nyssa aquatica. Amer. J. Bot.*, **77**, 413–19.

Huntley, B. and Birks, H. J. B. (1983) *An Atlas of Past and Present Pollen Maps for Europe*: *0–13000 Year Ago*, Cambridge Univ. Press, Cambridge.

Izhaki, I. and Safriel, U. N. (1990) The effects of some Mediterranean scrubland frugivores upon germination patterns. *J. Ecol.*, **78**, 56–65.

Janzen, D. H. (1971) Seed predation by animals. *Ann. Rev. Ecol. System.*, **2**, 465–92.

Johnson, W. C. and Webb, T. III. (1989) The role of bluejays (*Cyanocitta cristata* L.) in the postglacial dispersal of fagaceous trees in eastern North America. *J. Biogeog.*, **16**, 561–71.

Johnstone, I. M. (1986a) Plant invasion windows: a time-based classification of invasion potential. *Biol. Rev.*, **61**, 369–94.

Johnstone, I. M. (1986b) Macrophyte management: An integrated perspective. *New Zealand J. Marine Freshwater Res.*, **20**, 599–614.

Keddy, P. A. and Constabel, P. (1986) Germination of ten shoreline plants in relation to seed size, soil particle size and water level: An experimental study. *J. Ecol.*, **74**, 133–41.

Keddy, P. A. and Ellis, T. H. (1985) Seedling recruitment of 11 wetland species along a water level gradient: shared or distinct responses. *Canad. J. Bot.*, **63**, 1876–9.

Keeley, J. E. (1977) Seed production, seed populations in soil, and seedling production after fire for two congeneric pairs of sprouting and non-sprouting chaparral shrubs. *Ecology*, **58**, 820–9.

Keeley, J. E. and Zedler, P. H. (1978) Reproduction of chaparral shrubs after fire: A comparison of sprouting and seeding strategies. *Amer. Mid. Nat.*, **99**, 142–61.

Kellman, M. C. (1970) The viable seed content of some forest soil in coastal British Columbia. *Canad. J. Bot.*, **48**, 1383–5.

Kellman, M. C. (1978) Microdistribution of viable weed seed in two tropical soils. *J. Biogeog.*, **5**, 291–300.

Kemp, P. R. (1989) Seed banks and vegetation processes in deserts, in *Ecology of Soil Seed Banks* (eds M. A. Leck, V. T. Parker and R. L. Simpson), Academic Press, San Diego, pp. 257–81.

King, T. J. (1975) Inhibition of seed germination under leaf canopies in *Arenaria serpyllifolia*, *Veronica arvensis* and *Cerastium holosteoides*. *New Phytol.*, **75**, 870–90.

Klinkhamer, P. G., de Jong, T. J. and van der Meiden, E. (1988) Production, dispersal and predation of seeds in the biennial *Cirsium vulgare. J. Ecol.*, **76**, 403–14.

Krefting, L. W. and Roe, E. I. (1949) The role of some birds and mammals in seed germination. *Ecol. Monogr.*, **19**, 269–86.

Kurchenko, E. I. (1985) Coenopopulation structure of *Agrostis species*, in *The Population Structure of Vegetation* (ed. J. White), Dordrecht, Junk, pp. 207–224.

Leck, M. A. (1980) Germination in Barrow, Alaska, tundra soils cores. *Arctic Alpine Res.*, **12**, 343–9.

Leck, M. A. and Graveline, K. J. (1979) The seed bank of a freshwater tidal marsh. *Amer. J. Bot.*, **66**, 1006–15.

Leck, M. A., Parker, V. T. and Simpson, R. L. (1989) *Ecology of Soil Seed Banks*, Academic Press, San Diego.

Levin, D. A. and Kerster, H. W. (1974) Gene flow in seed plants. *Evolut. Biol.*, **7**, 139–230.

Livingston, R. B. (1972) Influence of birds, stones and soil on the establishment of pasture juniper, *Juniperus communis*, and red cedar, *J. virginiana*, in New England. *Ecology*, **53**, 1141–7.

Livingston, R. B. and Allessio, M. L. (1968) Buried viable seed in successional field and forest stands, Harvard Forest, Massachusetts. *Bull. Torrey Bot. Club*, **95**, 58–69.

Louda, S. M. (1989) Predation in the dynamics of seed regeneration, in *Ecology of Soil Seed Banks* (eds M. A. Leck, V. T. Parker and R. L. Simpson), Academic Press, San Diego, pp. 25–51.

Louda, S. M. and Zedler, P. H. (1985) Predation in insular plant dynamics: An experimental assessment of postdispersal fruit and seed survival, Enewak Atoll, Marshall Islands. *Amer. J. Bot.*, **72**, 438–45.

Lovett Doust, L. (1981) Population dynamics and local specialization in a clonal perennial (*Ranunculus repens*). I. The dynamics of ramets in contrasting habitats. *J. Ecol.*, **69**, 743–55.

Major, J. and Pyott, W. T. (1966) Buried viable seeds of two California bunchgrass sites and their bearing on the definition of a flora. *Vegetatio*, **13**, 253–82.

Marks, P. M. (1974) The role of pin cherry (*Prunus pensylvanica* L.) in the maintenance of stability in northern hardwood ecosystems. *Ecol. Monogr.*, **44**, 73–88.

Matlack, G. R. (1989) Secondary dispersal of seed across snow in *Betula lenta*, a gap-colonizing tree species. *J. Ecol.*, **77**, 853–69.

Mayer, A. M. and Poljakoff-Mayber, A. (1982) *The Germination of Seeds*, 3rd edn, Pergamon Press, New York.

McLeod, K. W. and McPherson, J. K. (1973) Factors limiting the distribution of *Salix nigra. Bull. Torrey Bot. Club*, **100**, 102–10.

McConnaughay, K. D. M. and Bazzaz, F. A. (1987) The relationship between gap size and performance of several colonizing annuals. *Ecology*, **68**, 411–16.

McDonnell, M. J. and Stiles, E. W. (1983) The structural complexity of old-field vegetation and the recruitment of bird-dispersed plant species. *Oecologia (Berlin)*, **56**, 109–16.

McEvoy, P. B. and Cox, C. S. (1987) Wind dispersal distances in dimorphic achenes of ragwort, *Senecio jacobaea. Ecology*, **68**, 2006–15.

McGraw, J. B. and Vavrek, M. C. (1989) The role of buried viable seeds in arctic and alpine plant communities, in *Ecology of Soil Seed Banks* (eds M. A. Leck, V. T. Parker and R. L. Simpson), Academic Press, San Diego, pp. 91–105.

Meredith, T. C. (1985) Factors affecting recruitment from the seed bank of sedge (*Cladium mariscus*) dominated communities at Wicken Fen, Cambridgeshire, England. *J. Biogeog*, **12**, 463–72.

Middleton, B. A., van der Valk, A. G., Mason, D. H., Williams, R. L. and Davis, C. B. (1991) Vegetation dynamics and seed banks of a monsoonal wetland overgrown with *Paspalum distichum* L. in northern India. *Aquatic Botany*, **40**, 239–59.

Mitchell, D. S. (1985) African aquatic weeds and their management, in *The*

Ecology and Management of African Wetland Vegetation (ed. P. Denny), Junk, Dordrecht, pp. 177–202.

Morris, W. F., Marks, P. L., Mohler, C. L., Rappaport, N. R., Welsey, F. R. and Moran, M. A. (1986) Seed dispersal and seedling emergence in an old field community in central New York (USA). *Oecologia (Berlin)*, **70**, 92–9.

Murray, D. R. (1986a) *Seed Dispersal*, Academic Press, Sydney.

Murray, D. R. (1986b) Seed dispersal by water, in *Seed Dispersal* (ed. D. R. Murray), Academic Press, Sydney, pp. 49–85.

Murray, K. G. (1988) Avian seed dispersal of three neotropical gap-dependent plants. *Ecol. Monogr.*, **68**, 271–98.

Nakagoshi, N. (1985) Buried viable seeds in temperate forests, in *The Population Structure of Vegetation* (ed. J. White), Junk, Dordrecht, pp. 551–70.

Niering, W. A. and Egler, F. E. (1955) A shrub community of *Viburnum lentago*, stable for twenty-five years. *Ecology*, **36**, 356–60.

Niering, W. A. and Goodwin, R. H. (1974) Creation of relatively stable shrublands with herbicides: Arresting 'succession' on rights-of-way and pastureland. *Ecology*, **55**, 784–95.

Niering, W. A., Whittaker, R. H. and Lowe, C. H. (1963) The saguaro: A population in relation to environment. *Science*, **142**, 15–23.

Nip-van der Voort, J., Hengeveld, R. and Haeck, J. (1979) Immigration rates of plant species in three Dutch polders. *J. Biogeog.*, **6**, 301–8.

Noble, I. R. and Slatyer, R. O. (1980) The use of vital attributes to predict successional changes in plant communities subject to recurrent disturbance. *Vegetatio*, **43**, 5–21.

Okubo, A. and Levin, S. A. (1989) A theoretical framework for data analysis of wind dispersal of seed and pollen. *Ecology*, **70**, 329–38.

Parker, V. T. and Kelly, V. R. (1989) Seed banks in California chaparral and other Mediterranean climate shrublands, in *Ecology of Soil Seed Banks* (eds M. A. Leck, V. T. Parker and R. L. Simpson), Academic Press, San Diego, pp. 231–55.

Peart, D. R. (1989a) Species interactions in a successional grassland. I. Seed rain and seedling recruitment. *J. Ecol.*, **77**, 236–51.

Peart, D. R. (1989b) Species interactions in a successional grassland. II. Colonization of vegetated sites. *J. Ecol.*, **77**, 252–66.

Peart, D. R. (1989c) Species interactions in a successional grassland. III. Effects of canopy gaps, gopher mounds and grazing on colonization. *J. Ecol.*, **77**, 267–89.

Peart, D. R. and Foin, T. C. (1985) Analysis and prediction of population and community change: a grassland case study, in *The Population Structure of Vegetation* (ed. J. White), Junk, Dordrecht, pp. 313–39.

Pickett, S. T. A. and McDonnell, M. J. (1989) Seed bank dynamics in temperate deciduous forest, in *Ecology of Soil Seed Banks* (eds M. A. Leck, V. T. Parker and R. L. Simpson), Academic Press, San Diego, pp. 123–47.

Pickett, S. T. A. and White, P. S. (1985) *The Ecology of Natural Disturbance and Patch Dynamics*, Academic Press, New York.

Platt, W. J. (1975) The colonization and formation of equilibrium plant species associations on badger disturbances in a tall grass prairie. *Ecol. Monogr.*, **45**, 285–305.

Platt, W. J. and Weis, I. M. (1977) Resource partitioning and competition within a guild of fugitive prairie plants. *Amer. Nat.*, **111**, 479–513.

Price, M. V. and Jenkins, S. H. (1986) Rodents as seed consumers and dispersers, in *Seed Dispersal* (ed. D. R. Murray), Academic Press, Sydney, pp. 191–235.

Price, M. V. and Reichman, O. J. (1987) Distribution of seeds in Sonoran Desert soils: implications for heteromyid rodent foraging. *Ecology*, **68**, 1797–1811.

Rabinowitz, D. (1981) Buried viable seeds in a North American tall-grass prairie: The resemblance of their abundance and composition to dispersing seeds. *Oikos*, **36**, 191–5.

Rabinowitz, D. and Rapp, J. K. (1980) Seed rain in a North American tall grass prairie. *J. Appl. Ecol.*, **17**, 793–802.

Rabinowitz, D. and Rapp, J. K. (1981) Dispersal abilities of seven sparse and common grasses from a Missouri prairie. *Amer. J. Bot.*, **68**, 616–24.

Rabotnov, T. A. (1978) On coenopopulations of plants reproducing by seeds, in *Structure and Functioning of Plant Populations* (eds A. H. J. Freysen and J. W. Woldendorp), North-Holland, Amsterdam, pp. 1–26.

Rabotnov, T. A. (1985) Dynamics of plant coenotic populations, in *The Population Structure of Vegetation* (ed. J. White), Junk, Dordrecht, pp. 121–42.

Reader, R. J. and Buck, J. (1986) Topographic variation in the abundance of *Hieracium floribundum*: relative importance of differential seed dispersal, seedling establishment, plant survival and reproduction. *J. Ecol.*, **74**, 815–22.

Reichman, O. J. (1979) Desert granivore foraging and its impact on seed densities and distribution. *Ecology*, **60**, 1085–92.

Reichman, O. J. (1984) Spatial and temporal variation of seed distributions in Sonoran Desert soils. *J. Biogeog.*, **11**, 1–11.

Rice, E. L. (1984) *Allelopathy*, 2nd edn, Academic Press, Orlando.

Rice, K. J. (1989) Impacts of seed banks on grassland community structure and population dynamics, in *Ecology of Soil Seed Banks* (eds M. A. Leck, V. T. Parker and R. L. Simpson), Academic Press, San Diego, pp. 211–30.

Ridley, H. N. (1930) *The dispersal of plants throughout the world*. Reeve, Ashford, Kent.

Roberts, H. A. (1981) Seed banks in soils. *Adv. Appl. Biol.*, **6**, 1–55.

Rozema, J. (1975) The influence of salinity, inundation and temperature on the germination of some halophytes and nonhalophytes. *Oecologia Plantarum*, **10**, 341–53.

Ryvarden, L. (1971) Studies in seed dispersal. I. Trapping of diaspores in the alpine zone at Finse, Norway. *Norwegian J. Bot.*, **18**, 215–26.

Sabiiti, E. N. and Wein, R. W. (1987) Fire and *Acacia* seeds: a hypothesis of colonization success. *J. Ecol.*, **74**, 937–46.

Saulei, S. M. and Swaine, M. D. (1988) Rain forest seed dynamics during succession at Gogol, Papua New Guinea. *J. Ecol.*, **76**, 1133–52.

Schenkeveld, A. J. and Verkaar, H. J. (1984) The ecology of shortlived forbs in chalk grasslands: Distribution of germinative seeds and its significance for seedling emergence. *J. Biogeog.*, **11**, 251–60.

Schneider, R. L. and Sharitz, R. R. (1988) Hydrochory and regeneration in a bald cypress-tupelo swamp forest. *Ecology*, **69**, 1055–63.

Schupp, E. W. (1990) Annual variation in seedfall, postdispersal predation, and recruitment of a neotropical tree. *Ecology*, **71**, 504–15.

Sculthorpe, C. D. (1967) *The Biology of Aquatic Vascular Plants*, Edward Arnold, London.

Sheldon, J. C. and Burrows, F. M. (1973) The dispersal effectiveness of the achene-pappus units of selected Compositae in steady winds with convection. *New Phytol.*, **72**, 665–75.

Silander, J. A. (1985) Microevolution in clonal plants, in *Population Biology and Evolution of Clonal Organisms* (eds J. B. C. Jackson, L. W. Buss and R. E. Cook), Yale University Press, New Haven, pp. 107–52.

Silvertown, J. W. (1980a) The evolutionary ecology of mast seeding in trees. *Biol. J. Linn. Soc.*, **14**, 235–50.

Silvertown, J. W. (1980b) Leaf-canopy-induced seed dormancy in a grassland flora. *New Phytol.*, **85**, 109–18.

Silvertown, J. (1987) *Introduction to Plant Population Ecology*, 2nd edn, Longman, Harlow.

Smith, A. J. (1975) Invasion and ecesis of bird-disseminated woody plants in a temperate forest sere. *Ecology*, **56**, 19–34.

Smith, B. H., Forman, P. D. and Boyd, A. E. (1989) Spatial patterns of seed dispersal and predation of two myrmecochorous forest herbs. *Ecology*, **70**, 1649–56.

Smith, J. M. B. (1990) Drift disseminules on Fijian beaches. *New Zealand J. Bot.*, **28**, 13–20.

Smith, J. M. B., Heatwole, H., Jones, M. and Waterhouse, B. M. (1990) Drift disseminules on cays of the Swain Reefs, Great Barrier Reef, Australia. *J. Biogeog.*, **17**, 5–17.

Sork, V. L. (1987) Effect of predation and light on seedling establishment in *Gustava superba*, *Ecology*, **68**, 1341–50.

Stamp, N. E. (1989) Seed dispersal of four sympatric grassland annual species of *Erodium*. *J. Ecol.*, **77**, 1005–20.

Steenberg, W. V. and Lowe, C. H. (1977) Ecology of the saguaro: II. Reproduction, germination, establishment, growth, and survival of the young plant. *National Parks Service Scientific Monograph Series No. 8*, US Government Printing Office, Washington, D.C

Steenberg, W. V. and Lowe, C. H. (1983) Ecology of saguaro: III. Growth and demography. *National Parks Service Scientific Monograph Series No. 17*, US Government Printing Office, Washington, D.C

Stephenson, A. G. (1981) Flower and fruit abortion: proximate causes and ultimate functions. *Ann. Rev. Ecol. System.*, **12**, 253–79.

Stieperaere, H. and Timmerman, C. (1983) Viable seeds in the soils of some parcels of reclaimed and unreclaimed heath in the Flemish district (northern Belgium). *Bull. Soc. Roy. Bot. Belgium*, **116**, 62–73.

Stiles, E. W. (1989) Fruits, seeds and dispersal agents, in *Plant–Animal Interactions* (ed. W. G. Abrahamson), McGraw-Hill, New York, pp. 87–122.

Symonides, E. (1985) Population structure of psammophyte vegetation, in *The Population Structure of Vegetation* (ed. J. White), Junk, Dordrecht, pp. 265–91.

Thompson, K. (1978) The occurrence of buried viable seeds in relation to environmental gradients. *J. Biogeog.*, **5**, 425–30.

Thompson, K. and Grime, J. P. (1979) Seasonal variation in seed banks of herbaceous species in ten contrasting habitats. *J. Ecol.*, **67**, 893–921.

Thompson, K. and Grime, J. P. (1983) A comparative study of germination responses to diurnally-fluctuating temperatures. *J. Appl. Ecol.*, **20**, 141–56.

Tiffney, B. H. (1986) Evolution of seed dispersal syndromes according to the fossil record, in *Seed Dispersal* (ed. D. R. Murray), Academic Press, Sydney, pp. 273–305.

Trapp, E. J. (1988) Dispersal of heteromorphic seeds in *Amphicarpaea bracteata* (Fabaceae), *Amer. J. Bot.*, **75**, 1535–9.

Tsuyuzaki, S. (1989) Analysis of revegetation dynamics on the volcano Usu, northern Japan, deforested by 1977–1978 eruptions. *Amer. J. Bot.*, **76**, 1468–77.

van der Pijl, L. (1982) *Principles of Dispersal in Higher Plants*, Springer-Verlag, New York.

van der Valk, A. G. (1974) Environmental factors controlling the distribution of forbs on coastal foredunes in Cape Hatteras National Seashore. *Canad. J. Bot.*, **52**, 1057–1073.

van der Valk, A. G. (1981) Succession in wetlands: A Gleasonian approach. *Ecology*, **62**, 688–96.

van der Valk, A. G. (1985) Vegetation dynamics of prairie glacial marshes, in *The Population Structure of Vegetation* (ed. J. White), Junk, Dordrecht, pp. 293–312

van der Valk, A. G. (1986a) Vegetation dynamics of freshwater wetlands; A selective review of the literature. *Arch. Hydrobiol. Beih. Ergebn. Limnol.*, **27**, 27–39.

van der Valk, A. G. (1986b) The impact of litter and annual plants on recruitment from the seed bank of a lacustrine wetland. *Aquatic Bot.*, **24**, 13–26.

van der Valk, A. G. and Davis, C. B. (1978) The role of seed banks in the vegetation dynamics of prairie glacial marshes. *Ecology*, **59**, 322–35.

van der Valk, A. G. and Pederson, R. L. (1989) Seed banks and the management and restoration of natural vegetation, in *The Ecology of Soil Seed Banks* (eds M. A. Leck, V. T. Parker and R. L. Simpson), Academic Press, San Diego, pp. 329–46.

van der Valk, A. G., Pederson, R. L. and Davis, C. B. (1992) Restoration and creation of freshwater wetlands using seed banks. *Wetland Ecol. Mgmt*, **1**, 191–7.

van der Valk, A. G., Swanson, S. D. and Nuss, R. F. (1983) The response of plant species to burial in three types of Alaskan wetlands. *Canad. J. Bot.*, **61**, 1150–64.

van der Valk, A. G. and Welling, C. H. (1988) The development of zonation in freshwater wetlands: an experimental approach, in *Diversity and Pattern in Plant Communities* (eds H. J. During, M. J. A. Werger and J. H. Willems), SPB Acad. Publ., The Hague, pp. 145–58.

van der Valk, A. G., Welling C. H. and Pederson, R. L. (1989) Vegetation change in a freshwater wetland: a test of *a priori* predictions, in *Freshwater Wetlands and Wildlife* (eds R. R. Sharitz and J. W. Gibbons), USDOE Office of Scientific and Technical Information, Oak Ridge, Tennessee, pp. 207–17.

Vazquez-Yanes, C. and Smith, H. (1982) Phytochrome control of seed germination in the tropical rain forest pioneer trees *Cecropia obtusifolia* and *Piper auritum* and its ecological significance. *New Phythol.*, **92**, 477–86.

Verkaar, H. J., Schenkeveld, A. J. and van der Klanshorst, M. P. (1983) On the ecology of short-lived forbs in chalk grasslands: Dispersal of seeds. *New Phytol.*, **95**, 335–44.

Vorontzova, L. I. and Zaugolnova, L. B. (1985) Population biology of steppe plants, in *The Population Structure of Vegetation* (ed. J. White), Junk, Dordrecht, pp. 143–78.

Walker, L. R., Zasada, J. C. and Chapin, F. C. III. (1986) The role of life history processes in primary succession on an Alaskan floodplain. *Ecology*, **67**, 1243–53.

Watt, A. S. (1923) On the ecology of British beechwoods with special reference to their regeneration. *J. Ecol.*, **11**, 1–48.

Webb, L. J., Tracey, J. G. and Williams, W. T. (1972) Regeneration and pattern in the subtropical rain forest. *J. Ecol.*, **60**, 675–95.

Welling, C. H., Pederson, R. L. and van der Valk, A. G. (1988a) Recruitment from the seed bank and the development of zonation of emergent vegetation during a drawdown in a prairie wetland. *J. Ecol.*, **76**, 483–96.

Welling, C. H., Pederson, R. L. and van der Valk, A. G. (1988b) Temporal patterns in recruitment from the seed bank during drawdowns in a prairie wetland. *J. Appl. Ecol.*, **25**, 999–1007.

Went, F. W. (1949) Ecology of desert plants. II. The effect of rain and temperature on germination and growth. *Ecology*, **30**, 1–13.

Westoby, M., Rice, B. and Howell, J. (1990) Seed size and plant growth form as factors in dispersal spectra. *Ecology*, **71**, 1307–15.

Whitmore, T. C. (1984) *Tropical Rain Forests of the Far East*, Clarendon Press, Oxford.

Whittaker, R. J., Bush, M. B. and Richards, K. (1989) Plant recolonization and vegetation succession on the Krakatau Islands, Indonesia. *Ecol. Monogr.*, **59**, 59–123.

Wienhold, C. E. and van der Valk, A. G. (1989) The impact of duration of drainage on the seed banks of northern prairie wetlands. *Canad. J. Bot.*, **67**, 1878–84.

Willson, M. F. (1983) *Plant Reproductive Ecology*, Wiley, New York.

Wood, D. M. and del Moral, R. (1987) Mechanisms of early primary succession in subalpine habitats on Mount St Helens. *Ecology*, **68**, 780–90.

Wood, D. M. and del Moral, R. (1988) Colonizing plants on the pumice plains, Mount St Helens, Washington. *Amer. J. Bot.*, **75**, 1228–37.

Yarranton, G. A. and Morrison, R. G. (1974) Spatial dynamics of a primary succession: Nucleation. *J. Ecol.*, **62**, 417–28.

Zammit, C. and Westoby, M. (1988) Pre-dispersal seed losses, and the survival of seeds and seedlings of two serotinous *Banksia* shrubs in burnt and unburnt heath. *J. Ecol.*, **76**, 200–14.

3 Community structure and ecosystem function[1]

Robert K. Peet

3.1 INTRODUCTION

3.1.1 Why structure and function?

In the introduction to his 1972 book on geographical ecology, Robert MacArthur wrote, 'To do science is to search for repeated patterns, not simply to accumulate facts. . . . Science should be general in its principles.' Most ecologists would probably agree with MacArthur's definition. However, much of the fascination and challenge of ecology, like biology in general, is that it is a discipline with a great wealth of facts, but relatively few general principles. If there is one universal, it is change, genetic change and evolution being among the best-known examples. Change is all pervasive in ecological systems. Indeed, scientists have been publishing reports of dramatic and predictable changes in the composition and structure of ecosystems since before the word 'ecosystem', or even the word 'ecology', had been coined. Not surprisingly, the search for a general theory of ecosystem change, or succession, has long been a major focus of ecological research.

For a theory of succession to be general, it must not depend on the particular species present. Consequently, much recent research on succession has focused on aspects of community structure and ecosystem function because they are species independent. Succession is a central organizing concept in ecology, in large part because of its perceived potential to reveal and explain general patterns. The objectives of this chapter are to summarize predictable changes in community structure and ecosystem function that occur during succession, and to examine mechanisms that might be responsible for those patterns.

3.1.2 Succession as a population process

Scientific progress has often resulted from taking a mechanistic or even reductionist approach wherein a process is broken into its component

[1] Botanical nomenclature follows Kartesz (1993).

Plant Succession: Theory and prediction Edited by David C. Glenn-Lewin, Robert K. Peet and Thomas T. Veblen © 1992 Chapman & Hall, London ISBN 0 412 26900 7

parts so as to determine how the workings of the larger system are consequences of lower-order processes. Many critical review papers have called for a reformulation of succession theory, and most have suggested such a mechanistic approach (e.g. Drury and Nisbet, 1973; Pickett, 1976; Peet and Christensen, 1980; Pickett *et al.*, 1987a; Tilman, 1985, 1988; Walker and Chapin, 1987). Collectively, these papers suggest that successional change at the community or ecosystem level can be understood, at least in part, as a consequence of population processes of the component species, which in turn might be understood as consequences of species attributes, such as life history and physiological characteristics.

Population processes critical for a mechanistic theory of succession include establishment of new individuals and their subsequent growth, reproduction, loss to predation, and death (Clements, 1916; Walker and Chapin, 1987; see also Chapter 2). The importance of these population processes in driving succession is illustrated below by studies conducted on abandoned agricultural land in southeastern North America, and in natural montane conifer forests in the Rocky Mountains. Oldfield succe

Succession on abandoned agricultural land in southeastern North America is well documented (e.g. Oosting, 1942; Keever, 1950; Christensen and Peet, 1981). On a typical site fast-growing annual weeds such as *Conyza canadensis* (horseweed) and *Ambrosia artemisiifolia* (ragweed) are quickly replaced by slower growing but more competitive herbaceous perennial forbs and grasses (primarily *Andropogon virginicus*). Within a few years numerous pine seedlings (*Pinus taeda, P. echinata*) become established and grow to overtop the forbs and grasses which largely die, apparently unable to compete for such resources as light, water and nutrients. As the pines grow larger, fewer fit in the available canopy space and light becomes progressively more limiting. Mortality is high among the overtopped individuals and density steadily declines. This highly competitive environment also precludes establishment of new pine seedlings and severely limits seedling establishment, even by those species tolerant of competitive environments. However, after 60 or 70 years a moderate number of small trees of *Acer* (maple), *Quercus* (oak), *Carya* (hickory) and other species occur beneath the canopy pines.

Canopy trees must eventually reach a size where space or light freed by deaths of large trees cannot be pre-empted by expansion of the crowns of the adjacent trees. During this period tree deaths produce relatively long-lasting gaps wherein resources are underexploited. In these gaps new seedlings are often able to become established, and pre-existing under-storey trees can grow rapidly. Most of the dominant canopy pines die 75–125 years after establishment and a new cohort, usually comprised of species more tolerant of low resource availability, assumes dominance. The new dominants are less synchronous in their establishment and the forest they comprise has many ages and sizes of trees represented.

Four phases of forest development can be identified in this example

(Bormann and Likens, 1979; Oliver, 1980; Oliver and Larson, 1990; Peet, 1981b; Peet and Christensen, 1987). The first is the *establishment phase* during which competitive pressures are low and numerous tree seedlings become established. The second is the *thinning phase* during which competition is intense, establishment of new seedlings is minimal and stem density steadily declines. The breakup of the original canopy and establishment or release of understorey trees marks a third phase, the *transition phase*. Eventually the forest reaches a relatively stable composition with a balance of small and large trees, and a balance of closed-canopy and gap environments. This is the *steady-state phase*.

This four-phase model of stand development appears to apply to most forest ecosystems, particularly where there is a single, relatively long-lived, post-disturbance dominant species, like *Pinus taeda* (loblolly pine) in southeastern USA or *Populus tremuloides* (aspen) in the upper Midwest. Similar developmental sequences can be seen in such diverse systems as montane (Daubenmire and Daubenmire, 1968; Day, 1972; Oliver, 1980; Peet, 1981b) and boreal (Bloomberg, 1950; Plochmann, 1956) *Picea-Abies* (spruce-fir) forests recovering from fire, low-diversity tropical forests recovering from agriculture or wind storm (Whitmore, 1984), Mediterranean-climate shrublands after fire (Schlesinger and Gill, 1978) and perhaps even native grasslands (Watt, 1947). However, even among strictly forest systems, considerable variation can be seen, particularly as influenced by patterns of plant establishment.

The coniferous forests of the Colorado Front Range, virtually all of which are subject to periodic fires, provide examples of succession patterns which result from variation in tree establishment rates (Peet, 1981b, 1988). The lower slopes of the mountains are covered by grass and shrublands, with *Pinus ponderosa* (Ponderosa pine) woodlands starting at about 1600 m. Middle-elevation forests are often dominated by *Pinus contorta* (lodgepole pine), while high-elevation forests are composed principally of *Picea engelmannii* (Engelmann spruce) and *Abies lasiocarpa* (subalpine fir). Upper treeline is reached at around 3500 m. Post-fire stand development of middle-elevation forests of the Front Range conforms well to the four-stage model of forest development (Peet, 1981b). The primary difference from the pattern found in southeastern USA forests is speed of recovery. The initial establishment phase in these Rocky Mountain forests often lasts 50 years or more (Peet, 1981b; Parker and Peet, 1984; Veblen, 1986; Veblen and Lorenz, 1986) and the transition phase may not appear for over 300 years, as compared to canopy closure within 5 to 10 years and initiation of the transition phase by 80 years in successional pine forests of the Southeast.

Although middle-elevation forests of the Rocky Mountains conform well to the four-phase model, extreme high- and low-elevation forests often do not (Whipple and Dix, 1979; Peet, 1981b; Parker and Peet, 1984; Veblen, 1986; Veblen and Lorenz, 1986). At high elevations the period of tree establishment is long relative to tree longevity. As a consequence, canopy closure, which initiates the thinning phase, sometimes does not occur until after the first trees to establish have aged and started to die. In such cases the thinning phase is largely skipped. Similar succession patterns appear to be common on extreme high-elevation sites world wide.

Low-elevation *Pinus ponderosa* stands behave in a still different manner. These semi-arid woodlands only rarely, perhaps at 30–50 year intervals, experience the combination of factors needed for tree regeneration (Peet, 1981b; Peet, unpublished data; White, 1985; also see Neilson, 1986). At least in Arizona this has been shown to consist of a drought period during which the vigour of the competing grasses is reduced, followed by a year with heavy seed production, and then several moist years without fire. As a consequence, seedling recruitment in these woodlands is low and often episodic (see Cooper, 1960; Potter and Green, 1964; Larsen and Schubert, 1969; Peet, 1981b; White, 1985). As in the extreme high-elevation forests, chronic low levels of recruitment result in the thinning phase being largely skipped. In addition, the final phase is not really a steady-state phase, but rather exhibits pulses of establishment alternating with periods of declining tree seedling and sapling density. Similar episodic establishment can be expected in most deserts and semi-arid shrub and woodlands (see Zedler, 1981), systems where the four-phase model is of little utility because of the minimal interplant competition for canopy space.

3.1.3 Succession as a consequence of environmental change

Although population processes and interactions can explain some successional patterns, many important successional trends resist such explanations. Thus far, we have assumed that the background environment and resource supply rates for a given site are constant. This is rarely if ever true. Changes in soil resources and the physical environment, while often minor on time scales appropriate for study of population processes, can be critical factors driving long-term successional changes.

During soil development, at least in moist environments, mineral nutrients essential to plant growth are steadily dissolved and transported out of the soil system, or become chemically fixed in forms unavailable to plants. These are replaced primarily by nutrients derived from rainfall, dust, or weathering of rock. As rock and soil weathering proceed, many nutrients become depleted and the ecosystem depends progressively more on atmospheric inputs. The final equilibrium levels for critical elements like phosphorus and calcium can be far short of those necessary to sustain the levels of plant growth or biomass found early in the succession. In addition, disturbances such as fire, agriculture or biomass harvest can remove nutrients or convert them to forms more vulnerable to loss from the system.

Climate varies on both long- and short-term cycles (Chapter 8). Conditions in one year or in one century might be appropriate for the establishment of one set of species, and the next year or next century might be appropriate for another. A Rocky Mountain forest site supporting a *Picea* (spruce) forest established in 1600 when global temperatures were the coolest in the last 4000 years (Bray, 1971) might well not regrow to *Picea* following deforestation today, but might instead regrow to *Pseudotsuga* (Douglas-fir). Similarly, Kullman (1987, 1988) has shown shifts in the recruitment and population dynamics of Swedish *Pinus sylvestris* forests in response to climatic changes of the past few centuries. Studies of stratified pollen deposits in lakes and bogs show numerous major changes in vegetation world wide over the 12 000 years since the last glacial retreat (e.g. Davis, 1981; Huntley and Birks, 1983). Certainly we cannot view the compositional endpoint of succession to be fixed and predictable over any long period of time (Chapter 8). In short, succession must be viewed, to quote H. C. Cowles (1901) '. . . as a variable approaching a variable'.

3.1.4 Primary and secondary succession

Historically, ecologists have adopted the term *primary succession* for successional seres where little prior modification of the environment by organisms (reaction, *sensu* Clements, 1916; e.g. Cooper, 1911), especially soil formation, is evident, though in practice the term has mostly been applied where the dominant mechanism of change is long-term modification of the environment or of supply rates of soil resources. For example, Tilman and Gleeson (Tilman 1988, 1990; Gleeson and Tilman, 1990) suggest that primary succession refers principally to 'soil-driven' processes. In contrast, the term *secondary succession* has been suggested for ecosystems recovering from disturbance where the impact of previous organisms remains clear (usually, soil remains), but in practice this term has been applied largely to ecosystems where population processes provide the dominant mechanism of change. Tilman and Gleeson (Tilman, 1988, 1990; Gleeson and Tilman, 1990) again support the conventional perspective, although they prefer to use the term 'transient dynamics' for the dominant mechanisms behind secondary succession.

The original meanings of the terms primary and secondary succession derive from the now long-rejected theories of Clements (1916). Further, many cases of succession are intermediate, and the terms have never been rigorously defined. Ecology would probably be better served if its practitioners were to classify successional seres as to the dominant processes or mechanisms of change (Pickett *et al.*, 1987a; Walker and Chapin, 1987), but unfortunately the critical mechanisms are often unknown. Further, the terms primary and secondary succession are in common usage and are likely to remain so. *The solution I propose, and which I take in this chapter, is to use primary succession to refer to successions where the dominant mechanisms involve long-term environ-mental change, and to use secondary succession to refer to successions where the dominant mechanisms are population processes.* Simple, transient, plant-mediated changes in soils, such as increases in total nitrogen, are intermediate in character and important for both primary and secondary succession.

3.2 DISTURBANCE

3.2.1 Problems of scale

Community change is often categorized as either episodic (i.e. discontinuous but recurrent, usually allogenic) or gradual (i.e. continuous, usually autogenic). Historically, ecologists have called the first type disturbance and the second type succession. However, as is usually the case when we try to categorize natural phenomena, what at first appear to be discrete classes really intergrade to form a continuum. In practice, the distinction between gradual community change (succession) and succession-initiating disturbance depends on the scale of investigation. The death of a tree in a forest is certainly a disturbance that locally redefines the environment for the community of tree seedlings, herbs and bryophytes beneath the tree, but it is only part of the normal succession or regeneration process for the community of trees. If we accept tree death as part of forest succession, what of the simultaneous death of groups of adjacent trees? Just as tree death can be viewed not only as disturbance, but also as part of the regeneration dynamics of a forest, so can forest fires and other larger and dramatic disturbances be viewed as recurrent events in the natural disturbance regime of a landscape. A landscape perspective requires that all disturbances, major or minor, be viewed as components of a dynamic ecosystem mosaic. Because it is not possible to separate fully disturbance from succession, both must be included in any general treatment of vegetation dynamics.

3.2.2 Natural disturbance regimes

Careful examination of any landscape will inevitably reveal that the vegetation cover of that landscape is not homogeneous, but rather is continuously varying with individual localities representing various states of succession following past disturbances. In some landscapes the scale of variation will be predominantly small and correspond to deaths of individual plants. In others the scale will be much larger, resulting from more catastrophic events such as wildfires or hurricanes (White, 1979).

Each ecosystem has a disturbance regime, which can be compared with the disturbance regimes of other ecosystems in that landscape or else-where. The mean interval between disturbances is one of the simplest, yet nevertheless most meaningful statistics for comparing such disturbance regimes (Pickett and White, 1985). In addition, a complete character-ization would include the variances associated with the return intervals, plus the means and variances of disturbance size, shape and intensity. These are not easy statistics to determine, and in no case have all of them been determined for even a simple landscape.

Among the best studied disturbance regimes are those associated with wildfire. Mean return intervals (MRI) for fire have been determined for numerous North American forests, woodlands and grasslands. In one of the first comprehensive studies of this sort, Heinselman (1973) used fire scars on trees plus tree ages to reconstruct the borders of major fires that occurred in the Boundary Waters Canoe Area of northern Minnesota, USA, over approximately a 300-year period. In this 215000-ha region, fire frequency was shown to depend on site conditions and to have changed dramatically during the study interval. Before European settle-ment, dry sites dominated by *Populus tremulodeis* (aspen) or *Pinus banksiana* (jack pine) burned with a MRI of roughly 50 years, whereas *Pinus resinosa* (red pine) and *P. strobus* (white pine) sites had MRIs of 150–350 years. Soon after European settlement in the area in the mid-1800s, the MRI dropped sharply owing to careless use of fire, as well as a proclivity towards arson among the settlers. However, following intro-duction of modern fire-suppression policies, natural fire was effectively eliminated from much of the area, the current MRI being nearly 2000 years (Heinselman, 1981).

Heinselman's results are not necessarily typical of northern forests or even those of northern Minnesota, nor should they be interpreted to mean that other disturbance events are unimportant. Further west in Minnesota, nearer the ecotone between forest and grassland, Frissell (1973) found that the pre-1900 MRI for fire was roughly 20 years. Here the typical forest fire was not the stand-destroying conflagration typical of the Boundary Waters area, but a creeping ground fire that kept the woods open and facilitated regeneration of light-demanding species. No formal studies have been made of fire frequency in the grasslands found immediately west of Frissell's study area, but historical evidence suggests that most tall-grass prairies near the forest border burned with a MRI of 2–5 years (see Curtis, 1959; Kucera, 1981; Collins and Barber, 1985).

Other forms of disturbance important in vegetation dynamics can also be characterized by MRIs and similar statistics. Canham and Louck's (1984) analysis of presettlement blowdowns of timber in northern Wisconsin, wherein they found a 1210 year MRI for blowdowns of 0.5 ha or larger, shows that windstorms can be locally important and require the same sort of analytical treatment as fire. Treefall gaps are ubiquitous in forest ecosystems and thus of widespread interest. The MRIs range between 50 and 200 years for most forest types, though scattered indi-vidual trees can have much greater longevities (Runkle, 1985). Relatively high turnover in grasslands has been reported to result from activities of ants, pocket gophers, prairie dogs, badgers and crayfish (e.g. Baxter and Hole, 1967; Grubb *et al.*, 1969; Platt, 1975). Among other important disturbance types, floods, avalanches, landslides, insect outbreaks, shifting cultivation and volcanism all can be important (see White, 1979). With this great diversity of disturbance types, the best that the vegetation scientist can hope to achieve is a characterization of the disturbance regimes for the most significant factors influencing his or her study area. Other disturbance types will be sufficiently infrequent that they can only be treated on a landscape or regional basis (e.g. Harmon *et al.*, 1983).

Disturbance regimes themselves are subject to change during succession. For example, fuel accumulation can increase the probability of fire, and mean tree size correlates with risk of catastrophic blowdown. Because species vary in their susceptibility to disturbances such as fire (Mutch, 1970) and wind (Webb, 1989), species replacements during succession can cause major changes in disturbance regimes (e.g. Jackson, 1968). Even changes in stand structure can be important. For example, in a study of moist tropical forest isolated by the creation of Lake Gatun during the construction of the Panama Canal, Brokaw (1982) found oldgrowth forest to have a treefall gap MRI of 114 years and a mean gap size of $277 \, m^2$. In contrast, in an adjacent second-growth forest he found a MRI of 159 years and a mean gap size of $198 \, m^2$. This example shows how a MRI can change during stand development, plus it demonstrates that just because a forest is of great age and has experienced little disturbance does not mean that it has a low tree turnover rate.

Disturbance regimes were first seriously studied when ecologists realized that because individual locations in a community undergo major changes owing to disturbance and the subsequent recovery processes, the steady-state for the ecosystem can be described effectively only by scaling up so that all of the recovery stages are represented in the observation area. However, even at such large scales we should be cautious about expecting equilibrium. For example, Baker (1989) re-examined Heinselman's (1973) data from the Boundary Waters Canoe Area and found that if he divided the region into three geographic regions, the regions had clear differences in their disturbance regimes. Further, normal climatic variation is often accompanied by changes in disturbance regimes, particularly wind and fire regimes (Clark, 1988, 1990a,b). Thus, while it is convenient and useful to characterize ecosystems in terms of their disturbance regimes, it must be remembered that any such characterization is at best an approximation subject to both spatial and temporal variation.

3.3   BIOMASS AND PRODUCTION

One of the most fundamental yet obvious observations about succession is that biomass is initially low or absent, and eventually reaches a higher level (e.g. Odum, 1969). Since biomass (and thus leaf area) is absent or very low early in succession, primary production must also be assumed to start low, and to increase during succession. However, it is not obvious whether to expect these increases to be monotonic, or even to parallel one another. Certainly numerous trends have been reported in the forest succession literature (Peet, 1981a; Shugart, 1984; Sprugel, 1985), but no consensus has emerged as to which trends are associated with which conditions. Even less information is available for non-forested terrestrial systems.

3.3.1 Biomass accumulation in forests

Numerous successional trends in forest biomass have been hypothesized. The simplest hypothesis is that biomass increases in a smooth, linear or logistic fashion towards a maximum at climax (Fig. 3.1a). This was implied but not explicitly stated in Odum's (1969) review, and is given as a general trend in Sprugel's (1985) review. High-elevation *Pinus albicaulis* (whitebark pine) forests in Montana (Forcella and Weaver, 1977), *Pinus banksiana* forests in New Brunswick (MacLean and Wein, 1976), *Abies balsamea* (balsam fir) forests in New Hampshire (Sprugel, 1984), coastal *Picea–Tsuga* (spruce–hemlock) in Oregon (Harcombe *et al.*, 1990) and *Quercus–Pinus* (oak–pine) forests on Long Island, New York (Holt and Woodwell, in Whittaker, 1975) all have been reported to be consistent with such a logistic biomass accretion model.

A second hypothesis or model of biomass accumulation (Fig. 3.1a) with wide support in the forestry literature is that biomass increases to a maximum, and then decreases to an intermediate level at climax (Loucks, 1970; Peet, 1981a). This pattern can be readily explained in terms of a population or gap-dynamics model of forest development (Bormann and Likens, 1979; Peet, 1981a,b; Shugart, 1984). In this perspective, a forest is viewed as a mosaic of patches with each patch big enough for from one to several canopy trees (see Chapter 4). At the beginning of secondary succession, all the patches contain only tree seedlings and thus little biomass. However, late in the thinning phase the great majority of the patches are occupied by large trees and average biomass is high. As these large trees die and are replaced by small trees, the average patch biomass drops. In steady-state or climax forests some patches are occupied by young trees and others by old trees with the consequence that the average patch biomass is at some intermediate level. Because, failing another disturbance, we cannot expect all the patches to again synchronously have large trees, the forest cannot be expected to again approach the peak biomass achieved late in the thinning phase.

A third hypothesis is that biomass increases to a maximum late in the thinning phase, but then declines dramatically with a subsequent recovery to the steady-state value, sometimes with a series of damped oscillations

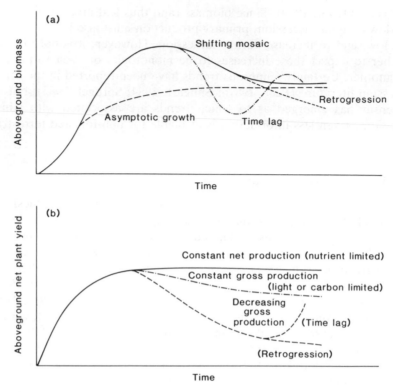

Figure 3.1 Successional changes in forest (a) biomass and (b) net plant yield as predicted by the various models described in the text. Biomass has been variously proposed to increase to an asymptote, and to increase to a peak associated with maturation of the initial post-disturbance cohort of trees and then decrease. The decrease has been proposed to be asymptotic, to oscillate around a steady state, and to steadily decline. Net plant yield has been proposed to increase to an asymptote (limitation by nutrient supply rate), to increase to a peak and then decrease such that gross production is constant (limitation by light or carbon compounds), and to increase to a maximum and then decrease sharply as a consequence of either changing population structure or soil fertility. (Modified from Peet, 1981a.)

around the asymptote (Fig. 3.1a). Even-aged coniferous forests in Finland (Ilvessalo, 1937; Sirén, 1955), Alberta (Plochmann, 1956) and Colorado (Peet, 1981b) are consistent with this hypothesis. The population or patch dynamics mechanism described above can also account for this pattern if we include some mechanism that assures that canopy tree death is relatively synchronous. The more synchronous the deaths, the more dramatic will be the drop and subsequent oscillations in biomass.

Forests of the Colorado Front Range contain examples of all three hypothesized trends (Peet, 1981a). Even-aged tree populations at middle elevations (e.g. *Pinus contorta* forests) typically exhibit a peak in biomass followed by damped oscillations as predicted by the third hypothesis. In contrast, sites near treeline with low rates of tree establishment have asymptotic biomass accumulation as predicted by the first hypothesis. Similarly, if canopy mortality is a long, gradual process that starts well before maximum biomass or longevity is achieved, biomass can be asymptotic, as is described by Harcombe *et al.* (1990) for the coastal *Picea–Tsuga* forests of Oregon. Sites with only moderately synchronous tree establishment and mortality tend to conform to the predictions of the second hypothesis, an overshoot and asymptotic decline.

Population processes do not provide the only mechanisms responsible for changes in biomass during succession. Changes in soil characteristics have long been recognized as important (see references in Miles, 1979; Finegan, 1984; Tilman, 1988). In the classical model of Clements (1916), early successional species improve soil conditions by increasing soil organic matter and accumulating soil nutrients. This presumably makes the site more favourable for plant growth, thereby allowing a different suite of species to invade and a higher level of biomass to accumulate. Sites with very thin soils such as rock outcrops (e.g. Shure and Ragsdale, 1977) or lava flows exhibit this pattern (e.g. Aplet and Vitousek, 1990; Rejmánek *et al.*, 1982; also see Jenny, 1980).

Studies of chronosequences have led to the conclusion that a more typical pattern is for essential nutrients to be leached from the soil, chemically fixed in unusable forms, or tied up in dead organic matter (Stevens and Walker, 1970). The lower biomass of many older boreal forests, particularly on cold, moist sites, is a direct result of nutrients becoming tied up in undecomposed litter (see Bloomberg, 1950; Strang, 1973; van Cleve and Yarie, 1986). The lower decomposition rates, in turn, can result from cooler soil temperatures under thick tree canopies, or from changes in the character of the litter (van Cleve and Yarie, 1986). Similarly, declining biomass in mature forests has been correlated with soil age and degree of leaching (see Fig. 3.2; e.g. Jenny *et al.*, 1969; Walker and Syers, 1976; J. Walker *et al.*, 1981; Westman and Whittaker, 1975).

3.3.2 Changes in forest primary production

Net ecosystem production (NEP) – defined as gross photosynthetic production less losses to respiration, herbivory and death – can alternatively be interpreted as the rate of accumulation of living biomass. Thus, the living biomass at some time t_1 can be calculated from NEP as its integral over the interval t_0 to t_1. In the simplest case, we can predict that during succession NEP will start near zero, increase to a peak at the time of maximal biomass accumulation early in the thinning phase, and then decrease to zero as biomass approaches an asymptote during the

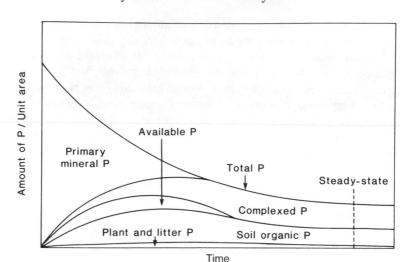

Figure 3.2 Changes in total phosphorus and its fractions during soil development. Early in soil development, phosphorus available to plants increases as P
is released from primary minerals through weathering. However, P is steadily
leached from the system, and much of what remains becomes complexed with
iron in forms unavailable to plants with the consequence that P eventually
becomes highly limiting for plant growth. (Modified from Walker and Syers,
1976.)

steady-state phase (Kira and Shidei, 1967; Odum, 1969). For any period
during which biomass declines, NEP will be negative, and if biomass
oscillates around some asymptotic level, NEP will oscillate around zero.

Net primary production (NPP), or net plant yield, is a measure of the
rate of production of plant material or harvestable biomass, and differs
from NEP in that material lost through death, shedding of plant parts, or
herbivory is not subtracted. The successional trends in NPP do not have
the neat one-to-one relationship to biomass found with NEP.

As with biomass, NPP has been observed or hypothesized to change in
a variety of ways during succession. The first and simplest proposed trend
is for NPP to increase in a logistic fashion to a constant level (Fig. 3.1b).
This is consistent with the classical Clementsian model where early
vegetation ameliorates the environment causing conditions to become
more favourable for plant growth. Consistent with this prediction are
reports by Holt and Woodwell (in Whittaker, 1975) of an uneven but
asymptotic increase in NPP for Long Island *Quercus–Pinus* forests, and
by Forcella and Weaver (1977) of a steady, asymptotic increase in net
primary production of *Pinus albicaulis* over an age sequence of 600 years.

The Clementsian model appears to apply well only where environmental conditions are extreme. The more typical pattern during succession is an initial but short recovery period during which leaf area
increases quickly and asymptotically to a maximum (Marks, 1974;

Sprugel, 1985), after which NPP is limited by resource availability rather than photosynthetic machinery. If, after leaf area is fully developed, NPP is limited by some soil resource with a relatively constant supply rate (typically nitrogen or phosphorus), NPP can be expected to remain constant (Fig. 3.1b). Several studies contain evidence that supports this model. In New England, deciduous forests recover from cutting to reach maximum NPP in as little as 4–5 years (Marks, 1974), after which NPP is relatively constant (see Sprugel, 1985). In Britain, *Pinus sylvestris* (scotch pine) plantations have been reported to reach constant NPP in 15–35 years, and *Betula* (birch) forests in 20–55 years. Using long-term permanent plots in established, mixed-age deciduous forests on the North Carolina piedmont, I have found NPP to be essentially constant for at least 50 years (Peet, 1981a), despite stand structural development and a steady increase in biomass.

An alternative proposal is that gross primary production, not NPP, should be limited by constant resource supply and thus should be constant. Such a view is consistent with the theoretical work of Odum and Pinkerton (1955), Odum (1969) and Margalef (1963), all of whom predict a shift in energy allocation from production to maintenance during succession. As already discussed, living biomass, and particularly structural biomass, accumulates during succession. Some workers have suggested that this added biomass incurs an added respiratory cost for the plants (Kira and Shidei, 1967; Odum, 1971; Horn, 1974; Waring and Schlesinger, 1985). If true, we should expect total respiration to increase during succession. Sprugel (1985) argues that the logic of increasing respiratory cost is flawed. Leaves have higher respiration rates than branches and stems, and the branch and wood respiration is largely confined to the surface cambial layers. At very least, wood surface rather than volume should be used to estimate respiration, and what few data we have suggest that wood surface area may actually decline during some successional sequences (Sprugel, 1984).

Whether gross primary production or net primary production remains constant following canopy development will likely depend on which resource is limiting growth. If the limiting resource is light, or more particularly energy or carbon compounds, then an increasing cost of respiration might cause NPP to decline to an asymptote below the early succession maximum. However, if the limiting resource is a nutrient, like nitrogen or phosphorus, then respiration will be much less of a drain on the limiting resource and NPP will likely be constant. If water is limiting, the result may be intermediate.

Numerous studies have shown that NPP can decline more rapidly during succession than can be accounted for by increased respiration. These studies include such diverse forest types as *Tsuga* (hemlock) in the Pacific Northwest, USA (Meyer, 1937), *Pinus* and *Picea* (pine and spruce) in the subtaiga of European CIS, *Tilia* (basswood) near

Moscow, *Quercus* (oak) near Voronezh (Rodin and Bazilevich, 1967) and *Eucalyptus* in Tasmania (Attiwill, 1979). The drop has been variously attributed to changes in forest structure (Peet, 1981a), tree senescence or declining vigour (Hellmers, 1964; Kira and Shidei, 1967) and nutrient availability (see Sprugel, 1985), but no consensus has emerged. In some cases a change in species composition may explain the decrease in NPP. For example, in southeastern USA early successional stands of species like *Pinus taeda* (loblolly pine) and *Liriodendron* (tulip popular) have NPPs as much as 50% higher than those of the climax *Quercus* (oak) forests of the region (Peet, 1981a). One possible explanation is provided by a theory of tree canopy architecture elaborated by Horn (1971, 1974) that predicts early successional species to have a 'multilayered' shape which is more efficient at harvesting light than the typically 'monolayered' shape of climax species.

Just as a depression in regeneration during the thinning phase of stand development can lead to stand breakup during the transition phase and an associated drop in biomass, so can the same process lead to a more transient drop in NPP. Whether the more gradual decline in NPP observed in many even-aged stands during the thinning phase can also be attributed to changes in stand structure is less clear. Some workers claim that NPP is independent of tree density (e.g. Mar:Moller, 1947, 1954; Baskerville, 1965) once equilibrium canopy leaf area is attained. In contrast, others (e.g. Assman, 1970) suggest there is a narrow optimal density range over which NPP is maximal. If such an optimum does exist, then we would expect NPP to reach a peak at that density and then decline steadily. More commonly, declining forest NPP is attributed to declining nutrient status (see section 3.4.1).

3.3.3 Biomass and production in non-forested ecosystems

Successional changes in biomass and production in natural, non-forested systems are little documented. It is likely that frequently burned shrublands such as the California chaparral exhibit patterns similar to those documented for forests. For example, Schlesinger and Gill (1978, 1980) have described for *Ceanothus* chaparral a stand development sequence that is very much like that found for even-aged development in forest stands. Nonetheless, they found biomass accumulation to be almost constant and clearly independent of plant density once the canopy had closed. Rundel and Parsons (1979) did report some evidence for stand senescence and declining biomass in very old *Adenostoma* chaparral, but their data are sketchy. In contrast, Westman (1982) found no evidence of stand senescence in California soft chaparral.

Most grassland species are sufficiently adapted to disturbances like fire and grazing that successional trends in above-ground biomass and production are not particularly pronounced. Further, grasslands often have as much as 75% of their biomass below ground where it is protected from most disturbances (Risser *et al.*, 1981; Svejcar, 1990). Except for arid sites where rootcrowns are easily damaged, above-ground production generally increases in grasslands the year following fire, often doubling (Kucera, 1981; Svejcar, 1990). Although it is possible that increases in above-ground production are achieved, in part, through reallocation of nutrients from below-ground to above-ground growth, below-ground production in prairies has rather consistently been shown to increase moderately following fire (Hadley and Kieckhefer, 1963; Kucera and Dahlman, 1968; Seastedt and Ramundo, 1990). When grassland is entirely destroyed as from ploughing and subsequent cultivation, biomass is reported to increase asymptotically, but with the above-ground component stabilizing much more quickly than the below-ground component (Gleeson and Tilman, 1990).

3.4 SOIL DEVELOPMENT AND NUTRIENT FLOWS

3.4.1 Pedogenesis and primary succession

Classical succession theory holds that soil development during primary succession leads to improved conditions for plant growth (Clements, 1916; Weaver and Clements, 1938). To quote Whittaker (1970), 'There is [according to classical theory] usually progressive development of soils with increasing depth, increasing organic content, and increasing differentiation of layers or horizons toward the mature soil of the final community.' However, important changes take place in soil nutrient availability during succession, and these changes are not consistently in the direction of increased fertility (Vitousek and White, 1981). While the general trend is towards increased nutrient availability early in primary succession, as is illustrated by such diverse systems as old fields on infertile sands in Minnesota (Tilman, 1987, 1988; Inouye *et al.*, 1987), lava flows in Hawaii (Vitousek *et al.*, 1983, 1988; Vitousek and Walker, 1987), and open land left by receding glaciers in Alaska (Bormann and Sidle, 1990), late in succession fertility often declines.

Walker and colleagues (Stevens and Walker, 1970; Adams and Walker, 1975; Walker and Syers, 1976) have summarized changes in the phosphorus (P) content of forest soils during primary succession (Fig. 3.2). Total P steadily declines as a consequence of leaching. At the same time, the distribution of the residual P among the component fractions changes. Early in succession most P is present in the form of primary minerals such as apatite, and thus is unavailable for use by plants. As these minerals weather, soluble P increases and is, in turn, converted into plant material or is complexed with iron and other materials to form stable, secondary minerals. Thus, the total amount of P available for plant growth increases during the period of significant weathering of primary minerals, but then drops as P is lost from the system or bound into recalcitrant mineral forms. Late in primary succession, the remaining P is largely tied up in secondary minerals and organic matter, leaving little available for plant uptake.

The time scale of the process of P loss described by Walker and Syers (1976) is long but variable, and depends on such factors as climate, the P adsorption capacity of the soil, and the initial amounts and types of P-containing primary minerals. However, the whole process is sufficiently slow that most secondary successional sequences can be assumed to have a constant supply rate of geological P.

Soil nitrogen (N) availability also changes during primary succession (e.g. Gorham et al., 1979; Inouye et al., 1987). Early in primary succession N is generally in low supply relative to P. The low N:P ratio early in primary succession favours the invasion of nitrogen fixers which then increase N availability (Walker and Syers, 1976; Gorham et al., 1979; Vitousek and Walker, 1987), though generally nitrogen fixers do not arrive until the second wave of invaders (Grubb, 1986). During the next phase of primary succession, both N and P are relatively abundant. A major disturbance such as fire or logging during this period can cause major losses of P and N, as described in section 3.4.2 below (see Vitousek and Reiners, 1975; Vitousek et al., 1982). As the N:P ratio increases, N fixation decreases. Then, as leaching and precipitation continue to remove P from the system, N will decrease in such a fashion as to maintain a relatively constant N:P ratio. Thus, very old systems will often have low levels of both P and N, but P will generally remain most limiting (Vitousek and White, 1981; Vitousek, 1982, 1984).

J. Walker and colleagues (1981) have described a chronosequence of sites on the Cooloola sand mass along the coast of southern Queensland, Australia. This chronosequence illustrates in particularly dramatic fashion how the stature of vegetation that can be supported by a particular site changes with soil age and degree of weathering. Soil surfaces at Cooloola range in age from a few thousand years to perhaps a half-million years. Because the soils are sandy, they are subject to relatively rapid nutrient loss through leaching. All of these forest types burn, and this likely increases the rates of nutrient loss (Kellman et al., 1985). On the oldest soils, the depth to the B horizon is in excess of 20 m. Along the chronosequence, both climax biomass and canopy height first increase to a peak around 20 000 years (Fig. 3.3(a)) and then decrease steadily so that old sites with soils in the range of 200 000–400 000 years (Fig. 3.3(b)) support only low, sclerophyll-scrub vegetation.

3.4.2 Biogeochemical consequences of secondary succession

The pool of mineral nutrients contained in biomass increases during succession as biomass accumulates. If we assume that over a time scale

Figure 3.3 Photographs contrasting the high and low stature and biomass of forests developed on dunes of (a) recent (*c.* 10 000–20 000 years) versus (b) ancient (*c.* 200 000–400 000 years) origin on the dune system at Cooloola, Queensland. The individual shown in the photographs is ≈1.65 m tall. (Photographs by author, see J. Walker *et al.*, 1981.)

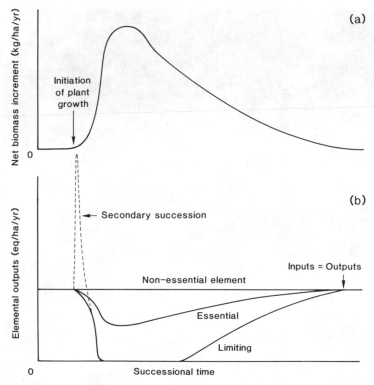

Figure 3.4 Trends in (a) net biomass increment and (b) nutrient loss (in hydrogen ion equivalents) from a forested ecosystem which experiences a major, biomass-removing disturbance. Essential nutrients are retained within the biomass as the forest grows, but as the steady state is reached, the nutrient outputs must equal the inputs. (Modified from Vitousek and Reiners, 1975.)

relevant to secondary succession soil nutrients have a relatively constant supply rate, then the availability of those nutrients ought to be inversely related to the rate at which they are being sequestered (removed from short-term cycling) in the accumulating biomass. This logic led Vitousek and Reiners (1975) to develop a relatively simple but highly predictive graphical model of changes in ecosystem nutrient dynamics during secondary forest succession (Fig. 3.4).

Assume that a nutrient is limiting to forest growth at a particular site, as is often the case with nitrogen or phosphorus. Over the long-term, the input of that nutrient into the ecosystem through such processes as atmospheric deposition and rock weathering must equal the losses, most of which will occur as dissolved material in stream flow. Vitousek and Reiners (1975) predicted that following a major disturbance such as fire or logging, nutrients will be released from decomposing, recently dead biomass. Because there are no plants to reabsorb them in their growth,

Vitousek and Reiners reasoned that nutrients will be lost from the system at levels far in excess of normal input. A particularly extreme example of the impact of disturbance on nutrient flux is provided by the experimental clearcut at the Hubbard Brook forest in New Hampshire. A 40-fold increase in stream nitrate concentration occurred the year following clearcutting (see Likens *et al.*, 1967; Bormann and Likens, 1979; cf. Swank, 1988).

In most cases the increase in nutrient availability and subsequent flush of nutrients from the system is short-lived. As the forest regrows, the nutrients are taken up by the growing plants and the loss rate drops below the input rate. In the case of a limiting nutrient, the plants take up most of the available forms of the nutrient with very little left to be lost. However, as biomass approaches steady-state levels, biomass increment and thus uptake of nutrient drops, while losses through death and decomposition increase. In a steady-state forest the rate of nutrient loss from combined living and dead biomass is roughly equal to uptake. Nutrients that are essential but less limiting to growth show similar but less dramatic trends in their availability (Fig. 3.4). Non-essential elements like silicon and sodium show little variation during stand development aside from that generated by erosion.

Vitousek and Reiners (1975; Vitousek, 1977) supported their nutrient flux model with data from New England forests. A series of watersheds with forests of various ages confirmed that rapidly growing forests were conservative of nutrients. Thus, Vitousek (1977), Bormann and Likens (1979) and others were able to conclude that thinning phase successional forests are very efficient in nutrient use and have extremely tight nutrient cycles, whereas older, transitional or steady-state forests are less efficient in nutrient retention.

Gorham *et al.* (1979) later extended the graphical model to incorporate the spectrum of variation represented by the primary–secondary succession continuum (Fig. 3.4). During primary succession the initial pulse of nutrient availability is necessarily absent. Otherwise, the model is much the same for primary and secondary succession, provided one assumes a constant input rate for nutrients.

The efficiency of nutrient use can also be examined in the context of the individual plant. Numerous indices of efficiency have been devised, but the basic concept is one of a ratio of biomass production (NPP) to amount of nutrient taken up from the soil. By any such measure, plants that have low nutrient concentrations or that recycle and retain nutrients efficiently should be the most nutrient efficient. Vitousek and colleagues (Vitousek, 1982, 1984; Vitousek *et al.*, 1982) calculated nutrient-use efficiencies for nitrogen, phosphorus and other elements from published data and found evidence of increased nutrient-use efficiency on sites low in nutrients, both among species (Vitousek, 1982, 1984) and within species (Birk and Vitousek, 1986; Saterson, 1985). When coupled with the observed changes in nutrient availability during secondary succession, these results lead to the prediction that nutrient-use efficiency of plants should be low early in secondary forest succession, increase to a maximum, and then decline with approach to steady-state conditions. The work of Birk and Vitousek (1986) on even-aged, coastal plain, USA pine forests supports these predictions.

3.5 COMMUNITY ORGANIZATION

3.5.1 Mechanisms of succession

In order to explain the considerable variation that exists in the succession process among different ecosystems and under different initial conditions, an organizational framework is needed. Ideally, such an organizational framework should reflect the causal mechanisms driving succession so that their relative importances might be understood.

One of the failings of most early studies of succession was a lack of consideration of the mechanisms responsible for succession, and few of the studies that did mention mechanisms examined alternative possibilities (Pickett *et al.*, 1987a). This lack of scientific rigour in early succession studies caught the attention of Connell and Slatyer (1977), who emphasized that successional mechanisms need to be 'stated in the form of hypotheses testable by controlled field experiments'.

In the same paper in which they urged experimental testing of mechanisms, Connell and Slatyer provided three alternative models of succession called facilitation, tolerance and inhibition. These models might more accurately be called scenarios in that they are projections of how and why a community might change during succession. Unfortunately, they are not readily distinguishable, and all three might well be important at the same time or at different times during a particular successional sequence, failings for which they have been firmly criticized (e.g. Peet and Christensen, 1980; Botkin, 1981; Quinn and Dunham, 1983; Grubb and Hopkins, 1986; Grubb, 1987; Pickett *et al.*, 1987a,b; Walker and Chapin, 1987). Although the models of Connell and Slatyer have not proved particularly useful for understanding succession, and they certainly do not and probably were never intended to provide the alternative testable hypotheses desired by their authors, they have been highly influential in reintroducing scientific rigour to studies of succession.

One result of the numerous published attempts to apply the Connell and Slatyer models is the recognition that the complexity of successional processes can only be understood if we allow for and examine the simultaneous and changing action and interaction of multiple mechanisms (Walker and Chapin, 1987; Pickett *et al.*, 1987a,b; Uhl, 1987). As Pickett *et al.* (1987a) wrote, 'Understanding how succession as a whole occurs would be best served by addressing the specific array of mechanisms and circumstances acting at a time and place rather than by dividing the suite into classes.'

In his classic treatise on succession, Clements himself (1916) described six processes (sets of interacting mechanisms) important in succession: nudation (or disturbance), migration, ecesis (or establishment), growth, reaction (or modification of the environment by the organisms) and stabilization. With the possible exception of stabilization, this set of processes remains broadly applicable for analysis of succession (Miles, 1979; MacMahon, 1981; Pickett *et al.*, 1987a). An alternative and more complete classification of mechanisms important in succession is provided by Pickett *et al.* (1987a,b), but virtually all of the component mechanisms they recognize fit into the five processes indentified by Clements.

Although a strict reductionist approach of studying separately all of the component mechanisms of successional change will probably, in the end, be necessary for maximizing predictive power, the complex and multivariate results are unlikely to be readily visualized by many ecologists. Only through use of individual-based simulation models (see Chapter 7) are the simultaneous consequences of the many possible interacting mechanisms likely to be understood (Huston and Smith, 1987; Tilman, 1988).

3.5.2 Two models of succession

One alternative means of organizing information on succession so as to enhance overall understanding is to provide reference points in the form of commonly encountered clusters of mechanisms and their consequences, which in the sense of Connell and Slatyer (1977) might be called models. In the introductory section of this chapter I defined primary succession as principally driven by mechanisms involving either permanent or allochthonous environmental change (usually soil and climate change, respectively), and secondary succession as driven primarily by mechanisms related to population interactions. When used in this way, primary and secondary succession can be called succession models. However, it is useful to further partition secondary succession into two recurrent clusters of successional mechanisms; these being a *gradient-in-time model* where biological characteristics of species can be used to explain their distributions along temporal gradients (Whittaker, 1953; Pickett, 1976), and a *competitive-sorting model* where population interactions, particularly competition, cause a temporal gradient in level of community organization and predictability (Margalef, 1963, 1968). Although these models are neither mutually exclusive nor all encompassing, it is possible to classify the autochthonous community change associated with secondary succession as more or less a consequence of the mechanisms implicit in each of them.

(a)   The gradient-in-time model

Each species has a unique set of physiological and life-history adaptations to the physical and biotic environment. Because species differ from each other, the success of species will change continuously along an environmental gradient. Further, natural selection will favour genetic variants that exploit resources and environments otherwise under-utilized. The expected result is that a graph showing the distributions of species along an environmental gradient might best be described as a sequence of overlapping unimodal curves.

Whittaker (1953, 1974) and Pickett (1976, 1982) have argued that a successional sequence can be viewed as simply another form of gradient. In much the same fashion as the differing adaptations of species result in their being arrayed as an overlapping series along a resource or physical environment gradient, so too might they be expected to form a successional sequence determined by physiological and life-history adaptations. The only differences are that the environmental changes occur in time rather than space, and some are brought about by the organisms themselves.

Any one successional phase is likely to be a short-lived or transient phenomenon, and there is a large element of chance in which species happen to be available to colonize a successional habitat. Consequently, we cannot expect species to have evolved in the context of the exact combination of species and environmental conditions that we might observe at any one time. Nonetheless, the predictable changes in community structure and competitive environment that occur during succession should have selected for species with predictable suites of adaptations for specific successional stages.

The gradient-in-time model of succession, with its emphasis on specialized niches, leads to a number of predictions. First, the successional sequence should be orderly in the sense that there are sets of allowable species for any point along the sequence (Lawton, 1987; Tilman, 1988; Gleeson and Tilman, 1990), but the exact species to expect for any given situation should be much less predictable. Second, the physiological and life-history traits represented should change in a predictable fashion. For example, the early succession species should have traits associated with r-selection in the sense of MacArthur and Wilson (1967) or R-selection in the sense of Grime (1979). That is, they should exhibit rapid growth, efficient dispersal, low investment in root and high investment in reproduction (Grime, 1979; Gleason and Tilman, 1990; Tilman and Cowan, 1989). In contrast, the climax community should have traits associated with K (MacArthur and Wilson, 1967) or C–S selection (Grime, 1979), such as slow growth, low investment in reproduction and efficient use of resources.

(b) The competitive-sorting model

The competitive-sorting model of succession initially developed from the observation made on marine phytoplankton by Margalef (1963, 1968) that the initial composition of a community is unpredictable. He suggested that this initial composition is largely a matter of chance, that because few individuals are present and competition is low, almost any species that can get to a site can grow and reproduce. However, as population sizes increase during succession, competition intensifies. With increasing competition, the poorer competitors are forced out and the composition of the community becomes more predictable. With intense competition, few species can invade, and those few that can will be those well adapted to the particular site. Thus, the competitive-sorting model is one of a gradient from low predictability early in succession to high predictability late in succession.

The competitive-sorting model leads to several predictions. First, for any given site there should be monotonic convergence towards the environmentally defined climax community. Second, increased competitive sorting should result in decreased niche width, or narrower species distributions along gradients. Third, as niche width declines, niche overlap should similarly decline. Fourth, the degree of community differentiation along an environmental gradient (i.e. beta diversity: Whittaker, 1965) should increase. Finally, physiological and life-history adaptations of major species should become increasingly predictable.

(c) Choice of models

The competitive-sorting and gradient-in-time models derive from generalizations based on very different organisms and environments. For this reason they should be viewed as idealized representations of extreme conditions rather than as competing models. In the case of succession among plankton from which the competitive-sorting model derives, propagules are widely distributed so almost any species could be present at the beginning, the environment is generally favourable, and there is relatively little amelioration or modification of the environment by the organisms. In contrast, for terrestrial vascular plants from which the gradient-in-time model derives, propagules are not as widely distributed, there is considerable variation in the environment, and the organisms present can significantly modify the environment, at least as it is perceived by the plants that occur there. The extreme gradient-in-time model might even apply to classical primary succession where the progression depends more on permanent environmental change.

3.5.3 Physical structure

During a typical successional sequence from an abandoned agricultural field to a forest, we can observe an increase in stature of the dominant plants from herbs to trees. In contrast, in grassland systems the above-ground changes are less dramatic than the progressive increase in allocation to roots and the resultant increase in root biomass (Gleeson and Tilman, 1990). Of course, other changes are more complex than simple changes in growth form. Leaf area and its distribution is one facet of structure that changes predictably. Where soils are fertile and establishment rapid, leaf area can reach its maximum in a few years in temperate forests (Marks, 1974), or even just a few months in moist tropical forests (Waring and Schlesinger, 1985). Once leaf area has reached its maximum, it tends to stay relatively constant except for periods of nearly synchronous tree mortality (Bormann and Likens, 1979). During the thinning phase, any stems that fail to keep up with the vertical growth of the dominants are quickly overtopped and die. As a consequence, leaf area tends to be concentrated at the top of the canopy (Peet and Christensen, 1987). This changes with forest maturation. Whereas a young, thinning-phase forest is probably uniform in dominant tree height with a continuous, even canopy, the horizontal structure of an older, steady-state canopy is patchy with numerous gaps of various sizes produced by tree deaths. Thus, the steady-state forest has much greater horizontal and vertical variance in leaf area distribution than the thinning phase.

The spatial distribution of plants also changes during succession. Initial seedling establishment is highly non-random, with successful seedlings being aggregated in localized 'safe-sites' (Grubb, 1977; Harper, 1977). Because closely spaced trees or shrubs experience greater competition for resources than plants far apart, there is a greater probability of death for a dominant or subdominant tree or shrub if it is near other dominants (Daniels, 1976; Lorimer, 1983; Weiner, 1984; Veblen, 1985; Kenkel et al., 1989; see also Chapter 4). The consequence is that during the thinning phase of stand development, surviving trees become progressively less aggregated and sometimes even regular in their spatial distribution (Cooper, 1961; Laessle, 1965; Kenkel, 1988; Kenkel et al., 1989). During the transition phase regeneration resumes, but in a patchy fashion with higher concentration in gaps caused by deaths of canopy trees. This pattern continues into the old-growth or mature forest, with the result that in these stands the small stems often show significant clumping and the large trees are little clumped or even somewhat regular in their spatial distribution (Christensen, 1977; Yeaton, 1978; Whipple, 1980; Good and Whipple, 1982).

3.5.4 Compositional convergence

The Clementsian (1916) model of succession predicts that early in succession there will be considerable between-site variation in vegetation because of differences in initial environmental conditions. However, as succession proceeds there should be convergence in the vegetation of these various sites towards a single regional climax vegetation. Few if any ecologists still think of all successional seres in a region eventually reaching the same climax, even over a time scale appropriate for geomorphological processes of the sort envisioned by Clements (1916) and Cowles (1901). However, the basic elements of convergence in Clements' model can be retained in a more modern formulation. Specifically, we can hypothesize composition to shift continuously towards a hypothetical, site-dictated climax community, and we can hypothesize that the differentiation among climax types in a region should be less than that among successional communities. That is, we can expect inter-site compositional differences to become smaller, even though complete convergence is unlikely.

The competitive-sorting model, like the Clementsian model, leads to the prediction of monotonic convergence on the site-specific climax composition. However, it also leads to the predictions that (1) there will be increasing differentiation of species along environmental gradients, which is to say that there will be greater compositional change along a given gradient, and (2) the compositional predictability of the community will increase. Both of these predictions distinguish the competitive-sorting model from the Clementsian model.

Christensen and I (Christensen and Peet, 1981, 1984) tested the hypothesis that composition converges monotonically towards a site-specific climax. For our study we focused on succession from old field to forest in the piedmont region of southeastern USA. We recognized five successional classes based primarily on time since field abandonment: 20–40-year-old successional pine, 40–60 year pine, 60–80 year pine, >80 year pine, and uneven-aged, near-steady-state hardwoods. In an initial study of mature hardwood forests of the region, we found that the site environmental variables with the highest correlation with compositional variation (as represented by ordination axes) were pH, soil calcium and soil magnesium. Soil pH, calcium and magnesium were also the site variables most useful for predicting composition within each of the four pre-climax classes.

The simplest of the several tests of convergence we used was a comparison of correlations between composition and environment for the five successional classes. In a second test we calculated mean species positions along the pH gradient for each of the five classes of stands, and then correlated species positions in each of the four pre-climax classes with species positions in the near-steady-state stands. In both tests correlations increased for both herbs (all leaf area <1 m tall) and trees (stems >1 cm diameter) to peaks in intermediate-aged pine stands, decreased in the oldest pine stands, and then increased to a maximum in the mature hardwood stands (Fig. 3.5(a)). These results show that convergence does not necessarily increase in a simple, monotonic fashion during succession.

In another test we used mean positions of species along a pH gradient in the near-steady-state stands as indicator values, and then calculated for each stand an expected soil pH. We then examined the correlations between observed and predicted pH for stands in each of the five successional classes. The correlation was very high in the near-steady-state hardwood stands, but was even higher in the intermediate-aged successional pine stands. In contrast, the correlation was low in the older pine stands (Fig. 3.5(b)). These results are consistent with those of the first two tests which demonstrated non-monotonic convergence, but in addition they suggest that the composition of steady-state stands may be less predictable with respect to soil factors than that of intermediate-aged pine stands.

The non-monotonic changes in convergence and compositional predictability during succession described above for piedmont pine forests can be explained as a result of corresponding non-monotonic variation in the intensity of competition. During the thinning and steady-state phases of succession when competition is likely to be most intense, composition is predictable, probably because competition severely limits the range of sites each of the various species can occupy. In contrast, early in succession in the establishment phase, and later when the canopy is breaking up during the transition phase, competition is probably of low intensity and those species available are thus capable of competing successfully over a broader range of sites. This is, in effect, a modification of the competitive-sorting model to accommodate the more complex population processes and interactions that occur among forest trees.

Although in our data the composition of the intermediate-aged successional pine stands appears more predictable than that of steady-state stands, we have no reason to expect that this is a general result. Whether steady-state communities will, in general, experience greater competitive pressure than occurs during the thinning phase will depend on the synchrony of establishment of the first cohort of trees, as well as their efficiencies of resource extraction relative to the species dominant at steady state. However, given similar efficiencies, the necessary occurrences of canopy gaps in the mature communities will probably result in a somewhat less intense competitive environment, and thus a somewhat lower compositional predictability.

The prediction derived from the competitive-sorting model, that early stages of succession should be heterogeneous owing to spatial or temporal variation in plant dispersal and establishment, has been born out by studies in several ecosystem types besides forests. For example, Inouye

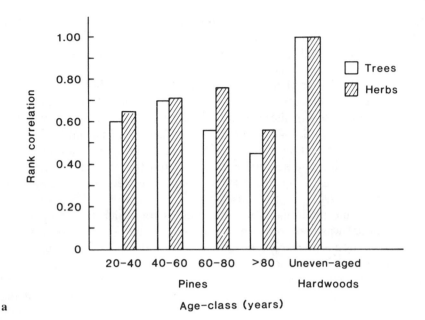

a

b

Figure 3.5   Convergence and changes in the predictability of species composition during secondary forest succession on the North Carolina piedmont. (a) Correlations between mean species positions on a pH gradient in the successional age-classes as compared to their positions in the near-climax, uneven-aged hardwoods. (b) Correlations between the pH of a site and the pH predicted for the site. Expectations were calculated using weighted average positions of species along the pH gradient in the uneven-aged hardwood stands. (From data in Christensen and Peet, 1984.)

et al. (1987) working with old fields in Minnesota, and Myster and Pickett (1990) working with old fields in New Jersey, both found small quadrats within fields to be highly variable early in succession, but to become much less variable with successional development. Van der Valk has described a conceptually similar phenomenon for wetlands, but where the variation is at the scale of the entire wetland and the variation depends on initial seedbank conditions in the wetland at the time of water drawdown (van der Valk, 1981; van der Valk and Davis, 1976, 1978). In a similar fashion, remains of past vegetation or historical events can greatly influence the initial colonization and succession of a site, but this impact declines with succession (Keever, 1950, 1979; Schafale and Christensen, 1986; Squires, 1989; Collins, 1990).

In the marine phytoplankton communities for which the competitive-sorting model was first conceived, dispersal can be assumed to be rapid. This assumption is probably less valid for communities of sessile terrestrial plants (Chapter 2). Chance factors in the distribution of propagules can result in the development of quite different communities (Myster and Pickett, 1990). For example, forests occurring in isolated valleys of the Bitterroot Mountains of Montana have local, idiosyncratic steady-state compositions, largely because of the absence of significant seed flow between valleys (McCune and Allen, 1985). In a more general sense, it can be shown mathematically that several ecologically equivalent species can coexist in a region for extended periods, simply because chance occurrences on isolated sites rarely bring the species into direct competition (Shmida and Ellner, 1984). Thus, the predictability of the steady-state composition of a particular vegetation can be expected to vary inversely with the insularity of the habitat.

3.5.5 Niche breadth and gradient length

The competitive-sorting model also provides testable predictions about successional changes in niche characteristics of species. Early in succession when competitive pressures are low, species should be widely distributed along environmental gradients. In contrast, late in succession when competitive pressures are high, species should be restricted to those portions of environmental gradients where they are best able to compete. This means that average niche breadth should decrease during succession. In addition, as niche width decreases during succession (and provided the total number of species along the gradient is constant), the length of the gradient as measured in units of compositional change (beta diversity) should increase. The Clementsian model, in contrast, leads to the prediction that gradient length should get shorter during succession as divergent communities on various sites converge towards the regional climax (Christensen and Peet, 1984; Peet and Christensen, 1988).

Several studies have shown that community differentiation along gradients increases during succession, thus conforming to the predictions derived from the competitive-sorting model. Working in the mountains of Norway, Matthews (1979) found that herb communities become progressively more differentiated along compositional gradients during primary succession following glacial retreat. Pineda *et al.* (1981a) studied pastures recovering from ploughing in central Spain and also found increasing differentiation with time. Christensen and I (1984; Peet and Christensen, 1988), as part of our study described above, calculated beta diversity for each of five age-classes of forest. The mature hardwoods had by far the greatest differentiation, and there was some suggestion of a decline in beta diversity in the oldest pine stands relative to stands still in the thinning phase.

Furthermore, niche widths of individual species have also been shown to change in a fashion consistent with the competitive-sorting model. A graphic illustration of this pattern is found in a comparison of the distributions of *Solidago* (goldenrod) species growing in mature prairie and in abandoned cropland. In this study Werner and Platt (1976) found relative niche breadth to be narrower and niche overlap less among the goldenrods on the mature prairie (Fig. 3.6). In community-wide studies incorporating all vascular plants, both Pineda *et al.* (1981b) and Christensen and Peet (1984) found decreases in average niche breadth during succession. Similar patterns have been described for such divergent communities as hot spring algae (Kullberg and Scheibe, 1989) and birds (May 1982).

Just as species composition changes during succession and becomes more predictable with increased competitive sorting, we should expect genetic variation within species to become restricted and the distribution of the remaining genetic variation to be more predictable. Unfortunately, few data currently exist on changes in genetic variation within species during succession (see Gray, 1987), but the data that are available are consistent with this prediction. The intense, density-dependent mortality associated with self-thinning has been shown to cause a decrease in genetic diversity in populations of various herbaceous species (Antonovics and Levin, 1980; Zangerl and Bazzaz, 1984). In an elegant set of experiments, Turkington and Aarssen (1984; Aarssen and Turkington, 1985) examined the association between genotypes of clover and of various grasses for four fields of different ages. They found stronger associations in the older fields. In another study Mulcahy (1975) examined *Acer saccharum* (sugar maple) seedlings in a mature forest and discovered a decrease in genetic variation of cohorts with increasing age, thus supporting the idea that genetic diversity decreases through time owing to competitive and environmental stress. In contrast, Schaal and Levin (1976) found plants of a large, approximately 30-year-old cohort of *Liatris cylindracea* (prairie blazingstar) had greater heterozygosity than the smaller cohorts of more recently established plants. The likely expla-

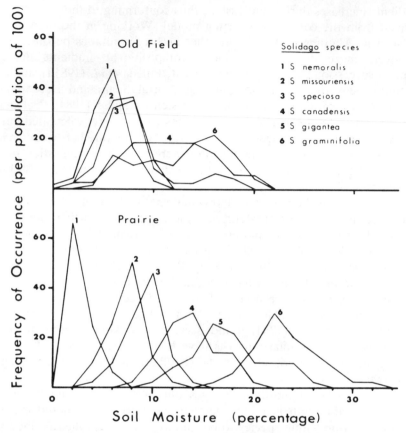

Figure 3.6 Frequency of occurrence (percentage) of plants of various *Solidago* (goldenrod) species along a soil moisture gradient (2% increments) in a mature prairie and in a grass-dominated old field. (From Werner and Platt, 1976.)

nation is that the large, genetically diverse cohort became established following a disturbance when competitive pressures were weak, though the authors stress the alternative explanation that there is some unexplained heterozygote advantage.

Competitive sorting is not the only reason to expect broader ecological niches early in succession. The gradient-in-time model suggests the alternative explanation that the early successional species have intrinsically broader niches than the late successional species. It can be argued that early successional species should be more generalized in their niche characteristics because the sites they are likely to occupy are less predictable than is the case for climax species. In addition, selection should have favoured such life-history characteristics as high allocation to reproduction, rapid reproduction, efficient dispersal and rapid growth, rather than habitat or niche specialization. In support of this hypothesis, Parrish and Bazzaz have shown that selected mature prairie species have narrower niches and less interspecific similarity than annual weeds with respect to rooting location (1976), pollination ecology (1979), nutrient requirements (1982a), and competition for resources (1982c). They also showed that a series of early successional tree species had broader niches along nutrient and water gradients than mature forest tree species seedlings (1982b). However, intrinsic species differences cannot be invoked for the high niche breadth early in succession reported by Christensen and Peet (1984). In that study the same set of species was used for each age-class.

3.5.6 Stability

Although succession is defined as ecosystem change, and thus implies instability, certain ecosystem characteristics related to stability can be examined in the context of succession. In particular, we can examine the rate of ecosystem change (or 'elasticity', *sensu* Westman and O'Leary, 1986), the length of the successional trajectory in terms of species turnover, and the predictability of that trajectory and especially its endpoint, the climax.

Rates of ecosystem change have been little studied and have never been subjected to a systematic, comparative review. Only D. Walker's (1970) application of stratigraphic data for determining rates of change along hydroseres in Britain stands out as exemplifying the sort of analysis needed. With respect to secondary succession, at best we are in a position to offer a few likely hypotheses. Bornkamm's (1981) work with herbaceous communities suggests that rates of change start high and steadily decline as species distributions more closely approach those expected in climax communities. For forest systems a similar result might be expected, except that a brief increase in rate of change might be expected to occur at the onset of the transition phase if death of the initial cohort of trees is relatively synchronous. We can predict that on favourable sites where most species become established early, the rate of change will be faster than on more extreme sites where succession will be limited by establishment rates. In addition, the rate of change will be linked to the longevity of the dominant plants. For example, the initial cohort of *Pinus taeda* on the piedmont of southeastern North America usually remains in the thinning phase for around 75 years and near-steady-state conditions are achieved in less than 200 years, whereas *Pinus contorta* in the Colorado Front Range takes closer to 275 years to reach the end of the thinning phase and in excess of 500 years to reach near-steady-state conditions.

As with rates of successional change, little is known about differences in lengths of (i.e. species turnover along) successional seres. Only a few generalizations can be made with certainty. The length of a successional trajectory will necessarily depend on the initial conditions. Secondary succession on favourable sites with ample migration can be expected to require little more than one episode of compositional sorting. Succession on extreme sites with little potential for biological amelioration of the environment, such as is the case for most deserts, can similarly be expected to have only modest compositional change (Zedler, 1981). In contrast, classical primary succession starting with a bare rock surface or a water-filled depression can be expected to have a long trajectory with many species coming and going during the process, as Walker (1970) has clearly demonstrated.

The predictability of the trajectory which succession follows is similar to the time sequence in degree of convergence expected. As explained in section 3.5.4, the predictability of community composition along a successional sere varies in response to changes in the intensity of competition. Like the predictability of the succession trajectory, the predictability of the endpoint or the climax composition has received little study. Various authors have discussed the possibility that for a given set of environmental conditions there might be a variety of alternative steady-state configurations (Connell and Sousa, 1983; Peterson, 1984). If true, then it is important to consider what conditions are required for the establishment of such alternative steady states, and what sort of disturbance would be needed to cause a shift to an alternative state.

Several apparent cases of alternative steady states in terrestrial plant communities have been linked to chance variation in the disturbance regime. A particularly graphic example is provided by the *Eucalyptus* forests of Tasmania described by Jackson (1968) in a paper that also illustrates changes that can occur in disturbance regimes during succession (Fig. 3.7). *Eucalyptus* forests are relatively flammable and typically burn 100–300 years after establishment. The trees become re-established from seed released by the original trees and the *Eucalyptus* forest regrows. However, if a site goes for a longer period (>350 years) without fire, the *Eycalyptus* die out without regenerating and no seed source remains, with the result that after the forest succeeds to the almost unburnable, mixed-species rainforest, there is little chance of returning to *Eucalyptus*. In contrast, if the *Eucalyptus* forest happens to burn before the trees have reached reproductive age, the *Eucalyptus* are killed without leaving seed for regeneration and the forest is replaced by a stable, highly flammable sclerophyll forest or heathland. Because of the high flammability of the new community, fires are frequent and the chance *Eucalyptus* tree that might establish has little possibility of growing to a sufficient size to regenerate before fire recurs.

Alternation between frequently burned, flammable communities and infrequently burned, non-flammable communities of the sort described by Jackson for Tasmania appears to be rather common. *Pinus contorta* forms a flammable community that can alternate with relatively less flammable

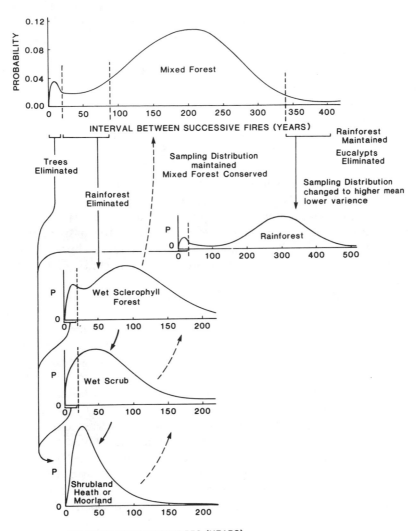

Figure 3.7 A diagrammatic model of the vegetation types that develop following destructive fires, and their associated fire-likelihood distributions. The model projects potential drift in composition and in fire likelihood that could result from chance occurrences of short or long intervals between successive destructive fires. Long intervals lead to elimination of the potentially dominant and moderately flammable *Eucalyptus* trees, thus lowering the fire risk for future generations of forest trees and maintaining a mixed-species rainforest. Short intervals lead to elimination of rainforest or of all forest species along with increased probability of subsequent fire. (Modified from Jackson, 1968.)

Populus tremuloides forests in the Rocky Mountain region (Parker and Parker, 1983; Peet, 1988). Flammable *Pinus palustris* alternates with less flammable *Quercus* spp. forest in Florida (Williamson and Black, 1981; Myers, 1985). Flammable *Aristida stricta* (wiregrass) savanna alternates with moist, non-flammable 'bay' forest (*Gordonia, Magnolia, Persea*) on the southeastern coastal plain of North America (Christensen, 1981; Frost *et al.*, 1986).

In semi-arid regions it is common for shrub-dominated communities to alternate with grass-dominated communities. Examples include South African *Acacia* savanna and shrubland (B. Walker, 1981; B. Walker *et al.*, 1981), grassland and desert scrub in southwestern North America (Humphrey, 1958, 1974) and grassland and *Prosopis* (mesquite) woodland in Texas and adjacent states (Buffington and Herbel, 1965). These alternative steady states can similarly be explained on the basis of fire and flammability. Grass is typically highly flammable and thus burns regularly, the fires preventing invasion of woody plants. In contrast, the sparse scrub and woodland systems are less flammable, plus the plants are sufficiently widely spaced that fire does not spread well.

Alternative explanations of multiple steady states have been offered based on Walter's (1971) hypothesis that in savanna systems woody plants must compete with more efficient grasses for the water present near the soil surface, but have nearly exclusive access to soil water at greater depths. Woody plants have difficulty becoming established in the presence of grasses and generally do poorly. However, if the grass layer is damaged such that woody plants can become established, they often remain at the expense of the grasses. This reduced success and associated regeneration failure of grasses has been variously attributed to increased leaching of critical nutrients from the grass rooting-zone to the shrub rooting-zone, decreased permeability of the upper soil horizon, funnelling of most precipitation to the bases of the woody plants via canopy interception and stem flow, and decreased light below the woody plant canopy (see B. Walker, 1981; B. Walker *et al.*, 1981).

3.5.7 Species diversity

Species diversity, or species richness, is a community attribute that reflects the combined influence of such processes as immigration, speciation, competition, predation and extinction. The assumed potential of species diversity measures to increase our understanding of these processes has inspired much research and has generated a massive literature. One major focus of this research has been the change in diversity during succession.

Plant species diversity repeatedly has been hypothesized (e.g. Odum, 1969; Harger and Tustin, 1973) and observed (Brünig, 1973; Nicholson and Monk, 1974; Peet, 1978) to increase to an asymptote over time.

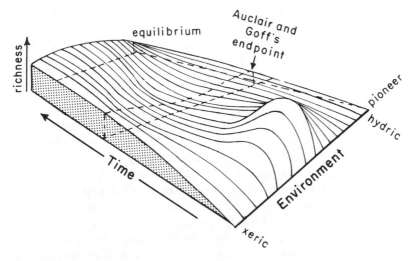

Figure 3.8 The pattern of change in species richness (species per 0.1 ha) of Rocky Mountain forests during succession, and the variation in this pattern associated with site moisture. This model represents a generalization of an earlier model by Auclair and Goff (1971) and can be applied with little modification to many boreal and montane conifer forest regions, as well as some temperate deciduous forests regions. (From Peet, 1978.)

Where colonization is slow and succession follows the gradient-in-time model, diversity usually does increase asymptotically towards an equilibrium representing the balance between immigration and extinction, a pattern well established for volcanic peaks like Paricutín (Rejmánek *et al.*, 1982) and Krakatau (Whittaker *et al.*, 1989).

A more frequently reported pattern has been for diversity to peak in late succession after most of the climax species have entered, and then to decrease as successional species are lost (Margalef, 1963, 1968; Loucks, 1970; Auclair and Goff, 1971; Whittaker, 1972; Horn, 1974, 1975; Bazzaz, 1975; Johnson *et al.*, 1976). Various other patterns have also been reported including a peak shortly after the succession initiating disturbance followed by a steady decline towards climax (Habeck, 1968; Peet, 1978), as well as an irregular increase to a peak at climax (Reiners *et al.*, 1971; Glenn-Lewin, 1980).

Part of the explanation for the conflicting observations and predictions is that diversity does not exhibit a simple and consistent response to any one factor. Change in diversity during succession depends on the environmental context. However, where a few environmental gradients dominate the vegetation pattern, it is often possible to interpret diversity as a region-specific multidimensional response (e.g. Fig. 3.8). Creation of a general model of successional change in diversity is much more difficult and perhaps impossible (see Glenn-Lewin, 1977, 1980; Whittaker, 1977; Yodzis, 1978).

The pattern of species diversity encountered in Rocky Mountain forests illustrates the type of regional interpretation that is often possible (Peet, 1978, 1988). On favourable sites diversity increases quickly to a high level because almost any species that arrives can grow. However, diversity declines dramatically during the thinning phase of forest development, probably due to competition for scarce resources and a dearth of tree species in the region. During the transition period when resources are less in demand, many species again invade and diversity increases to the climax level. On extreme sites at high elevations, or where soil moisture is retricted and forest development is more gradual and asymptotic in character, changes in diversity are also more gradual (see section 3.1.2). Owing to harsh conditions for establishment, few species invade early on such sites. However, gradual immigration of additional species plus some amelioration of environmental extremes by established plants results in a gradual increase to a maximum at climax. These various trends can be summarized as a three-dimensional response surface (Fig. 3.8). Most reports of diversity change in boreal and montane conifer forests are consistent with this response surface, as are many reports from temperate deciduous forest regions (e.g. Auclair and Goff, 1971; Peet, 1978).

Not all vegetation, or even all forest vegetation, fits neatly the model described above. For example, forests of the piedmont of southeastern North America show no neat pattern of change in species number during succession. However, species number in steady-state forests in this region correlates strongly with soil pH and soil cation content (Peet and Christensen, 1980). The successional trend that can be seen in these forests is one of an increasingly strong correlation between species number and soil pH during the thinning phase, a reduction in correlation during the transition phase, and finally an increase to a maximum in the steady-state phase (Peet and Christensen, 1988). This trend parallels closely the results described above for convergence and niche breadth, and is largely consistent with a competitive-sorting model of forest succession.

SUMMARY

1. Succession theory should transcend species changes and be applicable over a broad range of ecosystems with very different phylogenetic derivations. In this chapter quantifiable attributes of communities and ecosystems have been examined which show consistent trends during succession, yet which are independent of the particular species present and thus allow comparisons among dissimilar ecosystems. Where possible, trends in these attributes have been interpreted in terms of likely mechanisms.

2. Mortality can be interpreted as a necessary component of succession or as disturbance, depending on the scale of observation. Because succession and disturbance intergrade and cannot be distinguished consistently, a general theory of community dynamics should include both.

3. Major successional trends in forest biomass, production, tightness of nutrient cycles and species diversity can all be interpreted as reflecting underlying population processes. Early in forest succession when competition is low and plants are just becoming established, biomass and often diversity and production are low, and nutrient cycles are leaky. During the thinning phase competition is intense with existing trees increasing in size and many dying. Little tree regeneration occurs. As a consequence, biomass increases steadily, nutrient cycles are tight because available nutrients are being sequestered in the accumulating biomass, diversity is often low because of competition from the canopy trees plus spatial homogeneity, and production is relatively constant. After sufficient mortality has occurred that new gaps in the canopy can no longer be filled by expansion of remaining neighbours, forest processes shift dramatically. In this transition phase tree regeneration resumes or increases, biomass drops because of many deaths among the large trees, diversity increases perhaps due to increased resource availability and increased spatial heterogeneity, production at first declines due to tree death and then recovers, and nutrient loss slowly increases as biomass accumulation declines. Finally, in the steady-state or climax forest, and assuming a scale of observation larger than the typical disturbance event, biomass is roughly constant at an intermediate level, production is relatively constant, and nutrients lost from the system roughly balance their influx.

4. Two complementary models of secondary succession may be useful for understanding changes in community structure and ecosystem function. The *gradient-in-time model*, formulated originally by Whittaker (1953; also see Pickett, 1976, 1982), suggests that time should be treated in a fashion analogous to any other gradient. Along this gradient species are distributed in accordance with their physiologies and life-history traits. The *competitive-sorting model*, implicit in the work of Margalef (1963, 1968), suggests that communities start as largely random assemblages of species, but that with time competition leads to confinement of species to those environments for which they are best adapted. The two models can apply in various degrees to one successional sequence.

5. Community composition is most predictable and most similar to the ultimate climax composition after periods of intense competition. That is, convergence is greater where competition is more intense. Thus, composition is predictable in the thinning and climax phases of forest development, but is much less predictable in the establishment and

transition phases when competition is relaxed. The low predictability of composition early in succession can be attributed to broader species ranges or niche breadths, and the failure of species to occur in favourable habitats because of insufficient dispersal time and variation in initial conditions. Similarly, gradient length or community differentiation along gradients (beta diversity) is low early in succession when competition is low and niche breadths are high. During the thinning phase gradient length increases, and generally gradient length is longest in climax communities when competition is relatively high and most species have had time to disperse to favourable sites.

ACKNOWLEDGEMENTS

I thank R. B. Allen, N. L. Christensen, D. C. Glenn-Lewin, P. J. Grubb, O. L. Loucks, R. P. McIntosh, S. T. A. Pickett, T. T. Veblen and P. M. Vitousek for helpful comments on the manuscript. This work was supported by grants from the US National Science Foundation.

REFERENCES

Aarssen, L. W. and Turkington, R. (1985) Within-species diversity in natural populations of *Holcus lanatus*, *Lolium perenne* and *Trifolium repens* for four different-aged pastures. *J. Ecol.*, **73**, 869–86.

Adams, J. A. and Walker, T. W. (1975) Some properties of a chronotoposequence of soils in New Zealand. 2. Forms and amounts of phosphorus. *Geoderma*, **13**, 41–52.

Antonovics, J. and Levin, D. (1980) The ecological and genetic consequences of density-dependent regulation in plants. *Ann. Rev. Ecol. System.*, **11**, 411–52.

Aplet, G. H. and Vitousek, P. M. (1990) Vegetation response to age and climate gradients on windward Mauna Loa, Hawaii. *Bull. Ecol. Soc. Amer.* **71** (2), 78.

Assman, E. (1970) *The Principles of Forest Yield Study: Studies in the Organic Production, Structure Increment, and Yield of Forest Stands*, Pergamon Press, Oxford.

Attiwill, P. M. (1979) Nutrient cycling in a *Eucalyptus obliqua* (L.) Herit. Forest. III. Growth, biomass, and net primary production. *Austral. J. Bot.* **27**, 439–58.

Auclair, A. N. and Goff, F. G. (1971) Diversity relations of upland forests in the western Great Lakes area. *Amer. Nat.*, **105**, 499–528.

Baker, W. L. (1989) Landscape ecology and nature reserve design in the Boundary Waters Canoe Area, Minnesota. *Ecology*, **70**, 23–35.

Baskerville, G. L. (1965) Dry-matter production in immature balsam fir stands. *For. Sci. Monogr.*, **9**.

Baxter, F. P. and Hole, F. D. (1967) Ant (*Formica cinerea*) pedoturbation in a prairie soil. *Soil Science Soc. Amer., Proc.*, **31**, 425–8.

Bazzaz, F. A. (1975) Plant species diversity in old-field successional ecosystems in southern Illinois. *Ecology*, **56**, 485–8.

Birk, E. M. and Vitousek, P. M. (1986) Nitrogen availability and nitrogen use efficiency in loblolly pine stands. *Ecology*, **67**, 69–79.

Bloomberg, W. G. (1950) Fire and spruce. *Forestry Chron.*, **26**, 157–61.

Bormann, B. T. and Sidle, R. C. (1990) Changes in productivity and distribution of nutrients in a chronosequence at Glacier Bay National Park, Alaska. *J. Ecol.*, **78**, 561–78.

Bormann, F. H. and Likens, G. E. (1979) *Pattern and Process in a Forested Ecosystem*, Springer-Verlag, New York.

Bornkamm, R. (1981) Rates of change in vegetation during secondary succession. *Vegetatio*, **47**, 213–20.

Botkin, D. B. (1981) Causality and succession, in *Forest Succession: Concepts and Application* (eds D. C. West, H. H. Shugart, and D. B. Botkin), Springer-Verlag, New York, pp. 36–55.

Bray, J. R. (1971) Vegetational distribution, tree growth and crop success in relationship to recent climatic change. *Adv. Ecol. Res.*, **7**, 177–233.

Brokaw, N. V. L. (1982) Treefalls: frequency, timing and consequences, in *The Ecology of a Tropical Forest: Seasonal Rhythms and Long-Term Changes* (eds E. G. Leigh, A. S. Rand, and D. M. Windsor), Smithsonian, Washington, D. C., pp. 101–8.

Brünig, E. F. (1973) Species richness and stand diversity in relation to site and succession of forests in Sarawak and Brunei (Borneo). *Amazoniana*, **3**, 293–320.

Buffington, L. C. and Herbel, C. H. (1965) Vegetational changes on a semidesert grassland range. *Ecol. Monogr.*, **35**, 139–64.

Canham, C. D. and Loucks, O. L. (1984) Catastrophic windthrow in the pre-settlement forests of Wisconsin. *Ecology*, **65**, 803–9.

Christensen, N. L. (1977) Changes in structure, pattern and diversity associated with climax forest maturation in Piedmont, North Carolina. *Amer. Mid. Nat.*, **97**, 176–88.

Christensen, N. L. (1981) Fire regimes in southeastern ecosystems, in *Fire Regimes and Ecosystem Properties* (eds H. A. Mooney, T. M. Bonnicksen, N. L. Christensen, J. E. Lotan, and W. A. Reiners), USDA Forest Service, General Technical Report WO-26, pp. 112–36.

Christensen, N. L. and Peet, R. K. (1981) Secondary forest succession on the North Carolina piedmont, in *Forest Succession: Concepts and Applications* (eds D. C. West, H. H. Shugart, and D. B. Botkin), Springer-Verlag, New York, pp. 230–45.

Christensen, N. L. and Peet, R. K. (1984) Convergence during secondary forest succession. *J. Ecol.*, **72**, 25–36.

Clark, J. S. (1988) Effect of climate change on fire frequency in northwestern Minnesota. *Nature*, **334**, 233–5.

Clark, J. S. (1990a) Twentieth-century climate change, fire suppression, and forest production and decomposition in northwestern Minnesota. *Canad. J. For. Res.*, **20**, 219–32.

Clark, J. S. (1990b) Fire and climate change during the last 750 yr in northwestern Minnesota. *Ecol. Monogr.*, **60**, 135–59.

Clements, F. E. (1916) *Plant Succession: An Analysis of the Development of Vegetation*, Carnegie Institute Publication **242**, Washington, D. C.

Collins, S. L. (1990) Patterns of community structure during succession in tallgrass prairie. *Bull. Torrey. Bot. Club*, **117**, 397–408.

Collins, S. L. and Barber, S. C. (1985) Effects of disturbance on diversity in mid-grass prairie. *Vegetatio*, **64**, 87–94.

Connell, J. H. and Slatyer, R. O. (1977) Mechanisms of succession in natural communities and their role in community stability and organization. *Amer. Nat.*, **111**, 1119–44.

Connell, J. H. and Sousa, W. P. (1983) On the evidence needed to judge ecological stability or persistence. *Amer. Nat.*, **121**, 789–824.

Cooper, C. F. (1960) Changes in vegetation, structure, and growth of south-western pine forests since white settlement. *Ecol. Monogr.*, **30**, 129–64.

Cooper, C. F. (1961) Pattern in ponderosa pine forests. *Ecology*, **42**, 493–9.

Cooper, W. S. (1911) The climax forest of Isle Royale, Lake Superior, and its development. *Bot. Gazette*, **55**, 1–44.

Cowles, H. C. (1901) The physiographic ecology of Chicago and vicinity; a study of the origin, development, and classification of plant societies. *Bot. Gazette*, **31**, 73–108, 145–182.

Curtis, J. T. (1959) *The Vegetation of Wisconsin*, Univ. of Wisconsin Press, Madison.

Daniels, R. F. (1976) Simple competition indices and their correlation with annual loblolly pine tree growth. *Forest Sci.*, **22**, 454–6.

Daubenmire, R. and Daubenmire, J. B. (1968) *Forest Vegetation of Eastern Washington and Northern Idaho*, Washington Agricultural Experiment Station, Technical Bulletin 60.

Davis, M. B. (1981) Quaternary history and the stability of forest communities, in *Forest Succession: Concepts and Applications* (eds D. C. West, H. H. Shugart, and D. B. Botkin), Springer-Verlag, New York, pp. 132–53.

Day, R. J. (1972) Stand structure, succession, and use of southern Alberta's Rocky Mountain forest. *Ecology*, **53**, 474–8.

Drury, W. H. and Nisbet, I. C. T. (1973) Succession. *J. Arnold Arboretum*, **54**, 331–68.

Finegan, B. (1984) Forest succession. *Nature*, **312**, 109–14.

Forcella, F. and Weaver, T. (1977) Biomass and productivity of the subalpine *Pinus albicaulis–Vaccinium scoparium* association in Montana, USA. *Vegetatio*, **35**, 95–105.

Frissell, S. S. (1973) The importance of fire as a natural ecological factor in Itasca State Park, Minnesota. *Quat. Res.*, **3**, 397–407.

Frost, C. C., Walker, J. and Peet, R. K. (1986) Fire-dependent savannas and prairies of the Southeast: original extent, preservation status and management problems, in *Wilderness and Natural Areas in the Eastern United States* (eds D. L. Kulhavy and R. N. Conner), School of Forestry, Stephen F. Austin State University, Nacogdoches, pp. 348–57.

Gleeson, S. K. and Tilman, D. (1990) Allocation and the transient dynamics of succession on poor soils. *Ecology*, **71**, 1144–55.

Glenn-Lewin, D. C. (1977) Species diversity in North American temperate forests. *Vegetatio*, **33**, 153–62.

Glenn-Lewin, D. C. (1980) The individualistic nature of plant community development. *Vegetatio*, **43**, 141–6.

Good, B. J. and Whipple, S. A. (1982) Tree spatial patterns: South Carolina bottomland and swamp forests. *Bull. Torrey Bot. Club*, **109**, 529–36.

Gorham, E., Vitousek, P. M. and Reiners, W. A. (1979) The regulation of chemical budgets over the course of terrestrial ecosystem succession. *Ann. Rev. Ecol. System.*, **10**, 53–84.

Gray, A. J. (1987) Genetic change during succession in plants, in *Colonization, Succession, and Stability* (eds A. J. Gray, M. J. Crawley, and P. J. Edwards), Blackwell, Oxford, pp. 273–94.

Grime, J. P. (1979) *Plant Strategies and Vegetation Processes*, Wiley, Chichester.

Grubb, P. J. (1977) The maintenance of species-richness in plant communities: the importance of the regeneration niche. *Biol. Rev.*, **52**, 107–45.

Grubb, P. J. (1986) The ecology of establishment, in *Ecology and Design of Landscape* (eds A. D. Bradshaw, D. A. Goode, and E. Thorp), Blackwell, Oxford, pp. 83–97.

Grubb, P. J. (1987) Some generalizing ideas about colonization and succession in green plants and fungi, in *Colonization, Succession and Stability* (eds A. J. Gray, M. J. Crawley and P. J. Edwards), Blackwell, Oxford, pp. 81–102.

Grubb, P. J., Green, H. E. and Merrifield, R. C. J. (1969) The ecology of chalk heath: Its relevance to the calcicole–calcifuge and soil acidification problems. *J. Ecol.*, **57**, 175–212.

Grubb, P. J. and Hopkins. A. J. M. (1986) Resilience at the level of the plant community, in *Resilience in Mediterranean-Type Ecosystems* (eds B. Dell, A. J. M. Hopkins, and B. B. Lamont), Junk, The Hague, pp. 21–38.

Habeck, J. R. (1968) Forest succession in the Glacier Park cedar–hemlock forests. *Ecology*, **49**, 872–80.

Hadley, E. B. and Kieckhefer, B. J. (1963) Productivity of two prairie grasses in relation to fire frequency. *Ecology*, **44**, 389–95.

Harcombe, P. A., Harmon, M. E. and Greene, S. E. (1990) Changes in biomass and production over 53 years in a coastal *Picea sitchensis–Tsuga heterophylla* forest approaching maturity. *Canad. J. For. Res.*, **20**, 1602–10.

Harger, J. R. E. and Tustin, K. (1973) Succession and stability in biological communities. Part 1: Diversity. *Internat. J. Environ. Studies* **5**, 117–30.

Harmon, M. E., Bratton, S. P. and White, P. S. (1983) Disturbance and vegetation response in relation to environmental gradients in the Great Smoky Mountains. *Vegetatio*, **55**, 129–39.

Harper, J. L. (1977) *Population Biology of Plants*, Academic Press, New York.

Heinselman, M. L. (1973) Fire in the virgin forests of the Boundary Waters Canoe Area, Minnesota. *Quat. Res.*, **3**, 329–82.

Heinselman, M. L. (1981) Fire intensity and frequency as factors in the distribution and structure of northern ecosystems, in *Fire Regimes and Ecosystem Properties* (eds H. A. Mooney, T. M. Bonnicksen, N. L. Christensen, J. E. Lotan, and W. A. Reiners), USDA Forest Service, General Technical Report WO-26, pp. 7–57.

Hellmers, H. (1964) An evaluation of the photosynthetic efficiency of forests. *Quart. Rev. Biol.*, **39**, 249–57.

Horn, H. S. (1971) *The Adaptive Geometry of Trees*, Princeton Univ. Press, Princeton, New Jersey.

Horn, H. S. (1974) The ecology of secondary succession. *Ann. Rev. Ecol. System.*, **5**, 25–37.

Horn, H. S. (1975) Markovian properties of forest succession, in *Ecology and Evolution of Communities* (eds M. L. Cody and J. M. Diamond), Belknap Press, Cambridge, pp. 196–211.

Humphrey, R. R. (1958) The desert grassland – A history of vegetational change and an analysis of causes. *Botanical Review*, **24**, 193–252.

Humphrey, R. R. (1974) Fire in the deserts and desert grassland of North

144   Community structure and ecosystem function

America, in *Fire and Ecosystems* (eds T. T. Kozlowski and C. E. Ahlgren), Academic Press, New York, pp. 365–400.

Huntley, B. and Birks, H. J. B. (1983) *An Atlas of Past and Present Pollen Maps for Europe: 0–13000 Years Ago*. Cambridge Univ. Press, Cambridge.

Huston, M. and Smith, T. (1987) Plant succession: life history and competition. *Amer. Nat.*, **130**, 168–98.

Ilvessalo, Y. (1937) Perä-Pohjolan luonnon normaalien metsiköiden kasvu ja kehitys. *Metsätieteellisen Tutkimuslaitoken Julkaisuja*, **24** (2), 1–168.

Inouye, R. S., Huntley, N. J., Tilman, D., Tester, J. R., Stillwell, M. and Zinnel, K. C. (1987) Old-field succession on a Minnesota sand plain. *Ecology*, **68**, 12–26.

Jackson, W. D. (1968) Fire, air, water and earth – an elemental ecology of Tasmania. *Proc. Ecol. Soc. Austral.*, **3**, 9–16.

Jenny, H. (1980) *The Soil Resource: Origin and Behavior*, Springer-Verlag, New York.

Jenny, H., Arkley, R. J. and Schultz, A. M. (1969) The pygmy forest–podsol ecosystem and its dune associates of the Mendocino coast. *Madroño*, **20**, 60–74.

Johnson, W. C., Burgess, R. L. and Keammerer, W. R. (1976) Forest overstory vegetation and environment on the Missouri River floodplain in North Dakota. *Ecol. Monogr.*, **46**, 59–84.

Kartesz, J. T. (1993) *Synonymized Checklist of the Vascular Flora of the United States, Canada, and Greenland*, 2nd edn, Timber Press, Portland, Oregon (in press).

Keever, C. (1950) Causes of succession on old fields of the piedmont, North Carolina. *Ecol. Monogr.*, **20**, 231–50.

Keever, C. (1979) Mechanisms of plant succession on old fields of Lancaster County, Pennsylvania. *Bull. Torrey Bot. Club*, **106**, 299–308.

Kellman, M., Miyanishi, K. and Hiebert, P. (1985) Nutrient retention by savanna ecosystems. II. retention after fire. *J. Ecol.*, **73**, 953–62.

Kenkel, N. C. (1988) Patterns of self-thinning in jack pine: testing the random mortality hypothesis. *Ecology*, **69**, 1017–24.

Kenkel, N. C., Hoskins, J. A. and Hoskins, W. D. (1989) Local competition in a naturally established jack pine stand. *Canad. J. Bot.*, **67**, 2630–5.

Kira, T. and Shidei, T. (1967) Primary production and turnover of organic matter in different forest ecosystems of the western Pacific. *Japan. J. Ecol.*, **17**, 70–87.

Kucera, C. L. (1981) Grasslands and fire, in *Fire Regimes and Ecosystem Properties* (eds H. A. Mooney, T. M. Bonnicksen, N. L. Christensen, J. E. Lotan and W. A. Reiners), USDA Forest Service, General Technical Report WO-26, pp. 90–111.

Kucera, C. L. and Dahlman, R. C. (1968) Root-rhizome relationships in fire-treated stands of big bluestem, *Andropogon gerardii* Vitman. *Amer. Mid. Nat.*, **80**, 268–71.

Kullberg, R. G. and Scheibe, J. S. (1989) The effects of succession on niche breadth and overlap in a hot spring algal community. *Amer. Mid. Nat.*, **121**, 21–31.

Kullman, L. (1987) Little ice age decline of a cold marginal *Pinus sylvestris* forest in the Swedish Scandes. *New Phytol.*, **106**, 567–84.

Kullman, L. (1988) Short-term approach to tree-limit and thermal climate: evidence from *Pinus sylvestris* in the Swedish Scandes. *Ann. Botanici Fennici*, **25**, 219–27.

Laessle, A. M. (1965) Spacing and competition in natural stands of sand pine. *Ecology*, **46**, 65–72.

Larsen, M. M. and Schubert, G. H. (1969) *Root competition between ponderosa pine seedlings and grass*, USDA Forest Service, Research Paper RM-54.

Lawton, J. H. (1987) Are there assembly rules for successional communities?, in *Colonization, Succession and Stability* (eds A. J. Gray, M. J. Crawley and P. J. Edwards), Blackwell, Oxford, pp. 225–44.

Likens, G. E., Bormann, F. H., Johnson, N. M., Fisher, D. W. and Pierce, R. S. (1967) Effects of forest cutting and herbicide treatment on nutrient budgets in the Hubbard Brook watershed-ecosystem. *Ecol. Monogr.*, **40**, 23–47.

Lorimer, C. G. (1983) Tests of age-independent competition indices for individual trees in natural hardwood stands. *Forest Ecol. Mgmt*, **6**, 343–60.

Loucks, O. L. (1970) Evolution of diversity, efficiency, and community stability. *Amer. Zool.*, **10**, 17–25.

MacArthur, R. H. (1972) *Geographical Ecology*, Harper and Row, New York.

MacArthur, R. H. and Wilson, E. O. (1967) *Island Biogeography*, Princeton Univ. Press, Princeton, New Jersey.

MacLean, D. A. and Wein, R. W. (1976) Biomass of jack pine and mixed hardwood stands in northeastern New Brunswick. *Canad. J. For. Res.*, **6**, 441–7.

MacMahon, J. A. (1981) Succession processes: comparisons among biomes with special reference to probable roles and influences on animals, in *Forest Succession: Concepts and Applications* (eds D. C. West, H. H. Shugart and D. B. Botkin), Springer-Verlag, New York, pp. 277–304.

Mar:Moller, C. (1947) The effect of thinning, age, and site on foliage, increment and loss of dry matter. *J. Forestry*, **45**, 393–404.

Mar:Moller, C. (1954) The influence of thinning on volume increment, in *Thinning Problems and Practices in Denmark*, State Univ. College of Forestry, Syracuse, New York, pp. 5–44.

Margalef, R. (1963) On certain unifying principles in ecology. *Amer. Nat.*, **97**, 357–74.

Margalef, R. (1968) *Perspectives in Ecological Theory*, Univ. of Chicago Press, Chicago, Illinois.

Marks, P. L. (1974) The role of pin cherry (*Prunus pensylvanica* L.) in the maintenance of stability in northern hardwood ecosystems. *Ecol. Monogr.*, **44**, 73–88.

Matthews, J. A. (1979) A study of the variability of some successional and climax plant assemblage-types using multiple discriminant analysis. *J. Ecol.*, **67**, 255–71.

May, P. G. (1982) Secondary succession and breeding bird community structure: patterns of resource utilization. *Oecologia*, **55**, 208–16.

McCune, B. and Allen, T. F. H. (1985) Will similar forests develop on similar sites? *Canad. J. Bot.*, **63**, 367–76.

Meyer, W. H. (1937) *Yield of Even-aged Stands of Sitka Spruce and Western Hemlock*, USDA. Agricultural Technical Bulletin 544.

Miles, J. (1979) *Vegetation Dynamics*, Chapman and Hall, London.

Mulcahy, D. (1975) Differential mortality among cohorts in a population of *Acer saccharum* (Aceraceae) seedlings. *Amer. J. Bot.*, **62**, 422–6.

Mutch, R. W. (1970) Wildland fires and ecosystems – a hypothesis. *Ecology*, **51**, 1046–51.

Myers, R. L. (1985) Fire and the dynamic relationship between Florida sandhill and sand pine scrub vegetation. *Bull. Torrey Bot. Club*, **112**, 241–52.

Myster, R. W. and Pickett, S. T. A. (1990) Initial conditions, history and successional pathways in ten contrasting old fields. *Amer. Mid. Nat.*, **124**, 231–8.

Neilson, R. P. (1986) High-resolution climatic analysis and southwestern biogeography. *Science*, **232**, 27–34.

Nicholson, S. A. and Monk, C. D. (1974) Plant species diversity in old-field succession on the Georgia piedmont. *Ecology*, **55**, 1075–85.

Odum, E. P. (1969) The strategy of ecosystem development. *Science*, **164**, 262–270.

Odum, E. P. (1971) *Fundamentals of Ecology*, 3rd edn, Saunders, Philadelphia.

Odum, H. T. and Pinkerton, R. C. (1955) Time's speed regulator: the optimum efficiency for maximum power output in physical and biological systems. *Amer. Scient.*, **43**, 331–43.

Oliver, C. D. (1980) Forest development in North America following major disturbances. *Forest Ecol. Mgmt*, **3**, 153–68.

Oliver, C. D. and Larson, B. C. (1990) *Forest Stand Dynamics*, McGraw-Hill, New York.

Oosting, H. J. (1942) An ecological analysis of the plant communities of Piedmont, North Carolina. *Amer. Mid. Nat.*, **28**, 1–126.

Parker, A. J. and Parker, K. C. (1983) Comparative successional roles of trembling aspen and lodgepole pine in the southern Rocky Mountains. *Great Basin Nat.*, **43**, 447–55.

Parker, A. J. and Peet, R. K. (1984) Size and age structure of conifer forests. *Ecology*, **65**, 1685–9.

Parrish, J. A. D. and Bazzaz, F. A. (1976) Underground niche separation in successional plants. *Ecology*, **57**, 1281–8.

Parrish, J. A. D. and Bazzaz, F. A. (1979) Difference in pollination niche relationships in early and late successional plant communities. *Ecology*, **60**, 597–610.

Parrish, J. A. D. and Bazzaz, F. A. (1982a) Responses of plants from three successional communities to a nutrient gradient. *J. Ecol.*, **70**, 233–48.

Parrish, J. A. D. and Bazzaz, F. A. (1982b) Niche responses of early and late successional tree seedlings on three resource gradients. *Bull. Torrey Bot. Club*, **109**, 451–6.

Parrish, J. A. D. and Bazzaz, F. A. (1982c) Competitive interactions in plant communities of different successional ages. *Ecology*, **63**, 314–20.

Peet, R. K. (1978) Forest vegetation of the Colorado Front Range: Patterns of species diversity. *Vegetatio*, **37**, 65–78.

Peet, R. K. (1981a) Changes in biomass and production during secondary forest succession, in *Forest Succession: Concepts and Applications* (eds D. C. West, H. H. Shugart and D. B. Botkin), Springer-Verlag, New York, pp. 324–38.

Peet, R. K. (1981b) Forest vegetation of the Colorado Front Range: Composition and dynamics. *Vegetatio*, **45**, 3–75.

Peet, R. K. (1988) Forests of the Rocky Mountains, in *North American Terrestrial Vegetation* (eds M. G. Barbour and W. D. Billings), Cambridge Univ. Press, Cambridge, pp. 63–101.

Peet, R. K. and Christensen, N. L. (1980) Succession: a population process. *Vegetatio*, **43**, 131–40.

Peet, R. K. and Christensen, N. L. (1987) Competition and tree death. *Bio-Science*, **37**, 586–95.

Peet, R. K. and Christensen, N. L. (1988) Changes in species diversity during secondary forest succession on the North Carolina piedmont, in *Diversity and Pattern in Plant Communities* (eds H. J. During, M. J. A. Werger and J. H. Willems), SPB Acad. Publ., The Hague, pp. 233–45.

Peterson, C. H. (1984) Does a rigorous criterion for environmental identity preclude the existence of multiple stable points? *Amer. Nat.*, **124**, 127–33.

Pickett, S. T. A. (1976) Succession: an evolutionary interpretation. *Amer. Nat.*, **110**, 107–19.

Pickett, S. T. A. (1982) Population patterns through twenty years of oldfield succession. *Vegetatio*, **49**, 45–59.

Pickett, S. T. A., Collins, S. C. and Armesto, J. J. (1987a) Models, mechanisms and pathways of succession. *Bot. Rev.*, **53**, 335–71.

Pickett, S. T. A., Collins, S. C. and Armesto, J. J. (1987b) A hierarchical consideration of causes and mechanisms of succession. *Vegetatio*, **69**, 109–14.

Pickett, S. T. A. and White, P. S. (1985) *The Ecology of Natural Disturbance and Patch Dynamics*, Academic Press, Orlando.

Pineda, F. D., Nicolas, J. P., Pou, A. and Galiano, E. F. (1981a) Ecological succession in oligotrophic pastures of central Spain. *Vegetatio*, **44**, 165–76.

Pineda, F. D., Nicolas, J. P., Ruiz, M., Peco, B. and Bernaldez, F. G. (1981b) Succession, diversité et amplitude de niche dans les pâturages du centre de la péninsule ibérique. *Vegetatio*, **47**, 267–77.

Platt, W. J. (1975) The colonization and formation of equilibrium plant species associations on badger disturbances in a tallgrass prairie. *Ecol. Monogr.*, **45**, 285–305.

Plochmann, R. (1956) Bestockungsaufbau und Baumartenwandel nordischer Ürwalder dargestellt an Beispielen aus Nordwestalberta/Kanada. *Forstwiss. Forsch.*, **6**, 1–96.

Potter, L. D. and Green, D. L. (1964) Ecology of ponderosa pine in western North Dakota. *Ecology*, **45**, 10–23.

Quinn, J. F. and Dunham, A. E. (1983) On hypothesis testing in ecology and evolution. *Amer. Nat.*, **122**, 602–17.

Reiners, W. A., Worley, I. A. and Lawrence, D. B. (1971) Plant diversity in a chronosequence at Glacier Bay, Alaska. *Ecology*, **52**, 55–69.

Rejmánek, M., Haagerová, R. and Haager, J. (1982) Progress of plant succession on the Parícutin Volcano: 25 years after activity ceased. *Amer. Mid. Nat.*, **108**, 194–8.

Risser, P. G., Birney, E. C., Blocker, H. D., May, S. W., Parton W. J. and Wiens, J. A. (1981) *The True Prairie Ecosystem*, Hutchinson Ross, Stroudsburg.

Rodin, L. E. and Bazilevich, N. I. (1967) *Production and Mineral Cycling in Terrestrial Vegetation*, Oliver and Boyd, Edinburgh.

Rundel, P. W. and Parsons, D. J. (1979) Structural changes in Chamise (*Adenostoma fasciculatum*) along a fire-induced age gradient. *J. Range Mgmt*, **32**, 462–6.

Runkle, J. R. (1985) Disturbance regimes in temperate forests, in *The Ecology of Natural Disturbance and Patch Dynamics* (eds S. T. A. Pickett and P. S. White), Academic Press, New York, pp. 17–34.

Saterson, K. A. (1985) *Nitrogen Availability, Primary Production, and Nutrient*

Cycling during Secondary Succession in North Carolina Piedmont Forests, Doctoral Dissertation, Univ. of North Carolina, Chapel Hill, North Carolina.

Schafale, M. and Christensen, N. L. (1986) Vegetational variation among old fields in Piedmont North Carolina. *Bull. Torrey Bot. Club*, **113**, 413–20.

Schlesinger, W. H. and Gill, D. S. (1978) Demographic studies of the chaparral shrub, *Ceanothus megacarpus*, in the Santa Ynez Mountains, California. *Ecology*, **59**, 1256–63.

Schlesinger, W. H. and Gill, D. S. (1980) Biomass, production, and changes in the availability of light, water, and nutrients during the development of pure stands of the chaparral shrub, *Ceanothus megacarpus*, after fire. *Ecology*, **61**, 781–9.

Schaal, B. and Levin, D. A. (1976) The demographic genetics of *Liatris cylindricea. Amer. Nat.*, **110**, 191–206.

Seastedt, T. R. and Ramundo, R. A. (1990) The influence of fire on belowground processes of tallgrass prairie, in *Fire in North American Tallgrass Prairies* (eds S. L. Collins and L. L. Wallace), Univ. of Oklahoma Press, Norman, pp. 99–117.

Shmida, A. and Ellner, S. (1984) Coexistence of plant species with similar niches. *Vegetatio*, **58**, 29–55.

Shugart, H. H. (1984) *A Theory of Forest Dynamics*, Springer-Verlag, New York.

Shure, D. J. and Ragsdale, H. L. (1977) Patterns of primary succession on granite outcrop surfaces. *Ecology*, **58**, 993–1006.

Sirén, G. (1955) The development of spruce forest on raw humus sites in northern Finland and its ecology. *Acta Forestalia Fennica*, **62** (4), 1–363.

Sprugel, D. G. (1984) Density, biomass, productivity, and nutrient-cycling changes during stand development in wave-regenerated balsam fir forests. *Ecol. Monogr.*, **54**, 165–86.

Sprugel, D. G. (1985) Natural disturbance and ecosystem energetics, in *The Ecology of Natural Disturbance and Patch Dynamics* (eds S. T. A. Pickett and P. S. White), Academic Press, New York, pp. 335–52.

Squires, E. R. (1989) The effects of seasonal timing of disturbance on species composition in a first-year old field. *Bull. Torrey Bot. Club*, **116**, 356–63.

Stevens, P. R. and Walker, T. W. (1970) The chronosequence concept and soil formation. *Quart. Rev. Biol.*, **45**, 333–50.

Strang, R. M. (1973) Succession in unburned subarctic woodlands. *Canad. J. For. Res.*, **3**, 140–3.

Svejcar, T. J. (1990) Response of *Andropogon gerardii* to fire in the tallgrass prairie, in *Fire and North American Tallgrass Prairies* (eds S. L. Collins and L. L. Wallace), Univ. of Oklahoma Press, Norman, pp. 19–27.

Swank, W. T. (1988) Stream chemistry responses to disturbance, in *Forest Hydrology and Ecology at Coweeta* (eds W. T. Swank and D. A. Crossley), Springer-Verlag, New York, pp. 339–57.

Tilman, D. (1985) The resource ratio hypothesis of succession. *Amer. Nat.*, **125**, 827–52.

Tilman, D. (1987) Secondary succession and the pattern of plant dominance along experimental nitrogen gradients. *Ecol. Monogr.*, **57**, 189–214.

Tilman, D. (1988) *Plant Strategies and the Dynamics and Structure of Plant Communities*, Monographs in Population Biology **28**, Princeton Univ. Press, Princeton, New Jersey.

Tilman, D. (1990) Constraints and tradeoffs: toward a predictive theory of competition and succession. *Oikos*, **58**, 3–15.

Tilman, D. and Cowan, M. L. (1989) Growth of old field herbs on a nitrogen gradient. *Funct. Ecol.*, **3**, 425–38.

Turkington, R. and Aarssen, L. W. (1984). Local-scale differentiation as a result of competitive interactions, in *Perspectives on Plant Population Ecology* (eds R. Dirzo and J. Sarukhán), Sinauer Associates, Sunderland, Massachusetts, pp. 107–27.

Uhl, C. (1987) Factors controlling succession following slash-and-burn agriculture in Amazonia. *J. Ecol.*, **75**, 377–407.

van Cleve, K. and Yarie, J. (1986) Interaction of temperature, moisture, and soil chemistry in controlling nutrient cycling and ecosystem development in the taiga of Alaska, in *Forest Ecosystems in the Alaskan Taiga: A Synthesis of Structure and Function* (eds K. van Cleve, F. S. Chapin III, P. W. Flanagan, L. A. Viereck and C. T. Dyrness), Springer-Verlag, New York, pp. 160–89.

van der Valk, A. G. (1981) Succession in wetlands: a Gleasonian approach. *Ecology*, **62**, 688–96.

van der Valk, A. G. and Davis, C. B. (1976) The seed banks of prairie glacial marshes. *Canad. J. Bot.*, **54**, 1832–8.

van der Valk, A. G. and Davis, C. B. (1978) The role of seed banks in the vegetation dynamics of prairie glacial marshes. *Ecology*, **59**, 322–35.

Veblen, T. T. (1985) Stand dynamics in Chilean *Nothofagus* forests, in *The Ecology of Natural Disturbance and Patch Dynamics* (eds S. T. A. Pickett and P. S. White), Academic Press, Orlando, pp. 35–51.

Veblen, T. T. (1986) Age and size structure of subalpine forests in the Colorado Front Range. *Bull. Torrey. Bot. Club*, **113**, 225–40.

Veblen, T. T and Lorenz, D. C. (1986) Anthropogenic disturbance and recovery patterns in montane forests, Colorado Front Range. *Phys. Geog.*, **7**, 1–24.

Vitousek, P. M. (1977) The regulation of element concentrations in mountain streams in the northeastern United States. *Ecol. Monogr.*, **47**, 65–87.

Vitousek, P. M. (1982) Nutrient cycling and nutrient use efficiency. *Amer. Nat.*, **119**, 553–72.

Vitousek, P. M. (1984) Litterfall, nutrient cycling, and nutrient limitation in tropical forests. *Ecology*, **65**, 285–98.

Vitousek, P. M., Gosz, J. R., Grier, C. C., Melillo, J. M. and Reiners, W. A. (1982) A comparative analysis of potential nitrification and nitrate mobility in forest ecosystems. *Ecol. Monogr.*, **52**, 155–77.

Vitousek, P. M., Matson, P. A. and Turner, D. R. (1988) Elevational and age gradients in Hawaiian montane rainforests: foliar and soil nutrients. *Oecologica*, **77**, 565–70.

Vitousek, P. M. and Reiners, W. A. (1975) Ecosystem succession and nutrient retention: a hypothesis. *BioScience*, **25**, 376–81.

Vitousek, P. M., van Cleve, K. and Mueller-Dombois, D. (1983) Soil development and nitrogen turnover in montane rainforest soils on Hawaii. *Biotropica*, **15**, 268–74.

Vitousek, P. M. and Walker, L. R. (1987) Colonization, succession and resource availability: ecosystem-level interactions, in *Colonization, Succession and Stability* (eds A. J. Gray, M. J. Crawley and P. J. Edwards), Blackwell, Oxford, pp. 207–23.

Vitousek, P. M. and White, P. S. (1981) Process studies in succession, in *Forest Succession: Concepts and Applications* (eds D. C. West, H. H. Shugart and D. B. Botkin), Springer-Verlag, New York, pp. 267–76.

Walker, B. H. (1981) Is succession a viable concept in African savanna ecosystems?, in *Forest Succession: Concepts and Applications* (eds D. C. West, H. H. Shugart and D. B. Botkin), Springer-Verlag, New York, pp. 431–47.

Walker, B. H., Ludwig, D., Holling, C. S. and Peterman, R. M. (1981) Stability of semi-arid savanna grazing systems. *J. Ecol.*, **69**, 437–98.

Walker, D. (1970) Direction and rate in some British post-glacial hydroseres, in *Studies in the Vegetational History of the British Isles* (eds D. Walker and R. G. West), Cambridge Univ. Press, Cambridge, pp. 117–39.

Walker, J., Thompson, C. H., Fergus, J. F. and Tunstall, B. R. (1981) Plant succession and soil development in coastal sand dunes of subtropical eastern Australia, in *Forest Succession: Concepts and Applications* (eds D. C. West, H. H. Shugart and D. B. Botkin), Springer-Verlag, New York, pp. 107–31.

Walker, L. R. and Chapin, F. S. III. (1987) Interactions among processes controlling successional change. *Oikos*, **50**, 131–5.

Walker, T. W. and Syers, J. K. (1976) The fate of phosphorus during pedogenesis. *Geoderma*, **15**, 1–19.

Walter, H. (1971) *Ecology of Tropical and Subtropical Vegetation*, Oliver and Boyd, Edinburgh.

Waring, R. H. and Schlesinger, W. H. (1985) *Forest Ecosystems: Concepts and Management*. Academic Press, New York.

Watt, A. S. (1947) Pattern and process in the plant community. *J. Ecol.*, **35**, 1–22.

Weaver, J. E. and Clements, F. E. (1938) *Plant Ecology*, McGraw-Hill, New York.

Webb, S. L. (1989) Contrasting windstorm consequences in two forests, Itasca State Park, Minnesota. *Ecology*, **70**, 1167–80.

Weiner, J. (1984) Neighbourhood interference amongst *Pinus rigida* individuals. *J. Ecol.*, **72**, 183–95.

Werner, P. A. and Platt, W. J. (1976) Ecological relationships of co-occurring goldenrods (*Solidago*: Compositae). *Amer. Nat.*, **110**, 959–71.

Westman, W. E. (1982) Coastal sage scrub succession, in *Dynamics and Management of Mediterranean-type Ecosystems* (eds C. E. Conrad and W. C. Oechel), USDA Forest Service, General Technical Report PSW-58, pp. 91–9.

Westman, W. E. and O'Leary, J. F. (1986) Measures of resilience: the response of coastal sage scrub to fire. *Vegetatio*, **65**, 179–89.

Westman, W. E. and Whittaker, R. H. (1975) The pygmy forest region of northern California: studies on biomass and primary production. *J. Ecol.*, **63**, 493–520.

Whipple, S. A. (1980) Population dispersion patterns of trees in a southern Louisiana hardwood forest. *Bull. Torrey Bot. Club*, **107**, 71–6.

Whipple, S. A. and Dix, R. L. (1979) Age structure and successional dynamics of a Colorado subalpine forest. *Amer. Mid. Nat.*, **101**, 142–58.

White, A. S. (1985) Presettlement regeneration patterns in a southwestern ponderosa pine stand. *Ecology*, **66**, 589–94.

White, P. S. (1979) Pattern, process, and natural disturbance in vegetation. *Bot. Rev.*, **45**, 229–99.

Whitmore, T. C. (1984) *Tropical Rain Forests of the Far East*, 2nd edn, Oxford Univ. Press, Oxford.

Whittaker, R. H. (1953) A consideration of climax theory: the climax as a population and pattern. *Ecol. Monogr.*, **23**, 41–78.

Whittaker, R. H. (1965) Dominance and diversity in land plant communities. *Science*, **147**, 250–60.

Whittaker, R. H. (1970) *Communities and Ecosystems*, Macmillian, New York.

Whittaker, R. H. (1972) Evolution and measurement of species diversity. *Taxon*, **21**, 213–51.

Whittaker, R. H. (1974) Climax concepts and recognition, in *Vegetation Dynamics. Handbook of Vegetation Science* **8** (ed. R. Knapp), Junk, The Hague, pp. 139–54.

Whittaker, R. H. (1975) *Communities and Ecosystems*, 2nd edn, Macmillan, New York.

Whittaker, R. H. (1977) Evolution of species diversity in land communities. *Evolut. Biol.*, **10**, 1–67.

Whittaker, R. J., Bush, M. B. and Richards, K. (1989) Plant recolonization and vegetation succession on the Krakatau Islands, Indonesia. *Ecol. Monogr.*, **59**, 59–123.

Williamson, G. B. and Black, E. M. (1981) High temperature of forest fires under pines as a selective advantage over oaks. *Nature*, **293**, 643–4.

Yeaton, R. I. (1978) Competition and spacing in plant communities: Differential mortality of white pine (*Pinus strobus* L.) in a New England woodlot. *Amer. Mid. Nat.*, **100**, 285–93.

Yodzis, P. (1978) *Competition for Space and the Structure of Ecological Communities*, Springer-Verlag, Berlin.

Zangerl, A. R. and Bazzaz, F. A. (1984) Effects of short-term selection along environmental gradients on variation in populations of *Amaranthus retroflexus* and *Abutilon theophrasti*. *Ecology*, **65**, 207–17.

Zedler, P. H. (1981) Vegetation change in chaparral and desert communities in San Diego County, California, in *Forest Succession: Concepts and Applications* (eds D. C. West, H. H. Shugart and D. B. Botkin), Springer-Verlag, New York, pp. 406–30.

4 Regeneration dynamics

Thomas T. Veblen

4.1 INTRODUCTION

4.1.1 Stability concepts

Whereas the other chapters in this book treat successional change or
directional changes in the floristic composition of plant communities, this
chapter deals with the regeneration dynamics of plant communities which
do not appear to be experiencing successional change. Although the
concepts and mechanisms discussed here apply to a variety of types of
plant communities, this chapter will focus primarily on forests.

The traditional concept of the climax as formulated by Clements (1916)
emphasized continued dominance of a site by the same assemblage of
species as a result of the ability of climax species to establish and grow
under the influences of adults of the same species. In Clementsian successional theory, a self-reproducing climax state is regarded as the
culmination of progressive changes in species composition associated with
the sequential arrival of propagules to a site and plant-induced environmental changes (i.e. autogenic changes) which favour the reproduction of
species characteristic of later successional stages. Early criticisms of the
concept emphasized the importance of allogenic changes in the physical
environment (including gradual shifts in climate) that result in continued
successional change so that a climax is rarely attained (Jones, 1945;
Whittaker, 1953; Raup, 1957). Recent studies from a wide range of
habitat types emphasize the spatial and temporal frequencies of repeated
natural disturbance, such that successional development to a self-
reproducing climax state is no longer viewed as the norm (White, 1979;
Sousa, 1984). Nevertheless, relatively constant species compositions
appear to characterize some habitats at particular spatial and temporal
scales. Even in such habitats, however, repeated fine-scale disturbance is
viewed as an inherent feature of the regeneration dynamics of the plant
communities (Runkle, 1981; Bormann and Likens, 1979).

The degree of stability in community composition remains unresolved
for many plant communities. This is due both to variation in concepts of
stability and to the length of time required to assess the stability of most
plant communities. As applied to plant communities, two major view-
points on stability may be distinguished (see reviews by Holling, 1973;

Plant Succession: Theory and prediction Edited by David C. Glenn-Lewin, Robert K. Peet and Thomas
T. Veblen © 1992 Chapman & Hall, London ISBN 0 412 26900 7

Connell and Sousa, 1983). In the first, more quantitative viewpoint, constancy of numbers of individuals is emphasized. Accordingly, a stable equilibrium is defined as a particular state (e.g. population density) at which a population will remain or, if moved away from it, a state to which the population will return (Connell and Sousa, 1983). The second viewpoint emphasizes the presence or absence of species and does not require that a system remain at or return to a quantitatively defined equilibrium (Holling, 1973; Connell and Sousa, 1983). Thus, attention is shifted to how species populations in a particular habitat persist. Connell and Sousa (1983) term this qualitative viewpoint of stability *persistence*, to indicate that a species population does not become extinct during a given time period in a given area, or, if it does, it recolonizes the area within the time span required for one turnover of the population.

Given the dynamic nature of plant communities, it may not be reasonable to expect a quantitative equilibrium of population densities, particularly at small scales. Even over time periods shorter than the potential life spans of the species involved, thinning of initially dense populations is likely (see Chapter 3). Thus, qualitative persistence of species is a more appropriate criterion to apply to plant communities. Plant communities in which all the common plant species persist over time periods greater than that required for one turnover of their populations may be considered to be in *compositional equilibrium*. The concept of compositional equilibrium may be applied at a range of spatial scales. For example, for a species-rich forest in compositional equilibrium, the relative importances of species inevitably change at spatial scales corresponding to groups of a few trees (i.e. $<25\,m^2$) but may remain approximately constant at stand scales (thousands of square metres). Thus, scale of sampling is critical in the assessment of compositional equilibrium.

In communities in compositional equilibrium, fluctuations in the abundances of species populations may occur, but major changes in relative abundances and local extinctions are avoided. The degree of constancy in the relative abundance and dominance of different species will vary according to how narrowly the idea of compositional equilibrium is applied. For example, no variation in the relative abundances of species implies the occurrence of steady-state populations which requires a balance between establishment and mortality rates (Whittaker, 1975b). In practice, however, plant communities have been described as being in a steady state when the population age structures of the dominant species imply that they are all regenerating, even though there is usually no basis for quantitatively assessing their stabilities. This usage is essentially the same as the idea of compositional equilibrium, and is consistent with the traditional use of climax to denote relative stability of community composition.

4.1.2   The patch dynamics perspective

Appreciation of spatial scale is also essential for understanding the mechanisms by which different species persist in communities at or near compositional equilibrium. At a regional or landscape scale, spatial discontinuities in plant distributions often result in a mosaic of patches which differ in composition and/or structure (Wiens, 1976; Whittaker and Levin, 1977; Pickett and Thompson, 1978; Sousa, 1984). Some of these spatial discontinuities reflect patchiness in the physical environment. However, in many cases where background physical conditions remain relatively uniform, there are marked differences in opportunities for establishment and growth of plants, which result in relatively discrete patches, sometimes consisting of individuals of similar ages (e.g. Heinselman, 1973; Oliver, 1981; Nakashizuka and Numata, 1982a; Ogden, 1985a; Oliver and Larson, 1990). This patchwork mosaic may result from the influences of disturbances on the intensity of biological interactions and resource availability. This point of view is known as the patch dynamics perspective (Pickett and Thompson, 1978). Although it has been applied effectively to a wide range of ecosystem types from intertidal marine communities to temperate grasslands (Sousa, 1979, 1984; Pickett and White, 1985; see also Chapter 2), in this chapter it will be applied mainly to the regeneration dynamics of forests.

In most plant communities of mesic habitats, where densities may potentially be great enough to render space-related resources limiting, some individuals are usually capable of dominating the site for long time periods (e.g. for at least several hundreds of years for trees). While these individuals dominate the site, environmental conditions change slowly, but eventually some of the trees die and space-related resources (e.g. light, nutrients, soil moisture) become available to other individuals. Thus, during a short time period, known as the *gap phase*, relatively rapid change occurs as new individuals attain dominance (Watt, 1923, 1947; Bray, 1956). The gap phase is characteristic of both communities near compositional equilibrium and successional communities where the replacement species do not include the formerly dominant species. Gap-phase replacement occurs in many community types (White, 1979; Sousa, 1984) but is perhaps most easily understood and best studied in mesic forests. As described many years ago by Watt (1947), fine-scale treefalls sometimes control the regeneration dynamics of mesic temperate forests, creating a mosaic of gap, building and mature phases. These are structurally distinct phases which in their sequential development constitute a forest growth cycle (Watt, 1947; Whitmore, 1975, 1982). For a forest stand in compositional equilibrium, at a particular point the canopy composition may be continually changing but over the entire stand the pattern of gap-phase replacement maintains similar relative abundances of dominant species. This viewpoint focuses attention on the gap phase and provides the framework for the consideration of regeneration dynamics in this chapter.

4.1.3   Mechanisms of species coexistence

Explanations of coexistence of plant species in communities at or near compositional equilibrium rely upon resource partitioning, life-history differentiation, or disruption of biological interactions by fine-scale disturbances, or a combination of all three (Grubb, 1977; Pickett, 1980; Aarssen, 1983; Shmida and Ellner, 1984; Silvertown and Law, 1987; Denslow, 1985; Tilman, 1988).

In explanations based on resource partitioning, natural selection is believed to result in the evolution of species with differentiated niches, so that in equilibrium communities these species coexist without competing directly. The idea that two species occupying the same niche cannot coexist indefinitely is known as Gause's hypothesis or the principle of competitive exclusion. Whittaker (1975a) described the application of this principle as follows:

(1) If two species occupy the same niche in the same stable community, one will become extinct.
(2) No two species observed in a stable community are direct competitors limited by the same resources; the species differ in niche in ways that reduce competition between them.
(3) The community is a system of interacting niche-differentiated species populations that tend to complement one another, rather than directly competing, in their uses of the community's space, time, resources, and possible kinds of interactions.

For species of substantially different life-forms or patterns of seasonal development, complementarity of resource use patterns is easily envisaged (Grubb, 1977). However, for species of the same life-form and phenological behaviour in the same habitat, niche differentiation is a less attractive explanation of species coexistence. For plant species with similar modes of acquiring the same few essential resources, it is not clear how these resources can be differentially partitioned, particularly in species-rich communities (Grubb, 1977; Denslow, 1985).

Another explanation of species coexistence, known as the non-equilibrium viewpoint, suggests that disturbances truncate the process of competitive exclusion so that a compositional equilibrium with one species becoming extinct is never attained (Pickett, 1980). In this context, a *disturbance* is defined as 'a discrete, punctuated killing, displacement, or damaging of one or more individuals (or colonies) that directly or indirectly creates an opportunity for new individuals (or colonies) to become established' (Sousa, 1984). Both biotic events, such as herbivory or pathogen attack, and abiotic events, such as fine-scale windthrow, are frequent disturbances even in forests which appear to be at or near compositional equilibrium (Bormann and Likens, 1979; Runkle, 1982; Whitmore, 1982). According to the non-equilibrium viewpoint, most species, particularly the inferior competitors, grow to maturity when the control of the site by competitive dominants is disrupted by disturbance (Pickett, 1980). Thus, even in habitats where the background environmental conditions are uniform, periodic disturbances create spatial and temporal heterogeneity in resource availability which, in turn, permits species coexistence (see section 4.3.2).

Grubb's (1977, 1986) concept of the 'regeneration niche' as an explanation for species coexistence is consistent with both the ideas of resource partitioning and environmental heterogeneity induced by fine-scale disturbance. The *regeneration niche* of a species is 'an expression of the requirements for a high chance of success in the replacement of one mature individual by a new individual of the next generation . . .' (Grubb, 1977). According to this concept, in some communities important niche differences among coexisting species may be manifested only during the early stages of their life histories. In other words, the requirements for processes such as initiation of vegetative reproduction, propagule dispersal, germination, seedling establishment, and growth of juveniles may differ for species which have apparently undifferentiated needs as mature individuals. Changes in niche relationships as individuals grow may be the consequences of ontogenetic changes or the result of smaller individuals responding to the environment at a finer scale (Woods, 1984). The importance of micro-environmental influences on germination, establishment and early growth is also emphasized by concepts such as 'microsite mosaic' (Whittaker, 1975b) and 'safe site' (Harper, 1977). A common source of variation in these micro-environmental conditions in some communities is fine-scale disturbance.

In addition to non-equilibrium explanations of the coexistence of species which appear to have similar niches, explanations based on differences in life-history strategies have been proposed (Hutchinson, 1951; Shmida and Ellner, 1984; Silvertown and Law, 1987). Using a simple mathematical model, Shmida and Ellner (1984) predicted the coexistence of two species where one species emphasized fecundity and the other adult survivorship. Empirical support for coexistence based on differences in life-history traits comes from studies of regeneration and mortality patterns of *Picea–Abies* forests in the southern Appalachians and central Rockies (White *et al.*, 1985b; Veblen, 1986b). Although different species occur in the two areas, *Abies* is consistently the more abundant taxon in the seedling size class, implying greater fecundity, or at least a lower mortality rate for juveniles. In contrast, *Picea* is substantially longer lived and has a lower adult mortality rate. These differences in recruitment and mortality rates appear to contribute to the coexistence of these ecologically similar species in the same habitat.

4.2 RECOGNITION OF COMPOSITIONAL EQUILIBRIUM

Although it was often asserted in the older ecological literature that a particular community was in a climax state, rarely have detailed data on the populations of the constituent species been available to assess such conclusions. Since vegetation managers have traditionally accepted the idea of climax both in theory and in practice, the recognition of the climax state is of considerable practical importance. Rigorous proof of compositional equilibrium (or of steady-state populations) would require repeated measurements of a plant community over a time span equivalent to that needed for at least one turnover of all the species populations. Such observations have been made only for short-lived herbaceous plants (Connell and Sousa, 1983). In the case of forests dominated by long-lived individuals, it has not been feasible to document their species composition for time periods greater than a small fraction of the turnover period. Consequently, compositional equilibrium, as opposed to successional change, must be inferred from the past and present characteristics of the species populations. Attempts to infer the likelihood of species persistence based on demographic analysis of forest tree populations fall into three broad categories: (1) repeated censuses over multi-year periods; (2) inference of past and present demographic trends from forest structure, including dead as well as live trees; and (3) projection of future composition based on the spatial association between potential replacement individuals and canopy trees (i.e. by means of transition probabilities). Although discussed separately, these three approaches are not mutually exclusive.

4.2.1 Repeated censuses

There have been relatively few multi-year studies of rates of tree-seedling establishment, recruitment into succeeding age classes and direct measurement of mortality rates (e.g. Hett and Loucks, 1971, 1976; Sarukhán, 1978; Christy and Mack, 1984; Martínez-Ramos et al., 1988). Generally, these have revealed a pattern of a declining mortality rate with age for the dominant tree species. Too few demographic studies of tree populations have been conducted, however, to establish whether decreasing mortality rate with age is a general pattern. Such studies have a great potential for revealing differences in regeneration niches, and if continued long enough will indicate patterns of species replacement or persistence.

A partial solution to the problem of long-term demographic observation is the inclusion of older age groups in repeated censuses. One approach is to census repeatedly different same-aged populations representing a series of stages in the development of a species' population.

Such studies can reveal the major demographic features of older cohorts (e.g. Peet and Christensen, 1980, 1987). In forests with individuals of different ages and different species, repeated observations of the populations can also reveal important differences in recruitment and mortality patterns (e.g. Christensen, 1977; Lorimer, 1981; Harcombe and Marks, 1983; Lang and Knight, 1983; Connell *et al.*, 1984; Lieberman *et al.*, 1985; Harcombe, 1986, 1987). If we can assume that the differences observed over relatively short time spans will remain consistent over the long term, repeated censuses of forest populations provide an effective means of assessing the likelihood of species persistence. This is not always a safe assumption, given the likelihood that mortality rates will change due to climatic fluctuation, site changes associated with stand development, or selective disturbances such as disease or insect attack.

4.2.2   Inference from forest structure

An approach frequently used to distinguish successional trends from compositional equilibrium is based on analysis of forest stand structure. In this context, stand structure refers to the relative abundance and spatial dispersion of trees of different ages and/or sizes. In its simplest application, correspondence of the composition of seedling and sapling populations with the canopy composition has been taken as evidence of species persistence (i.e. the 'accordance' criterion of Braun, 1950). Interspecific differences in growth rates and recruitment from juvenile stages into the canopy (e.g. Lang and Knight, 1983; Connell *et al.*, 1984) may, however, confound predictions of abundances of canopy trees based on the relative abundances of seedlings and saplings.

Stand structural analysis based on both age and size data has often been used to reconstruct details of the history of stand development, infer past mortality rates, or assess quantitative population stability. Sometimes this method has been applied without consideration of its limitations. Given the difficulty of aging large numbers of trees by counting annual rings on increment cores, size (usually diameter at breast height) is sometimes used as a substitute for age. This substitution is unavoidable for many tropical trees which usually do not form annual rings. For most temperate tree species, however, the strength of the correlation of age with diameter can be determined. In contrast to Harper's (1977) often-cited statement that the correlation of age and size in multi-aged communities 'usually turns out to be very weak', for many tree species of both the northern and southern temperate zones correlation coefficients >0.8 or >0.9 are often reported (Spring *et al.*, 1974; Leak, 1975; Hett and Loucks, 1976; Lorimer, 1980; Donoso *et al.*, 1984; Veblen, 1985). However, the strength of the relationship between age and diameter is highly variable, depending on species and stage of stand development, and in each case the strength of the relationship must be investigated. Where the objective is to reconstruct the details of the history of stand development, size data alone are insufficient. However, size rather than age may be a better indicator of survival probability or reproductive capacity of trees and, hence, of their ability to contribute to the next generation of canopy trees (Harper, 1977). Size data may be sufficient, then, so long as it is only the future of the stand that is of concern.

When past mortality rates are inferred from static age structure, the individuals in successively older age classes are treated as if they were the survivors from initially equally sized cohorts. In other words, it is assumed that input into the population has remained constant over the time span represented by the oldest individuals, and that the age distribution is stationary (Caughley, 1977; Kellman, 1980). In reality, static age structure reflects past variability in both input to the population (i.e. seedling recruitment) and mortality (Harper, 1977). Nevertheless, a static age structure is often assumed to be indicative of the survivorship curves which would be observed if mortality in equal-sized cohorts was measured over time (e.g. Hett and Loucks, 1976). Given the annual variation in seed production, seed viability and seedling establishment characteristic of most tree species (Fowells, 1965; Grubb, 1977), constant input is unlikely. Consequently, tree mortality rates should not be inferred from static age structure.

Making similar assumptions, ideal age or size frequency distributions have been identified which allegedly represent stable tree populations (e.g. Liocourt, 1898; Meyer, 1952; Leak, 1964, Goff and West, 1975). When the numbers of individuals are plotted in successively larger size or older age classes, the distribution is regarded as 'balanced' if there are abundant individuals in the smallest or youngest class and progressively fewer individuals in successively larger or older classes (Meyer, 1952; Leak, 1964). Thus, as the oldest individuals die, the abundance of smaller or younger individuals allows for continued recruitment into larger or older classes so that the frequency distribution remains constant. When plotted with age or size as the horizontal axis, such a distribution is described by a smoothly declining curve which gradually approaches the horizontal axis towards greater age or size (i.e. a 'reverse J' curve in forestry terms). A negative exponential or a power function could mathematically describe the curve, depending on whether the rate of decline in successive age or size classes remains constant or decreases, respectively. The negative exponential model is $y = a\, e^{bx}$, where y is the number in any age or size class x, 'e' is the base of the natural logarithm, and a and b are constants (Hett and Loucks, 1976; Ross et al., 1982). Given a constant input of seedlings and a stationary age or size distribution, the negative exponential model implies a constant mortality rate. The power function model is $y = ax^b$, and given the same assumptions it implies declining mortality rates with age. Thus, these two models are analogous to Deevey Type II and Type III survivorship curves which result from constant mortality rates and declining mortality rates with age, respectively (Pearl and Miner, 1935; Deevey, 1947).

The power function model appears to be more consistent with the observed decline in mortality rate with age in seedling cohort studies, and with the idea that juveniles are more susceptible to lethal stresses than are mature trees (Hett and Loucks, 1976; Christy and Mack, 1984). Nevertheless, either the power function or the negative exponential model may best describe the observed size or age distributions of a specific tree population (e.g. Hett and Loucks, 1976; Ross et al., 1982).

While good fits of observed distributions clearly imply continuous recruitment of small or young trees into successively larger or older age classes, the same conclusion should be evident from qualitative inspection of the frequency distributions. Variation in survivorship curves of the same species occurring in different habitats suggests that such curves may be characteristic of a species in only a general way (Kellman, 1980; Butson et al., 1987). Without confirmation from long-term observations, neither the negative exponential nor the power function model should be used as a strict criterion of population stability. Age–frequency distributions that are dramatically different from the negative exponential and power function models (e.g. if young individuals are scarce) are clearly indicative of population instability. However, slight departures from expected distributions should not be interpreted as evidence of population instability or of past disturbances (e.g. Jackson and Faller, 1973; Johnson and Bell, 1975).

Inclusion of data on the relative abundances of dead standing trees of different species is often useful in determining if the species composition of a stand has changed over recent decades (e.g. Ogden, 1985b; Veblen, 1986a). Identification of fallen logs on the forest floor or partially buried logs may reveal trends in species composition initiated several centuries prior to the establishment of the current canopy dominants (e.g. Henry and Swan, 1974; Oliver and Stephens, 1977; Johnson and Fryer, 1989). When growth patterns of both live and dead trees are analysed by cross-dating and measuring tree-ring widths, detailed histories of stand development are possible (Lorimer, 1985; Foster, 1988).

Although stand structural analysis without supplemental data from repeated censuses cannot be used to determine precise past recruitment and mortality rates, it can be used effectively as a time-specific method to reconstruct the major features of stand history (e.g. disturbances) and to assess the chances of species persistence (e.g. Lorimer, 1977; Nakashizuka and Numata, 1982a; Glitzenstein et al., 1986; Read and Hill, 1988). Given that much of our current understanding of the regeneration dynamics of tree populations is inferred from stand structural analyses, substantial attention to the limitations of this methodology is warranted. The chances of obtaining meaningful results from stand structural analysis can be improved by considering the following recommendations, which are aimed at relating appropriate objectives to procedures.

Appropriate methods of stand structure analysis

(i) *Collect the maximum amount of age data feasible.*   Even though age determinations for large tree populations can be extremely time-consuming, such data are essential for reconstruction of the details of stand development. Also, in judging the likelihood that a species will persist at a site, it is extremely useful, if not essential, to know at least the major features of stand history (which requires age data).

(ii) *Avoid overestimating the precision of age data.*   Accurate total tree ages are surprisingly difficult to obtain. The ages of trees with rotten centres are sometimes estimated by extending ring counts to the centre based on the pattern of ring-width changes in the adjacent solid portion of the tree, or by computing age–diameter regressions (e.g. Lorimer, 1980; Butson *et al.*, 1987). The proportion of the total age data represented by such estimates has to be considered when deciding on the detail of interpretation which can be justified. Other errors result from false or missing rings, inaccurate ring counts on very old individuals, and increment cores with missing centres due to either asymmetrical growth or large tree diameter relative to increment borer length (Ogden, 1985a; Duncan, 1989). Ameliorative measures such as age determination by cross-dating (Fritts, 1976), using geometric techniques to estimate the number of missing rings on cores not quite reaching the pith (Duncan, 1989), or by counting rings on cut stumps or disks (e.g. Stewart, 1986) can reduce errors, but are not always practical.

A ring count reveals the age of the tree at the level of the bole at which it is cored, not the total age (i.e. time since germination). For species which can tolerate long periods of suppression as seedlings, the difference in age between a coring height of 30 or 40 cm and total age may be several decades or more (Morris, 1948; Veblen, 1986b). To correct for this, sometimes a sample of trees is cored both at ground level to approximate total tree age and at a more convenient coring height 30–40 cm above the ground; thus, an average difference in age at ground level and at the standard coring height may be computed and added to the other trees cored only at a standard coring height (e.g. Henry and Swan, 1974). Butt rot or greater frequency of partial rings near the base of highly stressed trees, however, may make coring at ground level impractical (Fritts, 1976). Another solution is to harvest seedlings of a given height, determine their mean or modal age, and add it to the number of rings counted on cores taken at the same height. Care must be taken that the harvested seedlings grew under the same conditions (i.e. open or closed canopy) as those under which the canopy trees established.

(iii)   Where tree size accounts for a significant proportion of the data, carefully investigate the relationship between size and age.   To test the strength of this relationship, a regression of age upon diameter must be performed for a sample of trees. The size of the sample should be as large as possible, but in practice usually consists of 20–60 trees per species (e.g. Leak, 1975; Lorimer, 1980). As large a proportion of the diameter range as possible must be included by uniformly sampling all size classes (e.g. <5 cm dbh, 5–10 cm dbh, etc.). Due to variations in tree growth rates at sites of strongly different edaphic or climatic characteristics, it is logical to test the relationship of age and diameter only for samples taken from relatively uniform habitats. Samples taken over steep physical gradients usually result in weak age–diameter relationships. Even within some homogeneous stands, the relationship between tree size and age may be so weak that no inferences about regeneration patterns can be made from size data alone (e.g. Veblen, 1986a).

(iv)   Sampling and interpretation must be conducted at scales appropriate to the objective of the study.   When the objective is to describe regeneration status over a larger area (i.e. tens of thousands of hectares) large numbers of small objectively located samples are usually used (e.g. Wardle and Guest, 1977). When stand structure and regeneration status of different species vary over the area sampled, interpretation of the data pooled from numerous dispersed samples becomes problematical. Given the spatial heterogeneity of large areas and the impracticality of sampling more than a tiny fraction of large areas, any attempt to identify more than the major trends of species replacement on the basis of broad-scale surveys is inappropriate. Because the stands may have markedly different structures and disturbance histories, pooling of structural data from many different sites may also confound age–diameter relationships and obscure the circumstances in which species regenerate. When the objectives are to assess compositional equilibrium in smaller tracts of forest and reveal the circumstances in which different species regenerate, sampling at more than one scale is often effective (e.g. Lorimer, 1980). In analysing regeneration processes, it is often useful to obtain data on spatial patterns of trees, either by mapping individual trees or by contiguously locating small quadrats which will allow examination of pattern at a range of spatial scales (e.g. Williamson, 1975; Veblen *et al.*, 1981; Read and Hill, 1988; Rebertus *et al.*, 1989). For example, fine-scale clumping of a species population may indicate regeneration following small treefalls, or non-random spatial associations of species populations may indicate important influences of one species on the regeneration of another.

(v)   Incorporate autecological information into the interpretation of stand structural data.   Knowledge of the approximate potential life spans of the species studied is essential in differentiating young invasive popu-

lations of long-lived species from relatively stable populations of short-lived species. Similarly, periodicity of seed production and patterns of seed storage and germination may be reflected in the age structure of mature populations. Mechanisms and effectiveness of vegetative reproduction are often critical when comparing the likelihood of persistence of different species. For example, parental subsidy of basal sprouts for species such as *Tilia americana* may result in relatively low mortality rates of saplings compared to other species (Woods, 1984). Also, for some tree species vegetative reproduction from aerial roots and stem bases appears to compensate for lack of regeneration by new seedling establishment (Veblen *et al.*, 1981; Johnston and Lacey, 1983).

(vi) Seek additional types of historical information on the stands studied.   For example, the relocation of historical landscape photographs can be effectively combined with quantitative analyses of stand structures (e.g. Kullman, 1986; Veblen and Lorenz, 1986, 1988). Similarly, data from land surveys sometimes allow comparisons of tree species abundances and sizes over time spans of many decades or even a century or more (e.g. Grimm, 1984).

4.2.3   The transition probability approach

A third approach to the prediction of change in forest composition is the computation of a table of transition probabilities based on the spatial association of canopy trees with potential replacement individuals (for details see Chapter 6). Waggoner and Stephens (1970) and Horn (1975, 1976) proposed a model of forest stand dynamics in which succession is viewed as tree-by-tree (or, in horizontal space, cell-by-cell) replacement, and the probability of each tree being replaced by another, of the same species or by a different species, is derived from the present state of the forest. A table of transition probabilities is best constructed on the basis of remeasurements of permanent plots. Where data from permanent plots are not available, the table of transition probabilities may be constructed based on the assumption that the recruitment of canopy trees is proportional to the local abundances of subcanopy populations. Thus, the transition probabilities are computed on the basis of relative abundances of subcanopy individuals occurring beneath canopy trees (e.g. Horn, 1975; Culver, 1981) or in gaps created by the fall of different canopy species (e.g. Barden, 1981). It is assumed that the transition probability of one species replacing another species depends only on the relative abundance of subcanopy individuals of the first species occurring beneath canopy trees or in gaps created by the second species.

   Although the transition probability approach has been applied effectively in numerous old-growth forests (Waggoner and Stephens, 1970; Horn, 1975, 1976; Barden, 1980, 1981; Runkle, 1981, 1984; Culver, 1981;

Ogden, 1983; Veblen, 1985; Taylor, 1990), its limitations need to be recognized. Use of the transition probability approach to predict future canopy composition is based on several assumptions, some of which are almost certainly not valid (Lippe *et al.*, 1985). Most importantly, it must be assumed that the transition probabilities remain constant over time. Due both to the effects of climate variation and changes in forest structure, transition probabilities are unlikely to remain constant. Similarly, where transition probabilities are estimated from relative abundances of saplings in the understorey, it is assumed that subcanopy individuals of different species have equal probabilities of replacing canopy trees (Horn, 1975; Usher, 1979). However, variations among species in survival and growth rates, both as seedlings and as larger subcanopy individuals, suggest that this assumption is generally not valid (Hett and Loucks, 1976; Abrell and Jackson, 1977; White *et al.*, 1985a; Veblen, 1986b). Where not already known (e.g. Fowells, 1965), major interspecific differences in growth and survival can be investigated by comparing height or radial increments (e.g. Hibbs, 1982; Kelty, 1986). Given the greater likelihood that taller individuals will reach the main canopy, predictions may alternatively be based on potential successors (i.e. the tallest subcanopy trees in each gap) instead of total abundances of saplings in gaps.

Predictions of future canopy compositions are also based on the assumption that regeneration or successional replacement is essentially a tree-by-tree replacement process resulting from small treefalls. While this assumption may be valid for some forests, in others, coarse-scale natural disturbances such as blowdown or wildfire often influence forest dynamics (e.g. Whitmore, 1974; Romme, 1982). In the absence of evidence supporting long-term site stability, the iterative use of transition probabilities over several tree generations to predict an equilibrium species composition is largely a theoretical exercise. An alternative application of the technique is to use it to test whether the present canopy composition represents compositional equilibrium (e.g. Veblen, 1985).

Many of the assumptions required in the transition probability approach are also necessary in traditional stand structure analysis, and the degree to which the transition probability approach represents a new framework for studying forest dynamics has been questioned (McIntosh, 1980). The danger of the approach is that it tends to treat forest changes as a statistical rather than an ecological process by largely ignoring interspecific differences in responses to the spatial heterogeneity of the environment associated with either canopy trees or treefall gaps (e.g. Denslow, 1980; Woods, 1984). Although explicit terms to account for variation in species' abilities to respond to treefall gaps have been suggested (e.g. Acevedo, 1981; White *et al.*, 1985a), only long-term observations can reveal the full range of variation in biological behaviour. A fundamental shortcoming lies in the implicit assumption that such a small set of variables (tree size and location) constitutes an adequate description of complex ecological processes. The technique is, nevertheless, useful for summarizing vegetational data, discovery of major trends in species composition, and the generation of hypotheses for long-term investigation (Ogden, 1985b).

4.3 THE REGENERATION NICHE AND SPECIES COEXISTENCE

4.3.1 Regeneration modes

Grubb's (1977) concept of the regeneration niche emphasizes that niche differences among coexisting plant species need only be manifested during the early stages of life histories. The concept was offered as an explanation for the coexistence of species with much the same life-form, phenology and habitat range. It emphasizes the importance of fine-scale environmental heterogeneity. Disturbances may favour coexistence by contributing to environmental heterogeneity through direct modification of microsites and indirectly by altering plant influences on microsites.

A first step in characterizing a regeneration niche is to describe a species' general mode of regeneration. A species' *regeneration mode* is its regeneration behaviour in relation to disturbance. Regeneration mode refers to the spatial scale at which regeneration occurs in relation to disturbance and usually can be inferred from the age structure and spatial patterns of tree populations. A continuum of regeneration modes may be segregated arbitrarily into three types: catastrophic, gap-phase and continuous.

(a) The catastrophic regeneration mode

Catastrophic regeneration is the establishment of most of a local population over a relatively short period of time following a stand-devastating disturbance and sudden release of resources. Many tree species with catastrophic modes of regeneration establish massively after disturbance by wildfire, volcanic ash deposition, landslides or blowdown (White, 1979). Their populations are initially even-aged, although the age range of the cohort is highly variable according to site conditions and propagule availability (e.g. Whipple and Dix, 1979; Franklin and Hemstrom, 1981; Peet, 1981). Patch size tends to be large, often exceeding a hectare or more, and, as the patch ages, self-thinning usually results in a tendency towards less clustered tree dispersion (e.g. Ford, 1975; Kenkel, 1988). The catastrophic mode of regeneration allows many shade-intolerant tree species to coexist with shade-tolerant species that are often less effective at colonizing severely disturbed sites (e.g. Whitmore, 1974; Ashton, 1981; Veblen *et al.*, 1981; Dunn *et al.*, 1983).

(b)   The gap-phase regeneration mode

The gap-phase mode of regeneration refers to trees attaining main canopy stature in small- to intermediate-sized canopy gaps which have resulted from the death of one tree or small groups of trees (Watt, 1947; Bray, 1956). Whereas the catastrophic regeneration mode is related mainly to exogenous disturbances, the fine-scale, gap-phase mode is a response to the endogenous treefalls which inevitably occur, both in old successional stands and in stands near compositional equilibrium (Bormann and Likens, 1979; Oliver, 1981). In many old-growth forests, regeneration in canopy gaps of this size range (*c.* 25–1000 m^2) creates a mosaic of patches of same-aged individuals. At a stand scale this may be reflected by intense clustering of small trees and less intense clustering in old patches as the result of self-thinning (e.g. Veblen *et al.*, 1981; Nakashizuka and Numata, 1982a; Stewart, 1986; Read and Hill, 1988). Where the canopy gaps are large enough to accommodate more than one canopy tree, clumping of large trees may also be observed. When a small area (i.e. <250 m^2) of a forest consisting of gap-phase species is sampled, the age distributions of individual species are likely to be sporadic (i.e. discontinuous), but at larger scales a continuously all-aged distribution should be apparent.

(c)   The continuous regeneration mode

The continuous regeneration mode refers to attainment of maturity in the absence of a disturbance-caused canopy opening. If the main canopy is not too dense, some shade-tolerant species may sometimes grow directly into it (Spurr and Barnes, 1980; Canham, 1989). This category also includes shade-tolerant understorey trees that grow slowly to adult size but rarely reach the upper part of the forest canopy. These understorey specialists are most common in tropical rainforests and apparently do not require gaps for either germination or growth to reproductive maturity (Whitmore, 1974; Denslow, 1980). This category also includes canopy tree species which reach the main canopy by establishing epiphytically and then growing as vines through even dense canopies in some southern temperate forests (Veblen *et al.*, 1981; Wardle, 1983). In contrast to the other two regeneration modes, the availability of the resources required for regeneration of these species is much more continuous. Compared to the other two modes, however, the continuous mode of regeneration appears to be rare.

   A particular regeneration mode may be characteristic of a species in only a general way, and may vary with forest type or stage of stand development (Veblen, 1989). For example, *Pinus contorta* in the central Rocky Mountains is well known as a species which regenerates mainly after wildfire (Stahelin, 1943; Daubenmire, 1943). Its catastrophic regeneration mode generally maintains its abundance in subalpine habitats where its range overlaps with the more shade-tolerant *Picea engelmannii* and *Abies lasiocarpa*, which, in the absence of fire, successionally replace it (Peet, 1981; Veblen, 1986a). However, in less favourable habitats, such as on particularly nutrient-poor soils and xeric sites, the species which otherwise would successionally replace it may be unable to grow. In such habitats, and also in Sierran subalpine forests in California, *Pinus contorta* regenerates in old-growth stands long unaffected by wildfire in a gap-phase mode, beneath small- to intermediate-sized treefall gaps (Whipple and Dix, 1979; Peet, 1981; Despain, 1983; Parker, 1986).

4.3.2   Regeneration in treefall gaps

As trees grow taller and develop more massive crowns, they become more susceptible to windthrow. As a tree ages, other factors also contribute to the increasing risk of windthrow, including decreasing physiological efficiency and the weakening effects of disease and insect attack (Brokaw, 1985b; Runkle, 1985; Worrall and Harrington, 1988). In some tropical rainforests heavy loads of epiphytes and lianas may also contribute to treefalls (Strong, 1977; Putz, 1984). For a given forest, the rate of treefalls results from the interaction of internal factors like tree architecture, stratification and health, with external factors such as type and intensity of precipitation and wind speeds and durations. Thus, although the characteristics of treefalls are not totally controlled by the plants themselves, small treefalls approximate the concept of endogenous disturbance more closely than any other type of common disturbance (Bormann and Likens, 1979).

Even in the absence of treefalls, there is substantial heterogeneity beneath a 'closed' forest canopy in terms of composition, height, thickness and foliage density, which may have important influences on understorey composition and subsequent responses to treefalls (Veblen *et al.*, 1979b; Lieberman *et al.*, 1989). In fact, the dichotomy between gap and non-gap sites in a forest is arbitrary, given the continuum of variation in light levels within a forest (Canham, 1989; Lieberman *et al.*, 1989). Nevertheless, the dichotomy between gap and non-gap sites is a useful point of departure for understanding regeneration dynamics. There are three general mechanisms by which treefalls temporarily increase the availability of resources for plant growth. The first is simply the decrease in rate of uptake and use of resources (such as solar radiation, soil moisture and soil nutrients) due to the loss of biomass, both in the canopy and the undergrowth (Vitousek and Denslow, 1986; Denslow *et al.*, 1990). The second is the decay and mineralization of nutrients previously held in organic matter. The latter results from both decomposition of the fallen trees and from higher rates of decomposition of soil organic matter associated with increased insolation and temperature at the soil surface

(Bormann and Likens, 1979). The third is simply the exposure of bare mineral soil where a thick litter layer previously may have impeded seedling establishment.

There are two general patterns of vegetation response to the transient pulses of resources associated with treefalls (Marks, 1974; Canham and Marks, 1985). The reorganization pattern involves accelerated growth of individuals already established at the time of the treefall. Examples include lateral encroachment of branches and roots from surrounding trees, vegetative spread of understorey species, epicormic branching of undamaged trees, sprouting and root suckering of damaged individuals, and release of previously suppressed seedlings and saplings (i.e. 'advance regeneration' in forestry). The new establishment pattern includes both establishment from dormant seeds already in place at the time of the treefall as well as from propagules dispersed to the site following the treefall. It also includes the establishment of tree seedlings that originate as canopy epiphytes and subsequently establish in gaps as trees fall in some tropical rainforests (Lawton and Putz, 1988). Within the reorganization and new establishment patterns, more detailed differences in responses to treefall gaps further differentiate the regeneration niches of tree species. The following treefall gap characteristics appear to be significant in the divergence of regeneration niches, and, thereby, may contribute to species coexistence in forests that are at or near compositional equilibrium.

(a)   Gap size, shape and orientation

Gap size is related both to tree sizes and the manner in which they fall. As expected, for gaps created by the fall of a single tree, gap area is positively correlated with tree size (Brokaw, 1982). The largest gaps result from multiple treefalls, when a large canopy tree knocks down its neighbours in a domino fashion. Relatively small gaps are created when a tree dies in a standing position and gradually collapses downwards. Limb falls generally create the smallest gaps. Small gaps have a greater perimeter-to-area ratio than larger gaps, which may significantly influence the pattern of vegetation response. In smaller gaps the relative importance of encroachment of lateral growth of branches and roots from the surrounding trees will be greater (Runkle, 1985). Similarly, in smaller gaps and in more irregularly shaped gaps, the greater number of nearby mature trees per unit gap area will favour more rapid propagule recruitment.

The microclimate of gaps varies as a function of latitude, gap size, and shape and orientation in relation to the height of the surrounding trees (Geiger, 1965; Lee, 1978; Poulson and Platt, 1989; Canham et al., 1990). In tropical forests maximum initial light intensities should be greater and should occur closer to the centre of the gap than in temperate forests

(Poulson and Platt, 1989). As gap size increases, so does intensity and duration of solar radiation, mean soil and air temperature, and temperature ranges; conversely, humidity decreases with size (Denslow, 1980; Runkle, 1985; Brokaw, 1985b; Moore and Vankat, 1986; Collins and Pickett, 1987). Gap shape and orientation also influence gap microclimate. For example, long narrow gaps allow less insolation at the ground surface than circular gaps of the same size (Tomanek, 1960 in Runkle, 1985). The shape and orientation of the gap with respect to the heights of surrounding trees also influence the amount of direct solar radiation. For example, in some temperate forest gaps there is a stronger relationship between sapling abundance and direct solar radiation during the growing season than between sapling abundance and either gap area or total solar radiation (Veblen et al., 1979a). Similarly, in an eastern deciduous forest in North America, gap orientation has been shown to influence light levels and ingrowth by surrounding trees (Poulson and Platt, 1989).

Given the strong influence of gap size on dispersal probabilities and gap microclimate, it is not surprising that many tree species are specialized according to the size of the gap in which they are likely to regenerate (Denslow, 1980, 1987; Runkle, 1982; Brokaw, 1985a, 1987; Brokaw and Scheiner, 1989; Whitmore, 1989). The mechanisms which determine the success of different tree species in gaps of varying sizes may involve differences in germination, vegetative reproduction or seedling establishment. For species such as *Prunus pensylvanica* in the northeast United States, and many tree species in neotropical rainforests, treefalls trigger massive germination of viable seeds stored in the soil (Marks, 1974; Vázquez-Yanes and Smith, 1982; Young et al., 1987; Raich and Khoon, 1990). Experimental studies have shown that the germination triggers involve the higher temperatures, light levels, or ratios of red/far-red light typical of gaps (Bazzaz and Pickett, 1980; Canham and Marks, 1985). In addition to being hormone-mediated responses to direct damage to trees, branching from epicormic buds or sprouting from stumps or roots may also be responses to environmental cues associated with treefall gap environments (Zimmerman and Brown, 1971). Seedlings of many tree species are differentiated by their response to amounts of solar radiation. For example, generally there is a direct relationship between the growth rate of a tree species in large gaps and the minimum gap size required for net growth of seedlings (Grime, 1966; Marks, 1975).

(b)   Time and periodicity of gap creation

The time of the year during which treefalls are most frequent varies greatly from one forest to another (Brokaw, 1985b). Since in most habitats the production of propagules is seasonal for the majority of tree species, time of gap creation may be an important influence on the availability of propagules to colonize gaps (Brokaw, 1982; Canham and Marks, 1985). The seasonality of gap creation may even be a selective force on the timing of dispersal or germination of gap-phase species (Brokaw, 1982). Propagule availability is less influenced by seasonal variation in the creation of gaps when dormant propagules are stored in the soil (e.g. *Prunus pensylvanica*) or on parent trees (e.g. *Eucalyptus regnans*) (Marks, 1974; Ashton, 1976).

The periodicity of treefall gaps may have major influences on the species composition of a forest (Denslow, 1980; Runkle, 1985). Inspection of growth ring patterns of canopy trees in temperate forests often reveals multiple releases and suppressions, implying several episodes of gap creation and closure (Henry and Swan, 1974; Oliver and Stephens, 1977; Nakashizuka and Numata, 1982a; Runkle, 1985; Veblen, 1985, 1986b; Canham, 1990). Often before a canopy gap can be closed, the fall of surrounding trees will expand the gap and subject the gap occupants to further environmental changes (Hartshorn, 1978; Runkle, 1984; Foster and Reiners, 1986). This pattern of periodic gap expansion may allow gap-phase species to reach the canopy in gaps which initially may have been too small. Many shade-tolerant tree species are capable of multiple episodes of release and suppression in response to treefall dynamics (Runkle, 1985; Veblen, 1985; Runkle and Yetter, 1987). Thus, after equal time periods, a site affected by a single large treefall may be dominated by shade-intolerant canopy trees, whereas a site affected by multiple small treefall gaps totalling the same area would be expected to be dominated by more shade-tolerant canopy trees.

Multiple treefall episodes affecting the same site confound the concept of *gap turnover rate*, which is the mean time between successive creations of gap areas at any one point in the forest (Brokaw, 1985b). The most reliable estimates of gap turnover rate are based on annual observations in permanent plots over a period of time (e.g. Hartshorn, 1978; Brokaw, 1982; Foster and Brokaw, 1982; Uhl, 1982). In these studies overlapping gaps may be taken into account. Less reliable estimates are derived from size and age frequencies of presently recognizable gaps, sometimes taking into account rates of gap closure by lateral encroachment and upward growth (e.g. Runkle, 1982; Nakashizuka, 1984; Veblen, 1985; Taylor, 1990). Barden (1989) showed that slight differences in definitions and sampling methods (e.g. defined entry height into the canopy, minimum diameter of trees capable of creating gaps, and transect samples versus full enumerations of gaps) may result in large differences in estimated turnover rates. Thus, estimates of gap turnover rate reflect both real differences among forests as well as variations in methodology.

For both tropical and temperate mesic forests with disturbance regimes dominated by small- to intermediate-sized gaps, mean gap turnover rates are *c.* 100 years; the range is from *c.* 50 to 575 years (Runkle, 1982; Nakashizuka, 1984; Brokaw, 1985b; Veblen, 1985; Foster and Reiners,

1986). Two observations reconcile these relatively short mean turnover times with average tree longevities of 300–600 years for many of the dominants of temperate forests (Runkle, 1985). First, multiple gap episodes occur at some sites before a first gap is created at other sites, resulting in a large range of possible turnover times for any given site. Second, many potentially dominant trees survive gap creation as sub-canopy trees a hundred or more years old. Despite the difficulties of making reliable measurements, turnover rates of large areas of forests dominated by frequent fine-scale disturbance do not appear to differ greatly from those dominated by coarse-scale exogenous disturbances (Runkle, 1985). Thus, the frequencies of gaps of different sizes may be of more significance in explaining community composition than mean turnover time (e.g. Brokaw, 1985a, 1987).

(c)   Influences of gap-creating species

Canopy trees of different species sometimes influence the species com-position of the nearby understorey vegetation through their effects on the establishment, growth and mortality of herbs and seedlings of woody plants (e.g. Veblen et al., 1979b; Hicks, 1980; Woods and Whittaker, 1981; MaGuire and Forman, 1983; Beatty, 1984; Augspurger, 1983a). If future development of vegetation in a gap depends on advance regeneration, or if the effects of the canopy tree persist after its death, then there may be a non-random spatial association of the species of the former canopy tree and the replacement species. These effects may be manifested as reciprocal replacement (i.e. alternation of species), which has been reported for numerous types of northern temperate forest (Fox, 1977; Woods, 1979, 1984; Runkle, 1984).

The causes of differential success of saplings and seedlings under certain canopy species or in gaps created by certain species are not clear. Microsite differences beneath canopy species may result from variations in the quantity and quality of litter, which, in turn, may affect availability of soil nutrients or result in allelopathic interactions (e.g. Zinke, 1962; Tubbs, 1973; Gauch and Stone, 1979; Beatty and Sholes, 1988; Boettcher and Kalisz, 1990). Canopy trees may also differentially alter understorey light levels or soil moisture (e.g. Anderson, 1964; Browne and Bourn, 1973). Microsite differences may result from interspecific variation in the pattern of canopy tree death. For example, species which tend to die standing create different microsites than those that are more prone to wind-snap or uprooting (Falinski, 1978; Putz et al., 1983; Beatty and Stone, 1986). In addition to seedling responses to direct modification of the microsite by the canopy tree, canopy influence may be mediated through indirect biological interactions. For example, canopy species may differentially influence the abundance and size of understorey plants which, in turn, may affect the establishment and growth of tree seedlings (e.g. Veblen *et al.*, 1979b; MaGuire and Forman, 1983). Also, Connell (1971) and Janzen (1970) suggested that in tropical rainforests the presence of host-specific pests and pathogens near canopy trees could decrease growth and survival of conspecific seedlings. According to the Janzen–Connell hypothesis for coexistence of tropical tree species, herbivores and pathogens will be concentrated near food plants and will cause high mortality in a cluster of offspring near their parents. Despite some negative tests (e.g. Hubbell, 1979; Connell *et al.*, 1984), several studies in tropical rainforests support the idea (e.g. Augspurger, 1983a,b; Clark and Clark, 1984).

(d)   Within-gap environmental heterogeneity

Divergence of regeneration niches may involve differential abilities to establish and grow in the spatially heterogeneous environment of a treefall gap. In an idealized gap created by the fall of a single canopy tree in a dense forest, the fallen tree trunk forms the axis of a dumb-bell-shaped area of damage to adjacent canopy trees and undergrowth (Orians, 1982; Brokaw, 1985b). Environmental conditions along the bole of the fallen tree are likely to be different from those near the crown, and both may be different from conditions near the upturned root mass. For example, the root zone usually includes a pit of exposed mineral soil and a mound of upturned soil and decaying roots which differ in their physical and chemical properties from each other and from the crown and trunk zones (e.g. Beatty, 1984; Vitousek and Denslow, 1986; Beatty and Scholes, 1988; Schaetzl *et al.*, 1989). The smothering effect and massive input of nutrients from the decaying foliage of the crown may create microsites different from those associated with the trunk or root zone. In most gaps, the fall of subcanopy trees and breakage of the canopy tree create a complex pattern of microrelief and scattered debris beneath irregularly shaped canopy openings. The aspects of the gap environment most likely to differentially influence the vegetation response include the nature of the surface, presence of understorey plants and presence of other kinds of organisms.

   Within a gap there is a mosaic of microsites which may affect seed lodgement, germination or survival. For example, tip up mounds often have a significantly different floristic composition than pits or the undisturbed forest floor (Beatty, 1984; Nakashizuka, 1989; Peterson *et al.*, 1990). Plant growth rates may also be different in different microhabitats within a gap (Uhl *et al.*, 1988). A 'nurse-log' syndrome is characteristic of many temperate rainforest species, such as those of northwestern North America (Franklin and Dyrness, 1973), New Zealand (Odgen, 1971) and southern Chile (Veblen *et al.*, 1981). In these instances, survival of seedlings is better on fallen logs and stumps than on the litter-covered forest floor. Growth to maturity following germination on logs requires establishment of firm root anchorage in the soil before the supporting log decays. Observation of large trees with logs still beneath them indicates that at least some seedlings survive and attain main canopy stature. Greater seedling survival on logs appears to result from the negative influences of tree leaf litter on seedling emergence and survival on the forest floor, and from reduced competition from herbs and mosses (Sydes and Grime, 1981; Christy and Mack, 1984; Harmon and Franklin, 1989), or it may involve differential availability of nutrients or moisture (Graham and Cromack, 1982). Use of logs as rodent trails may also influence dispersal and seedling survival (Falinski, 1978). The differential species composition of mounds versus pits is well known and may be explained by either contrasting soil properties, differential dispersal, and/or greater herbivory at one type of microsite (Hutnik, 1952; Falinski, 1978; Thompson, 1980; Putz, 1983; Beatty, 1984).

Occupation of treefall gaps by understorey species is often a critical influence on the rate and composition of canopy species regeneration (e.g. Ehrenfeld, 1980; Brokaw, 1983). The falling tree damages or kills some understorey plants by bending, breaking, uprooting or burial, but in many cases a large part of the undergrowth remains intact. Initially, the density of understorey species depends on their density prior to the treefall, so that falls occurring in patches of dense canopy and low light levels may result in gaps of greater resource availability for gap colonists. Sometimes the relative importance of understorey and canopy species' response to gaps is related to gap size, with larger gaps favouring understorey species (e.g. Huenneke 1983; Veblen 1985). A dramatic example of inhibition of the response of canopy species to treefall gaps involves the proliferation of bamboos in gaps in many tropical and temperate forests (Oshima, 1961; Shidei, 1974; Whitmore, 1975; Veblen, 1982; Nakashizuka and Numata, 1982b; Taylor and Zisheng, 1988). The interference from bamboos in gaps not only varies with their initial abundances beneath the forest canopy, but also with gap size, influences of herbivores on the bamboo, and timing of gap creation with respect to synchronous flowering and death of bamboo clones (Veblen, 1982, 1985; Nakaskizuka and Numata, 1982b).

Relatively little is known about the spatial heterogeneity of herbivores, pathogens, seed predators and animal vectors of dispersal within gaps. However, variation in the presence of these organisms is potentially important to the recruitment and survival of tree seedlings. For example, foraging of ants which may nest in or near logs is important in seed predation and dispersal (Ashton, 1979; Thompson, 1980). Similarly, seeds of fleshy fruits are sometimes left on fallen logs in the faeces of birds and mammals, and fleshy fruits may have a higher probability of being eaten in gaps than beneath a closed canopy (Thompson and Willson, 1978; Thompson, 1980). On the other hand, in some forests new gaps provide few resources to birds and due to exposure to predators are dangerous perching sites; thus, animal-mediated seedfall in gaps may be lower (Brokaw, 1986; Schupp *et al.*, 1989). Large upturned tree butts and relatively impenetrable masses of large logs sometimes provide safe sites for seedlings of species which otherwise are destroyed by large, ground-dwelling herbivores (Falinski, 1978). Micro-organisms associated with roots or with leaf litter may have a deleterious effect on tree seedlings (e.g. Florence, 1965) so that the depth of soil disturbance may sometimes be critical to seedling establishment and survival.

4.4   REGENERATION DYNAMICS AND LANDSCAPE PATTERNS

The emphasis in this chapter has been on regeneration processes occurring at the scale of gaps produced by the fall of individual trees or small groups of trees. Regeneration processes at these fine scales result in forest stands which consist of a mosaic of subunits (typically less than $1000\,m^2$) of varying structure and composition. Regeneration processes may also occur at the scale of an entire stand affecting areas of several hectares to many square kilometres. Thus, consideration of disturbance events which affect large areas is essential not only to an understanding of broad-scale vegetation patterns, but also to species coexistence at a regional scale (Forman and Godron, 1981; Urban *et al.*, 1987; Ricklefs, 1987).

In many forested landscapes a self-reproducing climax or compositional equilibrium may only rarely be attained and, if so, often at only relatively large spatial scales. Coarse-scale exogenous disturbances such as fire, mass movements, insect outbreaks and extensive blowdown are widespread and sometimes frequent enough to have a controlling influence on forest structure and composition at a landscape scale (e.g. Heinselman, 1973; Whitmore, 1974; Veblen and Ashton, 1978; Garwood *et al.*, 1979; Ashton, 1981; Romme, 1982; Canham and Loucks, 1984; Johnson and Fryer, 1987; Knight, 1987; Veblen *et al.*, 1991). Thus, at a stand scale, attainment of compositional equilibrium may be prevented by frequent exogenous disturbance. In landscapes dominated by coarse-scale disturbances, populations of early successional tree species may periodically become locally extinct. If common tree species persist at similar levels of abundance over large areas, however, the landscape as a whole may be in a state of *global equilibrium*, or 'shifting-mosaic steady-state' (Bormann and Likens, 1979). Assessment of compositional equilibrium, then, requires sampling of forest composition over an entire landscape. This usually involves a combination of ground sampling and mapping of vegetation units from aerial photographs (e.g. Heinselman, 1973; Johnson and Fryer, 1987).

At a landscape scale, the proportions of different stand types (differentiated by composition and/or structure) may be related to the nature and history of disturbances which have affected the landscape. The concept of *disturbance regime* is a means of organizing spatial and temporal information about a range of disturbances affecting a particular landscape (Sousa, 1984; White and Pickett, 1985; see also Chapters 1 and 3). The key descriptors of a disturbance regime are: (1) spatial distribution of occurrence, particularly in relation to environmental gradients; (2) frequency of occurrence; (3) size of the area disturbed; (4) mean return interval (i.e. the inverse of frequency); (5) predictability; (6) rotation period (the time required to disturb an area equivalent in size to the study area); (7) magnitude of the disturbance measured either directly as the intensity of the disturbing agent or indirectly as the impact on the vegetation (i.e. severity); and (8) the synergistic interactions of different kinds of disturbances. Available descriptions of disturbance regimes, however, are only partial and usually deal exclusively with a single type of disturbance such as endogenous treefall (e.g. Runkle, 1982) or fire (e.g. Romme, 1982). To understand the role of disturbance in creating broad landscape patterns, the full range of disturbances affecting a particular landscape must be considered (e.g. Harmon *et al.*, 1983).

SUMMARY

1. A plant species may be termed persistent if its population does not become extinct during a given time period in a given area, or, if it does, it recolonizes the area within the time span required for one turnover of its population. Plant communities in which all the common plant species are persistent may be judged to be in compositional equilibrium.
2. A critical consideration in the assessment of compositional equilibrium is the spatial scale at which it is investigated. In a vegetation mosaic consisting of small patches of different species composition, individual patches are often not in compositional equilibrium, even though equilibrium may exist at the spatial scale of the entire mosaic.
3. The patch dynamics perspective emphasizes the importance of periodic disturbance in creating opportunities for establishment of new individuals or growth of previously suppressed individuals by releasing the resources utilized by the former dominants of the site. Investigation at a wide range of spatial scales of the temporal and spatial characteristics of disturbances and plant population responses to disturbances is essential to an understanding of the regeneration dynamics and vegetation patterns in any landscape.
4. Recognition of compositional equilibrium in contrast to successional trends is methodologically difficult for long-lived organisms such as trees. The surest method, but also the least practical, requires repeated censuses over multi-year periods. Faster but less certain methods include stand structural analysis and computation of transition probabilities. Both methods must be applied with great caution

due to the unlikelihood of obtaining all the data required (e.g. many trees cannot be precisely aged) and the difficulty, without long-term studies, of verifying assumptions underlying these techniques.

5. In forest stands at or near compositional equilibrium, species persistence often involves a combination of resource partitioning and non-equilibrium mechanisms of coexistence. For many tree species, differences in resource needs and mode of acquiring resources may be manifested only during the early stages in their life histories, as emphasized by the regeneration niche concept. Fine-scale disturbance in the form of treefalls, however, also favours coexistence by truncating processes of competitive exclusion and by contributing to the spatial and temporal heterogeneity of resource availability.

6. The regeneration niches of tree species may be differentiated in relation to the heterogeneity of resource availability associated with: (1) gap size, shape and orientation; (2) seasonality and periodicity of gap creation; (3) the micro-environmental influences of the canopy tree which previously occupied the gap; (4) the nature of substrates within gaps; (5) abundance and composition of gap-occupying understorey plants; or (6) presence of animals and micro-organisms within gaps. While differences in regeneration niches are often critical to the coexistence of tree species, differences among mature trees in their influences on microsites and in their mortality patterns also contribute significantly to their coexistence.

ACKNOWLEDGEMENTS

For their helpful comments on this chapter, I thank D. H. Ashton, S. W. Beatty, B. R. Burns, A. Lara, D. C. Glenn-Lewin, Y. B. Linhart, M. G. Noble, R. K. Peet and A. J. Rebertus. This review benefited from research funded by the Council on Research and Creative Work of the Graduate School of the University of Colorado, the National Geographic Society, the National Science Foundation, the John Simon Guggenheim Memorial Foundation and the John D. and Catherine T. MacArthur Foundation.

REFERENCES

Aarssen, L. W. (1983) Ecological combining ability and competitive combining ability in plants: toward a general evolutionary theory of coexistence in systems of competition. *Amer. Nat.*, **122**, 707–31.

Abrell, D. B. and Jackson, M. T. (1977) A decade of change in an old-growth beech-maple forest in Indiana. *Amer. Mid. Nat.*, **98**, 22–32.

Acevedo, M. F. (1981) On Horn's Markovian model of forest dynamics with particular reference to tropical forests. *Theor. Pop. Biol.*, **19**, 230–50.

Anderson, M. C. (1964) Light relations of terrestrial plant communities and their measurement. *Biol. Rev.*, **39**, 425–86.

Ashton, D. H. (1976) Even-aged stands of *Eucalyptus regnans* F. Muell. in central Victoria. *Austral. J. Bot.*, **24**, 397–414.

Ashton, D. H. (1979) Seed harvesting by ants in forests of *Eucalyptus regnans* F. Muell. in Victoria. *Austral. J. Ecol.*, **4**, 265–77.

Ashton, D. H. (1981) Fire in tall open-forests (wet sclerophyll forests), in *Fire in the Australian Biota* (eds A. M. Gill and I. R. Noble), SCOPE, Canberra, pp. 339–66.

Augspurger, C. K. (1983a) Offspring recruitment around tropical trees: Changes in cohort distance with time. *Oikos*, **40**, 189–96.

Augspurger, C. K. (1983b) Seed dispersal of the tropical tree, *Platypodium elegans*, and the escape of its seedlings from fungal pathogens. *J. Ecol.*, **71**, 759–71.

Baker, W. L. (1989) Landscape ecology and nature reserve design in the Boundary Waters Canoe Area, Minnesota. *Ecology*, **70**, 23–35.

Barden, L. S. (1980) Tree replacement in a cove hardwood forest in the southern Appalachians. *Oikos*, **35**, 16–19.

Barden, L. S. (1981) Forest development in canopy gaps of a diverse hardwood forest of southern Appalachian Mountains. *Oikos*, **37**, 205–9.

Barden, L. S. (1989) Repeatability in forest gap research: studies in the Great Smoky Mountains. *Ecology*, **70**, 558–9.

Bazzaz, F. A. and Pickett, S. T. A. (1980) Physiological ecology of tropical succession: A comparative review. *Ann. Rev. Ecol. System.*, **11**, 287–310.

Beatty, S. W. (1984) Influence of microtopography and canopy species on spatial patterns of forest understory plants. *Ecology*, **65**, 1406–19.

Beatty, S. W. and Sholes, O. D. (1988) Leaf litter effect on plant species composition of deciduous forest treefall pits. *Canad. J. For. Res.*, **18**, 553–9.

Beatty, S. W. and Stone, E. L. (1986) The variety of soil microsites created by treefalls. *Canad. J. For. Res.*, **16**, 539–48.

Boettcher, S. E. and Kalisz, P. J. (1990) Single-tree influence on soil properties in the mountains of eastern Kentucky. *Ecology*, **71**, 1365–72.

Bormann, F. H. and Likens, G. E. (1979) *Pattern and Process in a Forested Ecosystem*, Springer-Verlag, New York.

Braun, E. L. (1950) *Deciduous Forests of Eastern North America*, McGraw-Hill, New York.

Bray, J. R. (1956) Gap phase replacement in a maple-basswood forest. *Ecology*, **37**, 598–600.

Brokaw, N. V. L. (1982) Treefalls: frequency, timing, and consequences, in *The Ecology of a Tropical Forest: Seasonal Rhythms and Longer-term Changes* (eds E. G. Leigh, Jr, A. S. Rand and D. W. Windsor), Smithsonian Institution Press, Washington, D.C., pp. 101–108.

Brokaw, N. V. L. (1983) Groundlayer dominance and apparent inhibition of tree regeneration by *Aechmea magdalenae* (Bromeliaceae) in a tropical forest. *Trop. Ecol.*, **24**, 194–200.

Brokaw, N. V. L. (1985a) Gap-phase regeneration in a tropical forest. *Ecology*, **66**, 682–7.

Brokaw, N. V. L. (1985b) Treefalls, regrowth, and community structure in tropical forests, in *The Ecology of Natural Disturbance and Patch Dynamics* (eds S. T. A. Pickett and P. S. White), Academic Press, Orlando, pp. 53–69.

Brokaw, N. V. L. (1986) Seed dispersal, gap colonization, and the case of

Cecropia insignis, in *Frugivory and Seed Dispersal* (eds A. Estrada and T. H. Gleming), Junk, Dordrecht, pp. 323–31.

Brokaw, N. V. L. (1987) Gap-phase regeneration of three pioneer tree species in a tropical forest. *J. Ecol.*, **75**, 9–19.

Brokaw, N. V. L. and Scheiner, S. M. (1989) Species composition in gaps and structure of a tropical forest. *Ecology*, **70**, 538–41.

Browne, J. H. Jr and Bourn, T. G. (1973) Patterns of soil moisture depletion in a mixed oak stand. *Forest Sci.*, **19**, 23–30.

Butson, R. G., Knowles, P. and Farmer, R. E. Jr (1987) Age and size structure of marginal, disjunct populations of *Pinus resinosa*. *J. Ecol.*, **75**, 685–92.

Canham, C. D. (1989) Different responses to gaps among shade-tolerant tree species. *Ecology*, **69**, 548–50.

Canham, C. D. (1990) Suppression and release during canopy recruitment in *Fagus grandifolia*. *Bull. Torrey Bot. Club*, **117**, 1–7.

Canham, C. D., Denslow, J. S., Platt, W. J., Runkle, J. R., Spies, T. A. and White, P. S. (1990) Light regimes beneath closed canopies and tree-fall gaps in temperate and tropical forests. *Canad. J. For. Res.*, **20**, 620–31.

Canham, C. D. and Loucks, O. L. (1984) Catastrophic windthrow in the pre-settlement forests of Wisconsin. *Ecology*, **65**, 803–9.

Canham, C. D. and Marks, P. L. (1985) The response of woody plants to disturbance: patterns of establishment and growth, in *The Ecology of Natural Disturbance and Patch Dynamics* (eds S. T. A. Pickett and P. S. White), Academic Press, Orlando, pp. 197–216.

Caughley, G. (1977) *Analysis of Vertebrate Populations*, Wiley, Chichester.

Christensen, J. L. (1977) Changes in structure, pattern and diversity associated with climax forest maturation in Piedmont, North Carolina. *Amer. Mid. Nat.*, **97**, 176–88.

Christy, E. J. and Mack, R. N. (1984) Variation in demography of juvenile *Tsuga heterophylla* across the substratum mosaic. *J. Ecol.*, **72**, 75–91.

Clark, D. A. and Clark, D. B. (1984) Spacing dynamics of a tropical rainforest tree: evaluation of the Janzen-Connell model. *Amer. Nat.*, **124**, 769–88.

Clements, F. E. (1916) *Plant Succession: An Analysis of the Development of Vegetation*, Carnegie Institution of Washington, Publ. 242.

Collins, B. S. and Pickett, S. T. A. (1987) Influence of canopy opening on the environment and herb layer in a northern hardwoods forest. *Vegetatio*, **70**, 3–10.

Connell, J. H. (1971) On the role of natural enemies in preventing competitive exclusion in some marine animals and in rainforest trees, in *Dynamics of Populations* (eds P. J. den Boer and G. Gradwell), Proc. Adv. Study Inst., Oosterbeek, 1970. Center for Agric. Publishing and Documentation, Wageningen, The Netherlands. pp. 298–312.

Connell, J. H. and Sousa, W. P. (1983) On the evidence needed to judge ecological stability or persistence. *Amer. Nat.*, **121**, 789–824.

Connell, J. H., Tracey, J. G. and Webb, L. J. (1984) Compensatory recruitment, growth, and mortality as factors maintaining forest tree diversity. *Ecol. Monogr.*, **54**, 141–64.

Culver, D. C. (1981) On using Horn's Markov succession model. *Amer. Nat.*, **117**, 572–4.

Daubenmire, R. F. (1943) Vegetational zonation in the Rocky Mountains. *Bot. Rev.*, **9**, 325–93.

Deevey, E. S. Jr (1947) Life tables for natural populations of animals. *Quart. Rev. Biol.*, **22**, 383–14.

de Liocourt, F. (1898) De l'aménagement des sapinières. *Bull. Soc. Forestiere Franche-Comte*, **4**, 396–409.

Denslow, J. S. (1980) Gap partitioning among tropical rain forest trees. *Biotropica* (Suppl.), **12**, 47–55.

Denslow, J. S. (1985) Disturbance-mediated coexistence of species, in *The Ecology of Natural Disturbance and Patch Dynamics* (eds S. T. A. Pickett and P. S. White), Academic Press, New York, pp. 307–23.

Denslow, J. S. (1987) Tropical rainforest gaps and tree species diversity. *Ann. Rev. Ecol. System.*, **18**, 431–51.

Denslow, J. S., Shultz, J. C., Vitousek, P. M. and Strain, B. R. (1990) Growth responses of tropical shrubs to treefall gap environments. *Ecology*, **71**, 165–79.

Despain, D. G. (1983) Nonpyrogenous climax lodgepole pine communities in Yellowstone National Park. *Ecology*, **64**, 231–4.

Donoso, C., Grez, R., Escobar, B. and Real, P. (1984) Estructura y dinámica de bosques del tipo forestal siempreverde en un sector de Chiloé insular. *Bosque*, **5**, 82–104.

Duncan, R. P. (1989) An evaluation of errors in tree age estimates based on increment cores in kahikatea (*Dacrycarpus dacrydioides*). *New Zealand Nat. Sci.*, **16**, 31–7.

Dunn, C. P., Guntenspergen, G. R. and Dorney, J. R. (1983) Catastrophic wind disturbance in an old-growth hemlock-hardwood forest, Wisconsin. *Canad. J. Bot.*, **61**, 211–17.

Ehrenfeld, J. G. (1980) Understory response to canopy gaps of varying size in a mature oak forest. *Bull. Torrey Bot. Club*, **107** (2), 29–41.

Falinski, J. G. (1978) Uprooted trees, their distribution and influence in the primeval forest biotope. *Vegetatio*, **38**, 175–83.

Florence, R. G. (1965) Decline of old-growth redwood forests in relation to some soil microbiological processes. *Ecology*, **46**, 52–64.

Ford, E. D. (1975) Competition and stand structure in some even-aged plant monocultures. *J. Ecol.*, **63**, 311–33.

Forman, R. T. T. and Godron, M. (1981) Patches and structural components for a landscape ecology. *BioScience*, **31**, 733–40.

Foster, D. R. (1988) Disturbance history, community organization and vegetation dynamics of the old-growth Pisgah forest, south-western New Hampshire, U.S.A. *J. Ecol.*, **76**, 105–34.

Foster, J. R. and Reiners, W. A. (1986) Size distribution and expansion of canopy gaps in a northern Appalachian spruce fir forest. *Vegetatio*, **68**, 109–14.

Foster, R. B. and Brokaw, N. V. L. (1982) Structure and history of the vegetation of Barro Colorado Island, in *The Ecology of a Neotropical Forest: Seasonal Rhythms and Longer-term Changes* (eds E. G. Leigh Jr, A. S. Rand and D. M. Windsor), Smithsonian Institution Press, Washington, D.C., pp. 67–81.

Fowells, H. A. (1965) *Silvics of Forest Trees of the United States*, US Government Printing Office, Washington, D.C.

Fox, J. F. (1977) Alternation and coexistence of tree species. *Amer. Nat.*, **111**, 69–89.

Franklin, J. F. and Dyrness, C. T. (1973) *Natural Vegetation of Oregon and Washington*, USDA Forest Service General Technical Report, PNW-8.

Franklin, J. F. and Hemstrom, M. F. (1981) Aspects of succession in the coniferous forests of the Pacific Northwest, in *Forest Succession: Concepts and Applications* (eds D. C. West, H. H. Shugart and D. B. Botkin), Springer-Verlag, New York, pp. 212–29.

Fritts, H. C. (1976) *Tree Rings and Climate*, Academic Press, New York.

Garwood, N. C., Janos, D. P. and Brokaw, N. (1979) Earthquake-caused land-slides: A major disturbance to tropical forests. *Science*, **205**, 997–9.

Gauch, H. G. and Stone, E. L. (1979) Vegetation and soil pattern in a mesophytic forest at Ithaca, New York. *Amer. Mid. Nat.*, **102**, 332–45.

Geiger, R. (1965) *The Climate Near the Ground*, Harvard Univ. Press, Cambridge.

Glitzenstein, J. S., Harcombe, P. A. and Streng, D. R. (1986) Disturbance, succession, and maintenance of species diversity in an east Texas forest. *Ecol. Monogr.*, **56**, 243–58.

Goff, F. G. and West, D. (1975) Canopy–understory interaction effects on forest population structure. *Forest Sci.*, **21**, 98–108.

Graham, R. L. and Cromack, K., Jr (1982) Mass, nutrient content, and decay rate of dead boles in rain forests of Olympic National Park. *Canad. J. For. Res.*, **12**, 511–21.

Grime, J. P. (1966) Shade avoidance and shade tolerance in flowering plants, in *Light as an Ecological Factor* (eds R. Bainbridge, G. C. Evans and O. Rackham), Blackwell, London, pp. 187–207.

Grimm, E. C. (1984) Fire and other factors controlling the Big Woods vegetation of Minnesota in the mid-nineteenth century. *Ecol. Mongr.*, **54**, 291–311.

Grubb, P. J. (1977) The maintenance of species richness in plant communities. The importance of the regeneration niche. *Biol. Rev.*, **52**, 107–45.

Grubb, P. J. (1986) Problems posed by sparse and patchily distributed species in species-rich plant communities, in *Community Ecology* (eds J. Diamond and T. J. Case), Harper and Row, New York, pp. 207–25.

Harcombe, P. A. (1986) Stand development in a 130-year-old spruce-hemlock forest based on age structure and 50 years of mortality data. *Forest Ecol. Mgmt*, **14**, 41–58.

Harcombe, P. A. (1987) Tree life tables. *BioScience*, **37**, 557–68.

Harcombe, P. A. and Marks, P. L. (1983) Five years of tree death in *Fagus-Magnolia* forest, southeast Texas (USA). *Oecologia*, **57**, 49–54.

Harmon, M. E., Bratton, S. P. and White, P. S. (1983) Disturbance and vegetation response in relation to environmental gradients in the Great Smoky Mountains. *Vegetatio*, **55**, 129–39.

Harmon, M. E. and Franklin, J. F. (1989) Tree seedlings on logs in *Picea–Tsuga* forests of Oregon and Washington. *Ecology*, **70**, 48–59.

Harper, J. L. (1977) *Population Biology of Plants*, Academic Press, London.

Hartshorn, G. S. (1978) Tree falls and tropical forest dynamics, in *Tropical Trees as Living Systems* (eds P. B. Tomlinson and M. H. Zimmerman), Cambridge Univ. Press, Cambridge, Mass., pp. 617–38.

Heinselman, M. L. (1973) Fire in the virgin forests of the Boundary Water Canoe Area, Minnesota. *Quat. Res.*, **3**, 329–82.

Henry, J. D. and Swan, J. M. A. (1974) Reconstructing forest history from live and dead plant material – an approach to the study of forest succession in southwest New Hampshire. *Ecology*, **55**, 772–83.

Hett, J. M. and Loucks, O. L. (1971) Sugar maple (*Acer saccharum* Marsh.) seedling mortality. *J. Ecol.*, **59**, 507–20.

Hett, J. M. and Loucks, O. L. (1976) Age structure models of balsam fir and eastern hemlock. *J. Ecol.*, **64**, 1029–44.

Hibbs, D. E. (1982) Gap dynamics in a hemlock-hardwood forest. *Canad. J. For. Res.*, **12**, 522–7.

Hicks, D. J. (1980) Intrastand distribution patterns of southern Appalachian cove forest herbaceous species. *Amer. Mid. Nat.*, **104**, 209–23.

Holling, C. S. (1973) Resilience and stability of ecological systems. *Ann. Rev. Ecol. System.*, **4**, 1–23.

Horn, H. S. (1975) Markovian processes of forest succession, in *Ecology and Evolution of Communities* (eds M. L. Cody and J. M. Diamond), Harvard Univ. Press, Cambridge, Mass., pp. 196–211.

Horn, H. S. (1976) Succession, in *Theoretical Ecology: Principles and Applications* (ed. R. M. May), Saunders, Philadelphia, pp. 187–204.

Hubbell, S. P. (1979) Tree dispersion, abundance and diversity in a tropical dry forest. *Science*, **203**, 1299–1309.

Huenneke, L. F. (1983) Understory response to gaps caused by the death of *Ulmus americana* in central New York. *Bull. Torrey Bot. Club*, **110**, 170–5.

Hutnik, R. J. (1952) Reproduction on windfalls in a northern hardwood stand. *J. Forestry*, **50**, 693–4.

Hutchinson, G. E. (1951) Copepedology for the ornithologist. *Ecology*, **32**, 571–7.

Jackson, M. T. and Faller, A. (1973) Structural analysis and dynamics of the plant communities of Wizard Island, Crater Lake National Park. *Ecol. Monogr.*, **43**, 441–61.

Janzen, D. H. (1970) Herbivores and the number of tree species in tropical forests. *Amer. Nat.*, **104**, 501–27.

Johnson, E. A. and Fryer, G. (1987) Historical vegetation change in the Kananaskis Valley, Canadian Rockies. *Canad. J. Bot.*, **65**, 853–8.

Johnson, E. A. and Fryer, G. (1989) Population dynamics in lodgepole pine–Engelmann spruce forests. *Ecology*, **70**, 1335–45.

Johnson, F. L. and Bell, D. T. (1975) Size-class structure of three streamside forests. *Amer. J. Bot.*, **62**, 81–5.

Johnston, R. D. and Lacey, C. J. (1983) Multi-stemmed trees in rainforest. *Austral. J. Bot.*, **31**, 189–95.

Jones, E. W. (1945) The structure and reproduction of the virgin forest of the north temperate zone. *New Phytol.*, **44**, 130–48.

Kellman, M. C. (1980) *Plant Geography*, 2nd edn, St Martin's, New York.

Kelty, M. T. (1986) Development patterns in two hemlock-hardwood stands in southern New England. *Canad. J. For. Res.*, **16**, 885–91.

Kenkel, N. C. (1988) Pattern of self-thinning in jack pine: testing the random mortality hypothesis. *Ecology*, **69**, 1017–24.

Knight, D. H. (1987) Parasites, lightning, and the vegetation mosaic in wilderness landscapes, in *Landscape Heterogeneity and Disturbance* (ed. M. G. Turner), Springer-Verlag, New York, pp. 59–83.

Kullman, L. (1986) Recent tree-limit history of *Picea abies* in the southern Swedish Scandes. *Canad. J. For. Res.*, **16**, 761–71.

Lang, G. E. and Knight, D. H. (1983) Tree growth, mortality, recruitment, and canopy gap formation during a 10-year period in a tropical moist forest. *Ecology*, **64**, 1075–80.

Lawton, R. O. and Putz, F. E. (1988) Natural disturbance and gap-phase regeneration in a wind-exposed tropical cloud forest. *Ecology*, **69**, 764–77.

Leak, W. B. (1964) An expression of diameter distribution for unbalanced, uneven-aged stands and forests. *Forest Sci.*, **10**, 39–50.

Leak, W. B. (1975) Age distribution in virgin red spruce and northern hardwoods. *Ecology*, **56**, 1451–4.

Lee, R. (1978) *Forest Microclimatology*, Columbia Univ. Press, New York.

Lieberman, D., Lieberman, M., Peralta, R. and Hartshorn, G. S. (1985) Mortality patterns and stand turnover rates in a wet tropical forest in Costa Rica. *J. Ecol.*, **73**, 915–24.

Lieberman, M., Lieberman, D. and Peralta, R. (1989) Forests are not just Swiss cheese: canopy stereogeometry of non-gaps in tropical forests. *Ecology*, **70**, 550–2.

Lippe, E., de Smidt, J. T. and Glenn-Lewin, D. C. (1985) Markov models and succession: a test from a heathland in The Netherlands. *J. Ecol.*, **73**, 775–91.

Lorimer, C. G. (1977) The presettlement forest and natural disturbance cycle of northeastern Maine. *Ecology*, **58**, 139–48.

Lorimer, C. G. (1980) Age structure and disturbance history of a southern Appalachian virgin forest. *Ecology*, **61**, 1169–84.

Lorimer, C. G. (1981) Survival and growth of understory trees in oak forests of the Hudson Highlands, New York. *Canad. J. For. Res.*, **11**, 689–95.

Lorimer, C. G. (1985) Methodological considerations in the analysis of forest disturbance history. *Canad. J. For. Res.*, **15**, 200–13.

MaGuire, D. A. and Forman, R. T. T. (1983) Herb cover effects on tree seedling patterns in a mature hemlock-hardwood forest. *Ecology*, **64**, 1367–80.

Marks, P. L. (1974) The role of pin cherry (*Prunus pensylvanica* L.) in the maintenance of stability in northern hardwood ecosystems. *Ecol. Monogr.*, **44**, 73–88.

Marks, P. L. (1975) On the relation between extension growth and successional status of deciduous trees of the northeastern United States. *Bull. Torrey Bot. Club*, **102**, 172–7.

Martínez-Ramos, M., Sarukhán, J. and Pinero, D. (1988) The demography of tropical trees in the context of forest gap dynamics, in *Plant Population Ecology* (eds D. J. Davy, M. J. Hutchings and A. R. Watkinson), Blackwell, Oxford, pp. 293–313.

McIntosh, R. P. (1980) The relationship between succession and the recovery process in ecosystems, in *The Recovery Process in Damaged Ecosystems* (ed. J. Cairns), Ann Arbor Science Publication, Ann Arbor, Mich, pp. 11–62.

Meyer, H. A. (1952) Structure, growth, and drain in balanced uneven-aged forests. *J. Forestry*, **50**, 85–92.

Moore, M. R. and Vankat, J. L. (1986) Responses of the herb layer to the gap dynamics of a mature beech maple forest. *Amer. Mid. Nat.*, **115**, 336–47.

Morris, R. F. (1948) How old is a balsam tree? *Forestry Chron.*, **24**, 106–10.

Nakashizuka, T. (1984) Regeneration process of climax beech (*Fagus crenata* Blume) forests. IV. Gap formation. *Japan. J. Ecol.*, **33**, 409–18.

Nakashizuka, T. (1989) Role of uprooting in composition and dynamics of an old-growth forest in Japan. *Ecology*, **70**, 1273–8.

Nakashizuka, T. and Numata, M. (1982a) Regeneration process of climax beech forests. I. Structure of a beech forest with the undergrowth of *Sasa*. *Japan. J. Ecol.*, **32**, 57–67.

Nakashizuka, T. and Numata, M. (1982b) Regeneration process of climax beech forests. II. Structure of a forest under the influences of grazing. *Japan. J. Ecol.*, **32**, 473–82.

Ogden, J. (1971) Studies on the vegetation of Mount Colenso New Zealand. 2. The population dynamics of Red Beech. *Proc. New Zealand Ecol. Soc.*, **18**, 66–75.

Ogden, J. (1983) Community matrix model predictions of future forest composition at Russell State Forest. *New Zealand J. Ecol.*, **6**, 71–7.

Ogden, J. (1985a) An introduction to plant demography with special reference to New Zealand trees. *New Zealand J. Bot.*, **23**, 751–72.

Ogden, J. (1985b) Past, present and future: Studies on the population dynamics of some long-lived trees, in *Studies on Plant Demography: A Festschrift for John L. Harper* (ed. J. White), Academic Press, London, pp. 3–16.

Oliver, C. D. (1981) Forest development in North America following major disturbances. *Forest Ecol. Mgmt*, **3**, 153–68.

Oliver, C. D. and Larson, B. C. (1990) *Forest Stand Dynamics*, McGraw-Hill, New York.

Oliver, C. D. and Stephens, E. P. (1977) Reconstruction of a mixed-species forest in central New England. *Ecology*, **32**, 84–103.

Orians, G. H. (1982) The influence of tree-falls in tropical forests on tree species richness. *Trop. Ecol.*, **23**, 255–79.

Oshima, Y. (1961) Ecological studies of *Sasa* communities. I. Productive structure of some of the *Sasa* communities in Japan. *Bot. Mag. Tokyo*, **74**, 199–210.

Parker, A. J. (1986) Persistence of lodgepole pine forests in the central Sierra Nevada. *Ecology*, **67**, 1560–7.

Pearl, R. and Miner, J. R. (1935) Experimental studies on the duration of life. XIV. The comparative mortality of certain lower organisms. *Quart. Rev. Biol.*, **10**, 67–79.

Peet, R. K. (1981) Forest vegetation of the Colorado Front Range: Composition and dynamics. *Vegetatio*, **45**, 3–75.

Peet, R. K. and Christensen, N. L. (1980) Succession: A population process. *Vegetatio*, **43**, 131–40.

Peet, R. K. and Christensen, N. L. (1987) Competition and tree death. *BioScience*, **37**, 586–95.

Peterson, C. J., Carson, W. P. McCarthy B. C. and Pickett, S. T. A. (1990) Microsite variation and soil dynamics within newly created treefall pits and mounds. *Oikos*, **58**, 39–46.

Pickett, S. T. A. (1980) Non-equilibrium coexistence of plants. *Bull. Torrey Bot. Club*, **107**, 238–48.

Pickett, S. T. A. and Thompson, J. N. (1978) Patch dynamics and the design of nature reserves. *Biol. Conserv.*, **13**, 27–37.

Pickett, S. T. A. and White, P. S. (eds) (1985) *The Ecology of Natural Disturbance and Patch Dynamics*, Academic Press, New York.

Poulson, T. L. and Platt, W. J. (1989) Gap light regimes influence canopy tree diversity. *Ecology*, **70**, 553–5.

Putz, F. E. (1983) Treefall pits and mounds, buried seeds, and the importance of soil disturbance to pioneer trees on Barro Colorado Island, Panama. *Ecology*, **64**, 1069–74.

Putz, F. E. (1984) How trees avoid and shed lianas. *Biotropica*, **16**, 19–23.

Putz, F. E., Coley, P. D., Lu, K. and Montalvo, A. (1983) Uprooting and snapping of trees: Structural determinants and ecological consequences. *Canad. J. For. Res.*, **13**, 1011–20.

Raich, J. W. and Khoon, G. W. (1990) Effects of canopy openings on tree seed germination in a Malaysian dipterocarp forest. *J. Trop. Ecol.*, **6**, 203–17.

Raup, H. H. (1957) Vegetational adjustment to the instability of site. *Proc. Paper 6th Tech. Meeting Int. Union Conserv. Naturalist Res. (Edinburgh)*, pp. 36–48.

Read, J. and Hill, R. S. (1988) The dynamics of some rainforest associations in Tasmania. *J. Ecol.*, **76**, 558–84.

Rebertus, A. J., Williamson, G. B. and Moser, E. B. (1989) Fire-induced changes in *Quercus laevis* spatial pattern in Florida Sandhills. *J. Ecol.*, **77**, 638–50.

Ricklefs, R. E. (1987) Community diversity: relative roles of local and regional processes. *Science*, **235**, 167–71.

Romme, W. H. (1982) Fire and landscape diversity in subalpine forests of Yellowstone National Park. *Ecol. Monogr.*, **52**, 199–221.

Ross, M. S., Sharik, T. L. and Smith, D. W. (1982) Age-structure relationships of tree species in an Appalachian oak forest in southwest Virginia. *Bull. Torrey Bot. Club*, **109**, 287–98.

Runkle, J. R. (1981) Gap regeneration in some old-growth forests of the eastern United States. *Ecology*, **62**, 1041–51.

Runkle, J. R. (1982) Patterns of disturbance in some old-growth mesic forests of eastern North America. *Ecology*, **63**, 1533–46.

Runkle, J. R. (1984) Development of woody vegetation in treefall gaps in a beech-sugar maple forest. *Holarctic Ecol.*, **7**, 157–64.

Runkle, J. R. (1985) Disturbance regimes in temperate forests, in *The Ecology of Natural Disturbance and Patch Dynamics* (eds S. T. A. Pickett and P. S. White), Academic Press, New York, pp. 17–33.

Runkle, J. R. and Yetter, T. C. (1987) Treefalls revisited: gap dynamics in the southern Appalachians. *Ecology*, **68**, 417–24.

Sarukhán, J. (1978) Studies on the demography of tropical trees, in *Tropical Trees as Living Systems* (eds P. B. Tomlinson and M. H. Zimmerman), Cambridge Univ. Press, Cambridge, Mass., pp. 163–84.

Schaetzl, R. J., Burns, S. F., Johnson, D. L. and Small, T. W. (1989) Tree uprooting: review of impacts on forest ecology. *Vegetatio*, **79**, 165–76.

Schupp, E. W., Howe, H. F., Augspurger, C. K. and Levey, D. J. (1989) Arrival and survival in tropical treefall gaps. *Ecology*, **70**, 562–4.

Shidei, T. (1974) Forest vegetation zones, in *The Flora and Vegetation of Japan* (ed. M. Numata), Elsevier, Amsterdam, pp. 87–124.

Shmida, A. and Ellner, S. (1984) Coexistence of plant species with similar niches. *Vegetatio*, **58**, 29–55.

Silvertown, J. and Law, R. (1987) Do plants need niches? Some recent developments in plant community ecology. *Trends Ecol. Evol.*, **2**, 24–6.

Sousa, W. P. (1979) Disturbance in marine intertidal boulder fields: the nonequilibrium maintenance of species diversity. *Ecology*, **60**, 1225–39.

Sousa, W. P. (1984) The role of disturbance in natural communities. *Ann. Rev.*

Ecol. System., **15**, 353–91.

Spring, P. E., Brewer, M. L., Brown, J. R. and Fanning, M. E. (1974) Population ecology of loblolly pine *Pinus taeda* in an old field community. *Oikos*, **25**, 1–6.

Spurr, S. H. and Barnes, B. V. (1980) *Forest Ecology*, 3rd edn, Wiley, New York.

Stahelin, R. (1943) Factors influencing the natural restocking of high altitude burns by coniferous trees in the central Rocky Mountains. *Ecology*, **24**, 19–30.

Stewart, G. H. (1986) Population dynamics of a montane conifer forest, western Cascade Range, Oregon, U.S.A. *Ecology*, **67**, 534–44.

Strong, D. R. (1977) Epiphyte loads, treefalls, and perennial forest disruption: A mechanism for maintaining higher tree species richness in the tropics without animals. *J. Biogeog.*, **4**, 215–18.

Sydes, C. and Grime, J. P. (1981) Effects of tree leaf litter on herbaceous vegetation in deciduous woodland. I. Field investigations. *J. Ecol.*, **69**, 237–48.

Taylor, A. H. (1990) Disturbance and persistence of sitka spruce (*Picea sitchensis* (Bong) Carr.) in coastal forests of the Pacific Northwest. *J. Biogeog.*, **17**, 47–58.

Taylor, A. H. and Zisheng, Q. (1988) Regeneration patterns in old-growth *Abies-Betula* forests in the Wolong Natural Reserve, Sichuan, China. *J. Ecol.*, **76**, 1204–18.

Thompson, J. N. (1980) Treefalls and colonization patterns of temperate forest herbs. *Amer. Mid. Nat.*, **104**, 176–84.

Thompson, J. N. and Willson, M. F. (1978) Disturbance and the dispersal of fleshy fruits. *Science*, **200**, 1161–3.

Tilman, D. (1988) *Plant Strategies and the Dynamics and Structure of Plant Communities*, Princeton Univ. Press, Princeton, New Jersey.

Tomanek, J. (1960) Mikroklimatische Verhaltnisse in Lochhiebe, in *Verh. Ganzstaatle Bioklimatol. Konf.*, 1958, pp. 297–313.

Tubbs, C. H. (1973) Allelopathic relationships between yellow birch and sugar maple seedlings. *Forest Sci.*, **19**, 139–45.

Uhl, C. (1982) Tree dynamics in a species rich tierra firme forest in Amazonia, Venezuela. *Acta Cientifica Venezolana*, **33**, 72–7.

Uhl, C., Clark, K., Dezzeo, N. and Maquirino, P. (1988) Vegetation dynamics in Amazonian treefall gaps. *Ecology*, **69**, 751–63.

Urban, D. L., O'Neill, R. V. and Shugart, H. H., Jr (1987) Landscape ecology. *BioScience*, **37**, 119–27.

Usher, M. B. (1979) Markovian approaches to ecological succession. *J. Animal Ecol.*, **48**, 413–26.

Vázquez-Yanes, C. and Smith, H. (1982) Phytochrome control of seed germination in the tropical rain forest pioneer trees *Cecropia obtusifolia* and *Piper auritum* and its ecological significance. *New Phytol.*, **92**, 477–85.

Veblen, T. T. (1982) Growth patterns of *Chusquea* bamboos in the understory of Chilean *Nothofagus* forests and their influences in forest dynamics. *Bull. Torrey Bot. Club*, **109**, 474–87.

Veblen, T. T. (1985) Forest development in tree-fall gaps in the temperate rain forests of Chile. *Nat. Geog. Res.*, **1**, 162–83.

Veblen, T. T. (1986a) Age and size structure of subalpine forests in the Colorado Front Range. *Bull. Torrey Bot. Club*, **113**, 225–40.

186   Regeneration dynamics

Veblen, T. T. (1986b) Treefalls and the coexistence of conifers in subalpine forests of the central Rockies. *Ecology*, **67**, 644–9.

Veblen, T. T. (1989) Tree regeneration responses to gaps along a transandean gradient. *Ecology*, **70**, 541–3.

Veblen, T. T. and Ashton, D. H. (1978) Catastrophic influences on the vegetation of the Valdivian Andes, Chile. *Vegetatio*, **36**, 149–67.

Veblen, T. T., Ashton, D. H. and Schlegel, F. M. (1979a) Tree regeneration strategies in a lowland *Nothofagus*-dominated forest in south-central Chile. *J. Biogeog.*, **6**, 329–40.

Veblen, T. T., Veblen, A. T. and Schlegel, F. M. (1979b) Understory patterns in mixed evergreen-deciduous *Nothofagus* forests in Chile. *J. Ecol.*, **67**, 809–23.

Veblen, T. T., Donoso Z. C., Schlegel, F. M. and Escobar R. B. (1981) Forest dynamics in south-central Chile. *J. Biogeog.*, **8**, 211–47.

Veblen, T. T., Hadley, K. S., Reid, M. S. and Rebertus, A. J. (1991) The response of subalpine forests to spruce beetle outbreak in Colorado. *Ecology*, **72**, 213–31.

Veblen, T. T. and Lorenz, D. C. (1986) Anthropogenic disturbance and recovery patterns in montane forests, Colorado Front Range. *Phys. Geog.*, **7**, 1–24.

Veblen, T. T. and Lorenz, D. C. (1988) Recent vegetation changes along the forest/steppe ecotone of northern Patagonia. *Ann. Ass. Amer. Geog.*, **78**, 93–111.

Vitousek, P. M. and Denslow, J. S. (1986) Nitrogen and phosphorous availability in treefall gaps of a lowland tropical rainforest. *J. Ecol.*, **74**, 1167–78.

Waggoner, P. E. and Stephens, G. R. (1970) Transition probabilities for a forest. *Nature*, **225**, 1160–1.

Wardle, J. and Guest, R. (1977) Forests of the Waitaki and Lake Hawaea catchments. *New Zealand J. For. Sci.*, **7**, 44–67.

Wardle, P. (1983) Temperate broad-leaved evergreen forests of New Zealand, in *Temperate Broad-Leaved Forests* (ed. J. D. Ovington), Elsevier, Amsterdam, pp. 33–71.

Watt, A. S. (1923) On the ecology of British beechwoods with special reference to their regeneration. Part I. The causes of failure of natural regeneration of the beech (*Fagus sylvatica* L.). *J. Ecol.*, **11**, 1–48.

Watt, A. S. (1947) Pattern and process in the plant community. *J. Ecol.*, **35**, 1–22.

Whipple, S. A. and Dix, R. L. (1979) Age structure and successional dynamics of a Colorado subalpine forest. *Amer. Mid. Nat.*, **101**, 142–58.

White, P. S. (1979) Pattern, process, and natural disturbance in vegetation. *Bot. Rev.*, **45**, 229–99.

White, P. S., Mackenzie, M. D. and Busing, R. T. (1985a) Natural disturbance and gap phase dynamics in southern Appalachian spruce–fir forests. *Canad. J. For. Res.*, **15**, 233–40.

White, P. S., Mackenzie, M. D. and Busing, R. T. (1985b) A critique on overstory/understory comparisons based on transition probability analysis of an old growth spruce-fir stand in the Appalachians. *Vegetatio*, **64**, 37–45.

White, P. S. and Pickett, S. T. A. (1985) Natural disturbance and patch dynamics: An introduction, in *The Ecology of Natural Disturbance and Patch Dynamics* (eds S. T. A. Pickett and P. S. White), Academic Press, New York, pp. 3–13.

Whitmore, T. C. (1974) *Change with Time and the Role of Cyclones in Tropical Rain Forest of Kolombangara, Solomon Islands*, Commonwealth Forestry Institute Paper, No. 46, University of Oxford.

Whitmore, T. C. (1975) *Tropical Rain Forests of the Far East*, Oxford Univ. Press, London.

Whitmore, T. C. (1982) On pattern and process in forests, in *The Plant Community as a Working Mechanism* (ed. E. I. Newman), Blackwell, Oxford, pp. 45–59.

Whitmore, T. C. (1989) Canopy gaps and the two major groups of forest trees. *Ecology*, **70**, 536–8.

Whittaker, R. H. (1953) A consideration of climax theory: The climax as a population and pattern. *Ecol. Monogr.*, **23**, 41–78.

Whittaker, R. H. (1975a) *Communities and Ecosystems*, 2nd edn, MacMillan, New York.

Whittaker, R. H. (1975b) The design and stability of plant communities, in *Unifying Concepts in Ecology* (eds W. H. van Dobben and R. H. Lowe-McConnell), Junk, The Hague, pp. 169–81.

Whittaker, R. H. and Levin, S. A. (1977) The role of mosaic phenomena in natural communities. *Theor. Pop. Biol.*, **12**, 117–39.

Wiens, J. A. (1976) Population responses to patchy environments. *Ann. Rev. Ecol. System.*, **7**, 81–120.

Williamson, G. B. (1975) Pattern and seral composition in an old-growth beech-maple forest. *Ecology*, **56**, 727–31.

Woods, K. D. (1979) Reciprocal replacement and the maintenance of codominance in a beech-maple forest. *Oikos*, **33**, 31–9.

Woods, K. D. (1984) Patterns of tree replacement: Canopy effects on understory pattern in hemlock–northern hardwood forests. *Vegetatio*, **56**, 87–107.

Woods, K. D. and Whittaker, R. H. (1981) Canopy–understory interaction and the internal dynamics of mature hardwood and hemlock–hardwood forests, in *Forest Succession: Concepts and Application* (eds D. C. West, H. H. Shugart and D. B. Botkin), Springer-Verlag, New York, pp. 305–23.

Worrall, J. J. and Harrington, T. C. (1988) Etiology of canopy gaps in spruce–fir forests at Crawford Notch, New Hampshire. *Canad. J. For. Res.*, **18**, 1463–9.

Young, K. R., Ewel, J. J. and Brown, B. J. (1987) Seed dynamics during forest succession in Costa Rica. *Vegetatio*, **71**, 157–73.

Zimmerman, M. H. and Brown, C. L. (1971) *Trees: Structure and Function*, Springer, New York.

Zinke, P. J. (1962) The pattern of influence of individual forest trees on soil properties. *Ecology*, **43**, 130–3.

5 From population dynamics to community dynamics: modelling succession as a species replacement process

Robert van Hulst

5.1 INTRODUCTION

Prediction of the future composition of a community is a problem of considerable theoretical and practical importance, and the literature on the topic is extensive. What is required is often a prediction of the proportional abundance (cover, biomass) of at least the dominant species over a limited time span, typically on the order of several years to several decades. This chapter focuses on simple, analytical models of successional processes, models which often are amenable to analytical solution. A contrasting approach is used in Chapter 7, where the emphasis is on computer simulation of multiple instances of simple and known processes.

In what follows I shall assume that the quantities of interest are the proportional abundances of each of the species present, written as a vector $\mathbf{p}$, the elements of which sum to unity. This can be contrasted to an approach which seeks to predict more integrated measures of community structure and function, such as overall standing crop, or energy or material turnover. I shall refer to the former approach as a *population* one, and to the latter as an *ecosystem* approach.

The objective in succession modelling at the population level is to describe the dynamics of $\mathbf{p}(t)$, given information on the present state of the community ($\mathbf{p}(0)$), on how the species influence one another, and on other factors influencing species abundances. Note that approximate predictions made on the basis of unformalized knowledge are often quite feasible: foresters commonly succeed in predicting the future course of succession of a woodlot. Note also that information on the *spatial* distribution of organisms may be required, either for practical (e.g. manage-

Plant Succession: Theory and prediction Edited by David C. Glenn-Lewin, Robert K. Peet and Thomas T. Veblen © 1992 Chapman & Hall, London ISBN 0 412 26900 7

ment) or for theoretical (effective prediction) reasons. We shall return to this point below.

To transform the present state of the community into a future state we also require an appropriate *dynamic model*. If such a model exists (it may not!), its repeated application will permit us to predict the future (or, by running the model backwards, to 'retrodict' the past) of any community that can be sufficiently well characterized – provided that environmental conditions that are not explicitly modelled remain the same.

There are at least two different ways in which one could look for an appropriate dynamic model. One might attempt to reconstruct the dynamic behaviour of the system, treated as a black box, by adopting the simplest model compatible with the data, and fitting model parameters to reproduce observed sequences of community change. When this has been accomplished, the model may be further validated by testing its predictions against other data sets. An example of this *phenomenological* approach is provided by the Markov chain models of succession discussed in Chapter 6 by Usher.

The second approach to modelling ecological succession attempts to derive a dynamic model from first principles in community or population ecology. The obvious attraction of this approach is that it may provide a theoretical underpinning for community dynamics: the phenomenological alternative can come very close to curve-fitting, given the large number of parameters that typically have to be estimated in even the simplest models. A theoretical foundation that would guide us in parameter estimation and, even more importantly, in model validation (or falsification!) would be highly desirable. Given our very incomplete understanding of community-level phenomena, it is not surprising that most modellers who have followed the second approach have attempted to derive a community dynamic model from the dynamics of the species composing the community.

At first sight, it is not at all clear that this will be generally feasible. Population dynamics and community dynamics occur on different time and spatial scales, and a description or model that is satisfactory at one scale level may be quite inappropriate at another level (see Allen and Starr, 1982; O'Neill *et al.*, 1987). It is quite conceivable, for example, that unordered, largely random fluxes of different populations will underlie a behaviour that is more or less predictable at the community level, but one which is not easily predictable from knowledge of the fates of the individual populations alone, just as the behaviour of a gas is not easily derived from the behaviour of its molecules. However, given the paucity in ecology of true community-level descriptors, it is tempting to turn to population dynamics in our effort to better understand vegetation dynamics. Our knowledge of the dynamics of plant populations has also greatly increased in the last decades, and we may be in a better position now to derive community dynamics from populations dynamics than we were in the early seventies when the first, largely phenomenological community dynamics models were published.

One further *caveat* is in order. Any dynamic model is based on a separation of state variables on the one hand and parameters and initial conditions on the other (Lewontin, 1974; van Hulst, 1980; see also Chapter 6). The state variables provide a sufficient description of the system, enabling us to predict its future states, provided we know the present state and the model parameters. A dynamic model is successful only to the extent that such a separation can be found. In the context of succession models, this means that we must be able to specify precisely what constitutes a sufficient description of a successional community. This, it turns out, is far more difficult than might be expected. For example, it is not clear whether the abundances (or biomasses, etc.) of the various higher plants present in a plant community form a sufficient description. Are we able to predict the future composition of the community if we know only its present composition and the way different species influence one another? I have argued before (van Hulst, 1980) that the answer to this question has to be negative in many cases: herbivores, various disturbing factors, migration, spatial patterns, and historical effects, all further complicate an already very knotty prediction problem (see also Chapter 1).

Despite the numerous potential pitfalls one might try to construct a partial predictive model by attempting to emulate in a more formal manner the intuition-based predictions of the expert, while recognizing the limits inherent in any partial model. This is precisely the aim of this chapter, and in what follows we shall compare some of the models of community change that have been proposed in the literature, as well as present a novel and perhaps more promising approach.

Much of what follows will deal with plant community dynamics: plants are so much easier to count than animals, and many of them live so much longer, that most of the work has dealt with them. Generalizing our conclusions to include animals is sometimes straightforward, but often also the differences in temporal and spatial scales involved conspire to create problems even less tractable than those of the dynamics of vegetation alone. The very persistence of most plants does create one important difficulty: it makes experimental study of community change very time consuming. Yet experimentation based on a proper theoretical framework is not impossible, as we shall see below.

5.2   A VARIETY OF MODELS

What factors influence the future vegetational composition of a site? At least six general categories of factors may be distinguished. Depending on which of these factors are judged to be of paramount importance in succession, one is led to different models.

1. The **present vegetation** (through species interactions, reproduction, environmental modification, etc.). If present vegetation is thought to be crucial, species interaction models will be appropriate. Examples include the linear differential equation models of van Hulst (1979) discussed below, or Horn's (1976) application of Markov models (see Chapter 6) in which shade tolerance is the key factor.

2. The present vegetation in the surrounding area (or **immigration of propagules** into the area). If the arrival of propagules on a site is considered to be a crucial factor, then one is led either to an island biogeographical approach, in which the accent shifts from the actual identity of the species present to such parameters as species diversity, community organization, or life-cycle strategies, or to a purely historical description of species changes, yielding explanation but no prediction.

3. The **past vegetation**. It is not clear how much 'memory' there is in the environment for past vegetation. This may depend on whether or not a bank of dormant seeds or rhizomes is present (see Chapter 2), and also on the importance of the modifying effect that plants have on their environment (soil acidification, organic matter build-up, increase in soil nitrate and ammonia, etc.). The influence of the past vegetation in preparing a site for future vegetation change was thought to be crucial by earlier students of succession (Clements, 1916; Cowles, 1899). Modern workers have tended to reject the suggestion that past vegetation is an important factor (for example, Drury and Nisbet, 1973; see, however, also Finegan, 1984). Even if the influence of past vegetation is long lasting, it remains possible that the present vegetational composition of a site contains all the information on past vegetation that is required for predicting future vegetational change.

4. **Present resource levels**. If resource availability (light, humidity, soil minerals) is thought to be of central importance in succession, Tilman's (1982, 1988) models, which are outlined below, may be the most appropriate. On the other hand, it is also possible that a site's resource status will be sufficiently reflected in its present vegetation, and that vegetation parameters alone constitute an adequate description.

5. **Disturbance levels**, **including herbivory**. These too may be reflected in the present vegetation of the site, although disturbance levels will often change more rapidly than the levels of most resources, and even vegetation composition. In recent years many authors have stressed the importance of disturbance as a factor in vegetation dynamics (e.g. Picket and White, 1985; Grime, 1979; Grubb, 1977; Miles, 1979; Walker, 1970; Connell and Slatyer, 1977; see also Chapter 1). In Grime's (1979, 1987) treatment, competitively superior species (i.e. species with the highest maximal growth rate of vegetative tissues; see also Grace, 1990) will tend to replace the early colonizers whenever disturbance does not remain excessive. It is, however, also conceivable that a plant, once established, has a strong competitive advantage over

newcomers (for a discussion of site pre-emption, see Chapter 2). In this case a model similar to that proposed by Sale (1978) for coral fish may be appropriate. This model, the 'lottery for living space model', will be outlined below.

6. **Stochastic factors** (e.g. climatic variability, supply fluctuations of resources, of propagules, etc.). The degree to which different models proposed for succession are able to incorporate environmental or demographic stochasticity varies widely. On the one hand, we have the purely deterministic differential (or difference) equation models which are only appropriate when the state of the vegetation does not depend crucially on stochastic factors. On the other hand, we encounter models which are inherently stochastic, the best known being the Markov models discussed by Usher in Chapter 6. The introduction of a stochastic component in a population dynamics model can give results that are often at odds with expectations derived from corresponding deterministic models (Tuljapurkar, 1989). In the absence of a theoretical framework within which to interpret the results of stochastic simulation studies in community dynamics, stochastic aspects remain, to some extent, an afterthought, rather than an essential part of vegetation modelling.

5.3   A SIMPLE MODEL INCORPORATING HISTORICAL EFFECTS

Let us suppose, with Clements (1916), that the past vegetation of a site determines its present vegetation. More precisely, assume that the abundance of species i in a successional site is a weighted sum of the past abundances of the n different species that once grew at that site:

$$\mathbf{x}_i(t) = \mathbf{x}_i(0) + \sum_j b_{ij} \int_0^t \mathbf{x}_j(s)\mathrm{d}s \quad (i,j = 1,2,\ldots,n) \tag{5.1}$$

where $\mathbf{x}_i(t)$ is the abundance of species i at time t and the b_{ij} are weights that summarize the effect of one unit of species i over one unit of time on the carrying capacity of the site for species j. Equation (5.1) is equivalent to

$$\mathrm{d}\mathbf{x}_i(t)/\mathrm{d}t = \sum_j b_{ij}\mathbf{x}_j(t) \quad (i,j = 1,2,\ldots,n)$$

with initial conditions $x_i(0)$, or, in vector notation

$$\dot{\mathbf{x}} = \mathbf{Bx} \tag{5.2}$$

where $\mathbf{x}$ is the vector of values $x_i(t)$, the dot signifies taking the derivative with respect to time, and $\mathbf{B}$ denotes the matrix of coefficients b_{ij}.

Suppose now that the elements of $\mathbf{x}$ are arranged in the order in which they appear in succession. Most of the non-zero entries of $\mathbf{B}$ will be

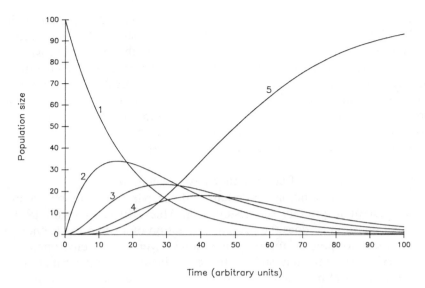

Figure 5.1   A differential equation model of succession. Shown are population sizes (in proportional terms) of five imaginary species as they replace one another in time (measured in imaginary units). The model used is equation (5.2), with the definition of **B** given in the text.

around or below the principal diagonal. Classical Clementsian succession theory states, moreover, that species alter their environment in such a way as to make it less favourable for themselves and more favourable for later successional species. In this case **B** will be a (nearly) lower triangular matrix with negative diagonal and positive entries below the diagonal. The solution of (5.2) with **B** of this form consists of roughly bell-shaped curves shifted along the time axis, thus mimicking the replacement of one species by another during succession (see Fig. 5.1). The conditions on **B** which will produce a non-negative and bounded solution are given in van Hulst (1979).

Figure 5.1 illustrates the behaviour of this model with **B** defined as follows:

$$\mathbf{B} = \begin{bmatrix} -0.06 & 0 & 0 & 0 & 0 \\ 0.06 & -0.07 & 0 & 0 & 0 \\ 0 & 0.07 & -0.08 & 0 & 0 \\ 0 & 0 & 0.08 & -0.09 & 0 \\ 0 & 0 & 0 & 0 & 0.09 \end{bmatrix}$$

An interesting prediction of this model with a **B** matrix as described above is that the course of succession should be extremely difficult to arrest. The very presence of populations will change the substratum so as to make it less and less suitable for the species involved (the negative terms b_{ii}). Furthermore, the early successional species 'prepare the path' for the later successional species, and the latter should become harder and harder to eliminate (the positive terms below the diagonal in **B**).

The above model can be easily modified to incorporate competitive (i.e. contemporaneous) effects, as well as the historical effects discussed above (see van Hulst, 1979). The rather exotic predictions both models produce (notably the irreversibility of succession) should warn us, however, that cumulative change in the environment of the kind described by Clements (1916) is unlikely ever to be an important cause of successional change.

It does not follow, of course, that past vegetation does not have any lasting effects. Vegetation-dependent soil 'preparation' can be an important factor in primary succession (e.g. Miles, 1979; Finegan, 1984). The real question is: does the present vegetation of a site provide enough information to predict future vegetation? Even if an environmental 'memory' exists, this may still be the case; increased organic matter or nitrate in the soil may, after all, show up in a vegetation of nitrophiles. Knowledge of the present state of the vegetation may then be sufficient to determine its future (although one can think of many other factors that may in fact prevent this).

5.4   TILMAN'S RESOURCE-BASED MODELS

Suppose that the distribution of plant species in space and in time is, ultimately, determined by resource levels. This may seem self-evident, but it is not necessarily true, except in a partial sense. While one will certainly not find a plant at a location where it cannot gather essential resources, it is not true either that a plant of a certain species will be present at every site where its essential resources are available. There are simply too many stochastic ecological factors that influence the occurrence of a species for such an assumption to be valid, and, perhaps more importantly, the time required for such a sorting process to run to completion may simply not be available. Nevertheless, it is certainly the case that major shifts in species abundance can often be traced to changes in resource levels (Tilman, 1982, 1985, 1987, 1988, 1990). The existence of well-documented (e.g. Fitter and Hay, 1987) differences in resource requirements and uptake rates between different species suggests that models of successional change should take into consideration resource requirements of competing species as well as the rates of supply of the limiting resources. Given our relative ignorance of both kinds of factors, as well as the extreme difficulty in elucidating them, this is a tall order indeed.

A qualitative picture of what is involved can, however, be gleaned from a graphical model presented in Tilman (1982, 1985). If one assumes that: (i) there are at least two limiting resources; (ii) that the potential inhabitants of a site have complementary resource requirements (i.e. the species with highest requirement for resource A has lowest requirement for resource B, etc.); (iii) that each species consumes relatively more of the resource that limits it; (iv) that resource availability not only constrains, but determines plant growth; and (v) that the system is left long enough under the same conditions that an equilibrium approach can be used, then Tilman's (1982, 1985) graphical model is appropriate. The model shows how the unequal consumption of two resources (e.g. phosphorus and silica in the case of phytoplankton), or the autogenous (i.e. vegetation-caused) accumulation of soil nutrients and decrease in light levels can cause shifts in species composition and in species richness.

If the rather stringent assumptions listed above are fulfilled (that is, when we *know* them to be fulfilled, as a result of experimental testing), this resource-supply-based model may provide a practical means for predicting future vegetational change. In general, though, this model is perhaps better seen as an interesting heuristic device: it provides yet another abstraction of successional change. By focusing on resource shifts, the model suggests that these by themselves suffice to cause vegetation to change in a way which is, in principle, predictable. As a closer look at some other models of succession will show, it is highly debatable whether endogenously caused resource shifts are ever the sole driving force for vegetation change. In fact, in a recent paper Tilman and Wedin (1991) argue that resource ratio competition is probably not an important driving force for early old-field succession. Note also, that resource-based models, such as Tilman's, are typically equilibrium models: they assume that the parameters which reflect the environmental effects and the species' reaction to these, remain constant for long enough periods to allow for a steady state (or at least a moving equilibrium) to be reached.

More recently, Tilman (1988, 1990) has extended his resource-based model to one more appropriate for size-structured populations of plants that have to allocate resources to both a system of leafed shoots used to harvest light, and a system of roots, used to forage for water and mineral nutrients. The new model, which is implemented in a computer simulation ('ALLOCATE'), is considerably more complex than his earlier models, and it possesses a much more sophisticated behavioural repertoire. Some of its implications with respect to allocation patterns in plants in various habitats have been criticized on physiological grounds (Shipley and Peters, 1990; Garnier, 1991). Nevertheless, it is clear already, that Tilman's work and that of his critics has done much to provide an empirical, ecophysiological basis for the previously mostly descriptive and mathematical work on vegetation succession. The fact that we still do not have a thorough enough understanding of the mechanisms of succession to allow us to predict effectively successional change, may mean simply that succession is too multifarious a phenomenon to be caught under a single net.

5.5 LOTTERIES FOR LIVING SPACE

Living space is not by itself a resource, but it provides access to essential resources (McConnaughay and Bazzaz, 1991). Thus, space acts as a surrogate for resources. Plants, being sedentary during most of their lives, are space limited in all but pioneer vegetation.

Spaces of sufficient size for colonization are normally made available by various forms of disturbance (Picket and White, 1985; Grime, 1979; see also Chapter 1). These spaces then become colonized, either by extension growth of neighbours or by seeds, or both (these processes are explained in detail in Chapter 2). It is reasonable to assume that the probability that a certain species will colonize a gap is, as a first approximation, equal to the proportional abundance of this species, weighted by average seed or propagule production. In particular, let us assume that there is no tendency for certain species to colonize preferentially gaps 'vacated' by certain other species, or adjacent to certain species. This implies that the (re-)colonization of empty sites is a spatially homogeneous random process. Let us also assume that, once a species has colonized such a gap, it has a high probability of holding it until it reaches reproductive age. It has, in fact, won the 'lottery for living space', and the only task that remains for it is to attempt to leave as many offspring as possible, some of whom may win a living space (see also the discussions of gap dynamics in Chapters 1, 2 and 4).

The above description may be appropriate, with suitable modifications, for certain coral fishes, as Sale (1978) has argued. Its application to plants seems limited to species colonizing only transient sites. While some such species obviously exist, it would seem that many, if not most species have evolved to some intermediate position on a continuum, which ranges from competitively ousting other plants from their sites, to leading a meager existence in habitats so poor in critical resources that gap ownership is not a guarantee of reproductive success (Grime, 1979). Nevertheless, evidence exists that suggests that for many species of plants, gap regeneration is crucial (Picket and White 1985; Grubb, 1977; Denslow, 1980; see also Chapters 2 and 4).

To the extent that the assumptions of the model are fulfilled, prediction of further vegetation change becomes relatively easy. Given an estimate of the frequency of each species in the seed 'rain' (easily obtainable with seed traps), of seed and seedling mortality, and knowledge of the rate of gap formation, one should be able to predict future vegetational composition.

The rate of gap formation will generally be related to the present vegetational composition (Denslow, 1980), so that the former may be expected to change over the course of succession. It is difficult to assume, however, that other factors that are quite unrelated to present vegetational composition, will not influence gap formation. So, even if the

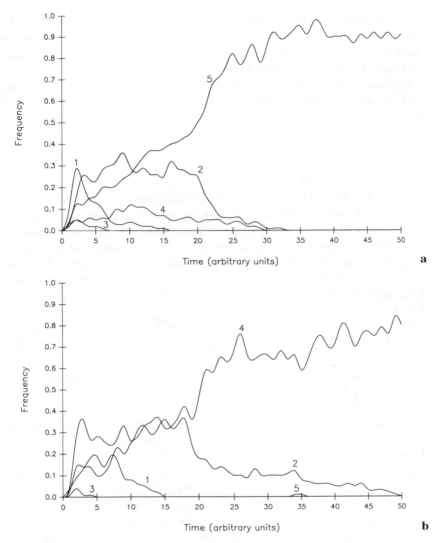

Figure 5.2 A spatial model of succession driven by a 'lottery for living space' mechanism. The abundances (in relative units) of each of 5 species are plotted against time. Figure 5.2(a) represents the output of the model without disturbance, while in Figure 5.2(b) disturbance has been added, as described in the text.

other assumptions apply (in particular, no competition for space and guaranteed reproductive success for every holder of a site), this model appears too schematic to be of practical value.

An example of a lottery model can be found in Worked Example 5.1 and Fig. 5.2. The chief importance of the 'lottery for living space' model lies in its emphasis on one factor, gap colonization, that by itself would be sufficient to drive successional change. Another property of the present model merits our attention: it is the first of the models discussed so far to incorporate explicitly a spatial structure (the basic resource is, after all, a spatial one). We shall proceed to examine this type of model in the next section, where I also present an example of a succession simulation using a lottery model.

Worked Example 5.1: A spatial model of succession driven by a 'lottery for living space' mechanism

On a 10 × 10 grid each of the 100 cells can contain one species, or the cell may be empty. Empty cells can be colonized from other cells on the grid, or from a common pool outside the grid. The seeds from all locally occurring species are assumed to have a constant abundance in the pool, and their dispersal is also assumed to be known. We can, therefore, calculate the probability that the seed of a certain species in the pool reaches a grid cell. To this probability is added the probability that a seed produced on the grid reaches the same grid cell. This is obtained by taking the total number of cells occupied by the species, weighted by a fecundity parameter for that species. Once a species has colonized an empty cell it is assumed to stay there until it dies. Each species has a fixed, species-specific probability of dying during a unit-time interval.

Using this model, and the parameters listed below, the trajectories plotted in Fig. 5.2(a) were obtained. Figure 5.2(b) is based on the same model and parameter values, except that there is an added disturbance: every unit of time there is a 20% chance of a disturbance that eliminates the plants in a randomly selected 20% of all cells. This clearly favours the species with high fecundity and highly dispersible seeds (i.e. the ruderal species in the simulation).

The parameters were as follows. Proportional pool abundances: 0.6, 0.5, 0.4, 0.3, 0.3; dispersabilities: 0.5, 0.4, 0.3, 0.2, 0.2; death rates: 0.5, 0.4, 0.3, 0.2, 0.1.

5.6   MODELS EXHIBITING SPATIAL STRUCTURE

So far we have mostly ignored the spatial dimension of communities. Only in the discussion of the lottery model above did we even consider the fact that all vascular plants and certain animals are immobile during their adult life, and therefore unable to mix and interact freely with all other members of the community. Yet, this fact should suggest caution in the application of models designed for animals or, say, chemical species.

The introduction of spatial structure into models of community change is far from straightforward, and it is not even clear that it would be worth while. Spatial patterns in vegetation usually do not represent equilibrium properties, but instead depend on the past history of the vegetation (Pickett, 1980). Conversely, the history of the vegetation (i.e. its successional path) may have been influenced by spatial patterns. There are also practical problems: information on the condition of each plant over an extended period is generally difficult to obtain.

For the purpose of prediction, spatial detail may not be required. It is, nevertheless, necessary to ask whether effective prediction of community composition alone is possible with models that do not incorporate the changing spatial distribution of plants. Even if the successional dynamics of a single microsite are perfectly Markovian (i.e. if there is no memory in the system), the dynamics of the vegetation of a larger site may not be Markovian at all. For this to happen it is sufficient that local processes depend crucially upon the nature of neighbouring vegetation (and thereby on the history of the site).

The difficulty of introducing spatial structure into models is that it increases the number of parameters that have to be estimated, and the number of variables that have to be measured. It is here that the curse of 'middle-number systems' (Allen and Starr, 1982) is truly felt – the systems we study have too many components to allow us to construct mechanics-style, single-parameter models, and not enough components to allow effective averaging of stochastic variations in the style of statistical mechanics.

While the incorporation of spatial effects in realistic, many-species models is likely to lead to models which are analytically intractable, a simulation approach may still be instructive, especially in studies with relatively few, well-studied species. Computer simulations of vegetation change driven by one's choice of a successional dynamic, are easily performed, even with large numbers of species, and they often produce remarkably realistic looking results (see Fig. 5.2). The production of life-like results alone, however, is not sufficient reason for accepting a model. A successful model should have both explanatory and predictive power, and it is not easy to see how models incorporating many variables and parameters can have either. A possible exception to this is the compound model which incorporates several well-tested partial models, as in some versions of FORET (Shugart, 1984; West et al., 1981; see also Chapter 7).

It is instructive to ask in what way spatial pattern can influence vegetation dynamics. There are at least three possibilities (not mutually exclusive): (i) through direct interactions among plants, for example competition; (ii) through indirect interactions, for example herbivore effects; and, most importantly, (iii) through modulation of local seed dispersal and/or clonal propagation. Of these three mechanisms, the last was directly treated in the lottery model discussed in the previous section, while all three can be included in tree-based simulation approaches such as FORET (Shugart, 1984).

Vegetation processes take place at a variety of different spatial and temporal scales (Allen and Starr, 1982). It seems reasonable to speculate that, as the spatial and temporal scale increases to that of landscape processes taking place over hundreds or even thousands of years, then the importance of spatial inhomogeneity will correspondingly decrease. Spatial inhomogeneity is, by definition, a local phenomenon. So it is especially in attempts at detailed prediction of local vegetation change, that spatial information will be required. Models addressing this type of prediction problem will thus tend to be ones with many parameters, and many state variables (the performance of the species at all sites). This virtually precludes analytical models, and the approach of choice here is probably computer simulation as discussed in Chapter 7.

5.7 INVASION MODELS

The process of invasion is basic to the study of population dynamics: pioneer species invade open areas, and later successional species invade existing vegetation. To be sure, almost all species that will appear later in a site may be present from the start (the 'initial floristic composition' hypothesis of Egler, 1954). Successful invasion, however, involves growth of a seed or propagule into an adult plant, and subsequent multiplication. Even if successful invasion requires the availability of small gaps in the vegetation cover, the invading species will generally be surrounded by plants of other species which may be aggressively foraging for limiting resources. What determines the fate of such an invading species? The question has so far received surprisingly little attention. It is easy to see, however, that the answer will depend on the answers to two further questions: what makes a species a good invader (in a certain community type); and, what features confer a resistance to invasion (by certain species) on communities (cf. Crawley, 1987)? Let us examine these two questions in turn.

The population aspects of invasive ability have been fairly well studied (e.g. Harper, 1977; Werner, 1976; Brown and Marshall, 1981; Salisbury, 1961; Baker, 1972; Levin, 1984). Common attributes of successful invaders include: high reproductive rate, effective seed dispersal, high seed survival and germinability, effective space pre-emption, and aggressive resource acquisition (Werner, 1976; Grime, 1979). Champion invaders include some of the world's worst (i.e. most successful) weeds (Holm *et al.*, 1977), but also phanerogamic parasites and hemiparasites (van Hulst *et al.*, 1987).

On the other hand, it should be emphasized that an aggressive foraging strategy is not always an essential or even an advantageous trait for invading species. Many communities are characterized by extremely low levels of one or more critical resources. Primary successions typically start out this way, while secondary successions, even in their early stages, are characterized by resource levels that are higher than usual (see Chapter 1). Under conditions of extreme resource shortages, any species with an aggressive mode of resource acquisition is likely to devote too great a share of its limited resources to foraging, and thus to perish. Sun plants often etiolate under a canopy, and die, whereas shade plants conserve their resources and (often) live under these conditions (Fitter and Hay, 1987). This suggests that special consideration should be given to community resistance to invasion, as well as to resource levels. Here our present knowledge is much more limited.

MacArthur (1972) and Turelli (1980, 1981) have studied invasion of competitors in communities structured by competitive relationships. The probability of successful establishment of a rare invader can be expected to depend on at least three factors: (i) the levels of competition the invader faces; (ii) demographic stochasticity (random variation among individuals in offspring production); and (iii) environmental stochasticity (random variation in the population's per capita growth rate). This builds on an earlier observation of MacArthur and Levins (1967), that the success of an invader depends on the amount of overlap between its niche and those of the members of the established community.

Turelli (1981) has demonstrated in his model systems that environmental stochasticity has relatively little effect on the probability that a rare invader will become established. In a separate study (Turelli, 1980) of the joint effect of competition, and demographic and environmental stochasticity, he concluded that P, the probability of successful establishment, will decrease if a, the level of interspecific competition, n, the number of residents, or γ, the level of environmental variation, increases. Small values of γ (<0.2) have very little effect. P increases as the initial population size of the invader, or r, the growth rate of the invader, increase.

Turelli's conclusions are based on varying one parameter at a time, and on the specific models employed (Lotka–Volterra competition model with stochasticity). If it can be assumed that rare invaders in multispecies communities generally face competition from many local residents, then demographic fluctuations and competitive pressure (and, to a lesser extent, environmental stochasticity) will lead to very small invasion probabilities, and hence much less close packing of competitors than might be expected otherwise. However, the question remains whether a simple niche-partitioning model, originally developed for communities of freely moving animals, is at all appropriate for plants which exhibit modular growth, pronounced ecological differences between juvenile and adult individuals, limited mobility, and, most importantly, which may not even divide up their environments solely or primarily by simple resource partitioning. Considering the non-substitutability of plant resources, the specialized foraging and herbivore avoidance adaptations that are so common in plants, may be much more important than a preoccupation with resource partitioning *per se* (see, for example, Fitter and Hay, 1987; but also Aarssen, 1983; and Turkington and Aarssen, 1984).

The intricacies of mechanism inherent in invasion processes can be avoided if one adopts a more general approach to invasion. Any mechanistic model of the process, it seems, would have to make far too great a demand on both our limited knowledge of the factors governing plant growth and our limited ability to obtain the required data at the scale of a community rather than that of a single plant. In the next section we turn to such a general (or macroscopic) approach, which, nevertheless, remains firmly embedded in population biological theory so as not to become a completely phenomenological description of vegetation change.

There exist profound differences in invasive ability among different plant species. Some species are chiefly or solely *gap* colonizers: they require a minimum size gap for successful colonization (e.g. Gross, 1980; Picket and White, 1985). Such gaps are formed at different rates and in different sizes in vegetation. Other species are able to invade very small openings which nearly always exist (e.g. van Hulst *et al.*, 1984).

Some vegetation is rather resistant to invasion. This is normally due to the near absence of gaps, as in the case of swaths of many rhizomatous, strongly tillering grasses. This suggests that the resistance to invasion of a community may be approximately proportional to the fraction of open soil in that community. However, resistance to invasion also depends frequently on the nature of the resident species (see van Hulst *et al.*, 1987). Conversely, the invasive ability of a species will depend on the growth rate of a small population when surrounded by plants of other species. Both factors, rate of establishment as a function of gap size, and initial growth rate as a function of the surrounding vegetation, can be experimentally determined for different combinations of invaders and established species.

Many of the models that have up to now been employed in succession modelling have been dynamically simple linear (or linearized) models such as systems of linear differential equations, or time homogeneous Markov chains. In all these models the eventual vegetation composition is independent of the initial conditions. This, it would seem, is far from realistic. Whether through external driving factors or through the nature of the governing dynamics, actual successions exhibit an extraordinary dependence on initial conditions. This is most clearly exhibited by the experiments of Schmidt (1981). In an impressive series of experiments Schmidt has studied secondary succession for over ten years on old fields that were given various initial and management treatments. Vegetation responded quite differently to the different preparations of the field (plowing, heat sterilization, chemical sterilization) and also to the different types of management (control, mulching, rotovator treatment, mowing with different frequencies).

The dependence of vegetation change on initial conditions suggests that non-linear models are required. The initial attraction of Markov chain and linear differential equation models, that they often exhibit a unique equilibrium which will eventually be reached from any starting point, can now be seen to have been based on too narrow a view of ecological succession (Connell and Slatyer, 1977; see also Chapter 6).

5.8 A GAME THEORETIC INVASION MODEL

To develop a non-linear model (cf. van Hulst, 1980), let us assume that we are given a vector $\mathbf{x}$ containing the proportion of an area free of vegetation and the proportions of the area covered by each of the n species of interest. We therefore have $\Sigma_i x_i = 1$. Assume also that we are given a matrix $\mathbf{A}$ $(a_{ij}, i,j = 1,2,\ldots,n)$ in which a_{ij} is proportional to the growth rate of population i when surrounded by species j. Then the growth rate of population i amidst community $\mathbf{x}$ will be proportional to $\Sigma_j a_{ij} x_j = (\mathbf{Ax})_i$. The proportional growth rates of the species forming a 'microcommunity' $\mathbf{x}$ amidst a larger community $\mathbf{y}$ will be equal to $\mathbf{x'Ay}$. Assume also that the following dynamic applies: the proportional growth rate of x_i equals the growth rate of species i if surrounded by community $\mathbf{x}$, minus the weighted average of the growth rates of all species in $\mathbf{x}$, or, in symbols:

$$\frac{x_i(t + \delta)}{x_i(t)} = (\mathbf{Ax})_i - \mathbf{x'Ay},$$

where $t + \delta$ is some increment of time after t. Hence

$$\mathrm{d}x_i/\mathrm{d}t = x_i[(\mathbf{Ax})_i - \mathbf{x'Ay}] \tag{5.3}$$

With these definitions, the following results of Žeeman (1980) can be applied. Let us define a *non-invadable community* to be a vector $\mathbf{x}$ such that $\mathbf{x}$ cannot be invaded successfully by any one or more of the available species (not all elements of $\mathbf{x}$ have to be positive, but none may be negative). The notion of the non-invadable community is thus directly analogous to the *evolutionarily stable strategy* of Maynard Smith (1982). We define a *local climax* to be the endpoint of a successional sequence in the absence of perturbations and changes in the species pool. Mathematically, a local climax is an attractor. Zeeman (1980) derives the following results (converted here to our terminology):

(i)   A local climax may be open to invasion; a non-invadable community is always a local climax.

(ii) Given a species assemblage S containing s species. If S can form a non-invadable community with all s species present, that community is unique and it forms the local climax of any mixed community with s species. An s species local climax is not necessarily a global climax since there may be other local climaxes with fewer than s species.

Since the dynamic of equation (5.3) is no longer linear (it is, in fact, cubic), the present model is far richer than models based on a linear dynamic. In particular, different starting configurations do not necessarily lead to the same local climax; a community can lose species, or even exhibit cyclical behaviour. There may be several alternative stable points.

Another important property of the model is that the notion of structural stability (Thom, 1975) can be introduced. A system is structurally stable if small changes in the parameters induce qualitatively similar trajectories (in technical terms: topologically equivalent flows). Several results on structural stability of the model under consideration are given by Zeeman (1980). Shipley (1987) has shown that the equations (5.3) are strictly equivalent to the Lotka–Volterra competition equations, with proportional abundances.

I present some examples of the model's behaviour below. First, however, several more conventions need to be explained. The only parameters of the model are the coefficients a_{ij} of $\mathbf{A}$, the pay-off matrix of game theory. It can be shown that adding a constant to a column of $\mathbf{A}$ (=increasing the pay-off to all strategies equally) does not change the flow induced by $\mathbf{A}$. The diagonal elements a_{ii} are, therefore, conveniently set equal to 0. As pointed out above, an element of $\mathbf{A}$ can be interpreted in our context as the proportional growth rate of a species i when surrounded by species j. If we agree to set $a_{ii} = 0$, then $a_{ij} > 0$ indicates relatively better growth when surrounded by plants of a different species than when surrounded by conspecifics (as in Trenbath's (1974) 'over-yielding' mixtures), while $a_{ij} < 0$ indicates an 'underyielding' mixture (with the proviso, that one of the species is much less abundant than the other).

If x_1 represents the proportion of unvegetated soil as suggested above, then the first row of the $\mathbf{A}$ matrix represents the rates of growth of bare soil in different populations, and the first column of $\mathbf{A}$ represents the rates of growth of the different species in patches of bare soil.

For two and three species communities, the dynamics are relatively simple and the different topological classes have been classified by Zeeman (1980). For two species $\mathbf{A}$ has the form

$$\mathbf{A} = \begin{bmatrix} 0 & a \\ b & 0 \end{bmatrix}$$

For $a, b > 0$ a 2-species climax exists, given by

$$\mathbf{x}' = (a/(a + b), b/(a + b)).$$

In all other cases either species 1 or 2 will eventually become extinct. For 3 species there are 19 topologically different classes of model behaviour (Zeeman, 1980). If the determinant of $\mathbf{A}$ is positive, 3-species, mixed communities are possible. The 19 classes include ones with several local climaxes (in one case as many as 3), repellors and saddle-points. If $n = 4$, the behaviour becomes even more complex: small parameter changes, for example, may yield a smooth transition from a stable equilibrium to cyclical behaviour.

Computer simulations, both with 'actual' data (invasive behaviour inferred from tables in Schmidt, 1981) and randomly generated data are shown in Fig. 5.3. It can be seen that sequences simulating successional species replacement are easily generated. Even an artificial 'community' with a_{ij} values generated randomly ($-1 \leq a_{ij} \leq 1$) usually generates shifts in abundance of the different species and either a local climax (not necessarily with the full species complement) or an uninvadable community (given the species complement employed), illustrating once more that succession-like species replacement curves are easily generated with a variety of models.

As a next step one might attempt to derive a_{ij} values from population biological data obtained experimentally. One could, for example, introduce seeds or plants of certain species in different communities and monitor their fates. This type of experiment is standard in plant population biology (e.g. Pacala and Silander, 1990; Harper, 1977) and, when employed systematically, it should permit estimation of a_{ij} values and further testing of the game theory model.

Worked Example 5.2: A succession model based on game theory

In all graphs (Fig. 5.3) the abscissa represents time (arbitrary units), and the ordinate indicates the proportional abundance of each species.

Figures 5.3(a) and (b) illustrate how different starting configurations can give rise to quite different outcomes. The invasion matrix **A** in both figures is as follows:

$$\mathbf{A} = \begin{bmatrix} 0 & 6 & -4 \\ -3 & 0 & 5 \\ -1 & 3 & 0 \end{bmatrix}$$

With the starting configuration in 5.3(a) species 1 wins, while with that in 5.3(b) an equilibrium results. In our terminology, the community consisting of species 1 by itself is a *non-invasible community*, whereas the equilibrium with all three species present is a *local climax*.

Figure 5.3(c) illustrates cycling; the matrix **A** is:

$$\mathbf{A} = \begin{bmatrix} 0 & 1 & -0.1 & 0 \\ 0 & 0 & 1 & -0.1 \\ -0.1 & 0 & 0 & 1 \\ 1 & -0.1 & 0 & 0 \end{bmatrix}$$

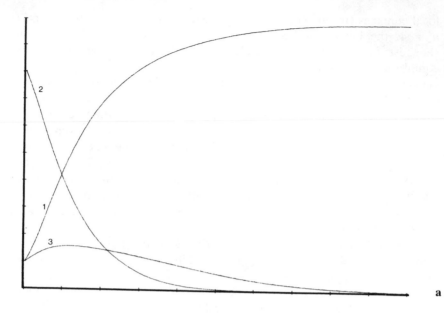

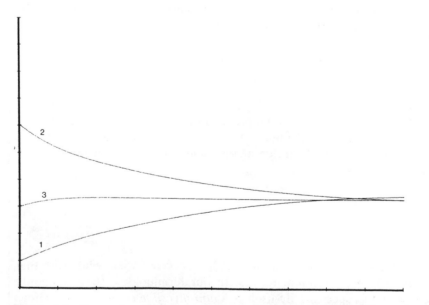

Figure 5.3 Succession driven by a game theory model. In all graphs the proportional abundances of the species are plotted against time (arbitrary units). See the text for further explanation. Figure 5.3(h) represents an attempt to mimic the process of species replacement in one of Schmidt's (1981) experimental successions (his table 5.1E). The species are: (1) Open space, (2) *Senecio vulgaris*, (3) *Conyza canadensis*, (4) *Epilobium tetragonum*, (5) *Poa trivialis*, (6) *Taraxacum officinale*, (7) *Solidago canadensis*, (8) *Tussilago farfara*, (9) *Dactylis glomerata*, (10) *Salix caprea*. The time scale is in unstandardized units.

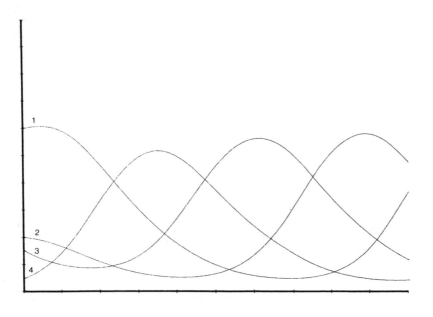

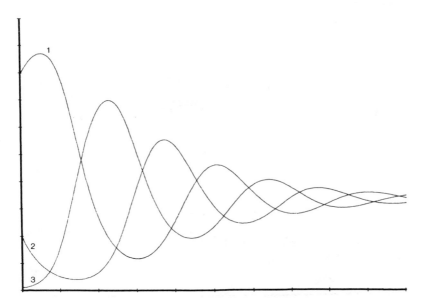

Figure 5.3   Continued

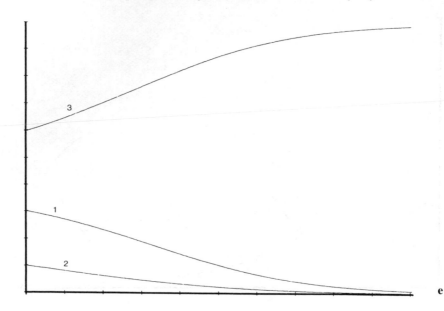

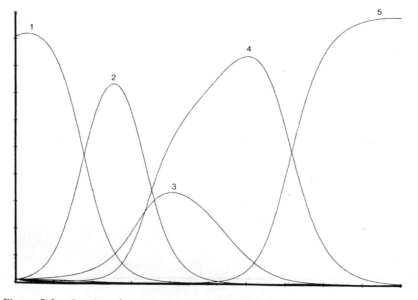

Figure 5.3   Continued

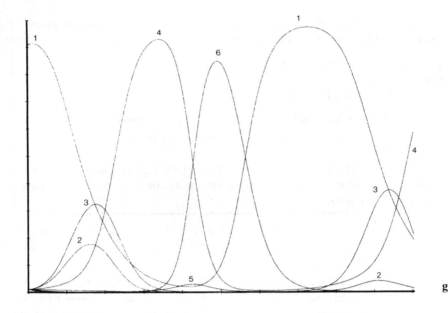

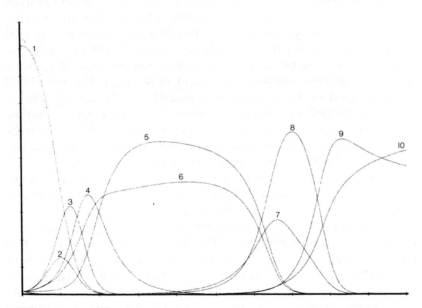

Figure 5.3   Continued

Figures 5.3(d) and (e) show how a change in the invasion matrix may alter the outcome. The matrix **A** is:

$$\mathbf{A} = \begin{bmatrix} 0 & 1 + \varepsilon & -1 \\ -1 & 0 & 1 + \varepsilon \\ 1 + \varepsilon & -1 & 0 \end{bmatrix}$$

with $\varepsilon = 2$ in 5.3(d) and $\varepsilon = -2$ in 5.3(e). The former results in stable oscillations to equilibrium, the latter to the exclusion of the two other species by species 2.

Figures 5.3(f) and (g) illustrate the trajectories resulting from two random invasion matrices. The community in Fig. 5.3(g) exhibits cyclical behaviour.

SUMMARY

1. Plant succession is perhaps too multifarious a phenomenon to be captured by any single mathematical model. I have shown that the general pattern of successional species replacement can be faithfully reproduced by a variety of simple models. These models are mutually exclusive in that they focus on different driving factors. It may, in fact, be argued that such partial models can only highlight one aspect of one particular type of successional replacement process. If resource shifts over successional time are considerable, then Tilman's (1982) resource-based model (or a related model) may be appropriate. A model incorporating environmental memory of past vegetation will be appropriate if early successional plants exert a lasting influence on the site, as seems to be the case in certain primary successions (Finegan, 1984). Succession in fertile communities where competition is likely to be intense should perhaps be modelled with an invasion model incorporating competition, whereas succession in high-disturbance areas seems to be well modelled by lottery models.

2. It is important to remember that the modelling of vegetation dynamics is still an immature discipline. The range of models proposed to date is exceedingly limited and most models proposed have been mathematically rather unsophisticated. Some recent attempts to use more sophisticated models have not dealt with vegetation but with sessile animals (e.g. the non-equilibrium model developed by Antonelli and Voorhees (1983) for coral communities), or have focused on population strategies, rather than on community development (e.g. Levin *et al.*, 1984). One may, therefore, hope that some progress can be achieved in succession modelling once mathematical models of sufficient sophistication are being used.

3. A more fundamental difficulty that bars the way to a better understanding of community dynamics is our profound ignorance of

community-level phenomena: the extent of community integration, the rules of community assembly (if such rules exist), the role of chance and chaotic dynamics, as well as that of intraspecific differences in species interactions, the effects of spatial and temporal scale, and one could go on. Both our theoretical and our practical understanding of such phenomena is extremely limited, and most community-level work in plant ecology is still largely descriptive. Given the current reductionist intellectual climate, the tendency exists to seek an explanation for community-level phenomena only at the level of the populations constituting the community. It is essential, however, to remember that a rapid increase in complexity will result from even a moderate increase in the number of interacting entities (e.g. Levins and Lewontin, 1980), and appropriate mathematical models are indispensable if one wishes to deal with this complexity.

4. At some point the lack of suitable community-level descriptors will also be felt: our present attempts to describe the fate of each individual species of vascular plant or tree in an area (but ignoring animals) may prove to be misguided. Yet, in practical applications of community ecology, just such a description is often required (Shugart, 1984; Finegan, 1984). This has led, for example in forest ecology, to a proliferation of phenomenological models of vegetation change. However, because many different models produce similar predictions and because large numbers of parameters have to be estimated, this approach may yield prediction (at least in the short term), with little explanation. The alternative approach, stressed in this paper, namely to derive community dynamics from population dynamics, clearly suffers from another ill: the models produced are so specifically geared to particular assumptions that their explanatory power may be considerable but their predictive power nil. The game theoretic invasion model proposed above may turn out to be sufficiently grounded in population biology and yet sufficiently general to avoid both the Scylla of phenomenological curve fitting and the Charybdis of a mechanistic particularism.

REFERENCES

Aarssen, L. W. (1983) Ecological combining ability in plants: toward a general evolutionary theory of coexistence in systems of competition. *Amer. Nat.*, **122**, 707–31.

Allen, T. F. H. and Starr, T. B. (1982) *Hierarchy: Perspectives for Ecological Complexity*, Univ. of Chicago Press, Chicago.

Antonelli, P. L. and Voorhees, B. H. (1983) Non-linear growth mechanics – I. Volterra-Hamilton systems. *Bull. Math. Biol.*, **45**, 103–16.

Baker, H. G. (1972) Migration of weeds, in *Taxonomy, Phytogeography and Evolution* (ed. D. H. Valentine), Academic Press, New York, pp. 327–47.

Brown, A. H. D. and Marshall, D. R. (1981) Evolutionary changes accompanying colonization in plants, in *Evolution Today, Proc. Sec. Internat. Congr. Syst. Evol. Biol.* (eds G. G. E. Scudder and J. L. Reveal), pp. 351–63.

Clements, F. E. (1916) *Plant Succession*. Carnegie Inst. Washington, Pub. 242.

Connell, J. H. and Slatyer, R. O. (1977) Mechanisms of succession in natural communities and their role in community stability and organization. *Amer. Nat.*, **111**, 1119–44.

Cowles, H. C. (1899) The ecological relations of the vegetation on the sand dunes of Lake Michigan. *Bot. Gazette*, **27**, 95–117.

Crawley, M. J. (1987) What makes a community invasible?, in *Colonization, Succession and Stability* (eds A. J. Gray, M. J. Crawley and P. J. Edward), Blackwell, Oxford, pp. 429–453.

Denslow, J. S. (1980) Gap partitioning among tropical forest trees. *Biotropica*, **12** (Suppl.), 47–55.

Drury, W. H. and Nisbet, I. C. T. (1973) Succession. *J. Arnold Arboretum*, **54**, 331–68.

Egler, F. E. (1954) Vegetation science concepts: I. Initial floristic composition, a factor in old field vegetation development. *Vegetatio*, **4**, 412–17.

Finegan, B. (1984) Forest succession. *Nature*, **312**, 109–14.

Fitter, A. H. and Hay, R. K. M. (1987) *Environmental Physiology of Plants*, 2nd edn, Academic Press, London.

Garnier, E. (1991) Resource capture, biomass allocation and growth in herbaceous plants. *TREE*, **6**, 126–31.

Grace, J. (1990) On the relationship between plant traits and competitive ability, in *Perspection on plant competition* (eds J. B. Grace and D. Tilman), Academic Press, NY, pp. 51–65.

Grime, J. P (1979) *Plant Strategies and Vegetation Processes*, Wiley, Chichester.

Grime, J. P. (1987) Dominant and subordinate components of plant communities: Implications for succession, stability, and diversity, in *Colonization, Succession and Stability* (eds A. J. Gray, M. J. Crawley and P. J. Edwards), Blackwell, Oxford, pp. 413–28.

Gross, K. L. (1980) Colonization by *Verbascum thapsus* (Mullein) of an old-field in Michigan: experiments on the effects of vegetation. *J. Ecol.*, **68**, 919–27.

Grubb, P. J. (1977) The maintenance of species richness in plant communities. The importance of the regeneration niche. *Biol. Rev.*, **52**, 107–45.

Harper, J. L. (1977) *Population Biology of Plants*, Academic Press, London.

Holm, L. G., Plucknett, D. L., Pancho, J. V. and Herberger, J. P. (1977) *The World's Worst Weeds: Distribution and Biology*, Univ. Press of Hawaii, Honolulu.

Horn, H. S. (1976) Succession, in *Theoretical Ecology: Principles and Applications* (ed R. M. May), Saunders, Philadelphia, pp. 187–204.

Levin, D. A. (1984) Immigration in plants: An exercise in the subjunctive, in *Perspectives on Plant Population Ecology* (eds R. Dirzo and J. Sarukhan), Sinauer Associates, Sunderland, Mass., pp. 242–60.

Levin, J. A., Cohen, D. and Hastings, A. (1984) Dispersal strategies in patchy environments. *Theor. Pop. Biol.*, **26**, 165–91.

Levins, R. and Lewontin, R. C. (1980) Dialectics and reductionism in ecology. *Synthese*, **43**, 47–78.

Lewontin, R. C. (1974) *The Genetic Basis of Evolutionary Change*, Columbia Univ. Press, New York.

MacArthur, R. (1972) *Geographical Ecology*, Harper & Row, New York.

MacArthur, R. H. and Levins, R. (1967) The limiting similarity, convergence and divergence of coexisting species. *Amer. Nat.*, **101**, 377–85.

May, R. M. (1973) *Stability and Complexity in Model Ecosystems*, Princeton Univ. Press, Princeton, N.J.

Maynard Smith, J. (1982) *Evolution and the Theory of Games*, Cambridge Univ. Press, Cambridge.

McConnaughay, K. D. and Bazzas, F. A. (1991) Is physical space a soil resource? *Ecology*, **72**, 94–103.

Miles, J. (1979) *Vegetation Dynamics*, Chapman and Hall, London.

O'Neill, R. V., DeAngelis, D. L., Waide, J. B. and Allen, T. F. H. (1987) *A Hierarchical Concept of Ecosystems*, Princeton Univ. Press, Princeton, N.J.

Pacala, S. W. and Silander, J. A. (1990) Field tests of neighborhood population dynamic models of two annual weed species. *Ecol. Monogr.*, **60**, 113–34.

Pickett, S. T. A. (1980) Non-equilibrium coexistence of plants. *Bull. Torrey Bot. Club*, **107**, 238–48.

Pickett, S. T. A. and White P. S. (eds) (1985) *The Ecology of Natural Disturbance and Patch Dynamics*, Academic Press, Orlando.

Sale, P. F. (1978) Coexistence of coral reef fishes – a lottery for living space. *Env. Biol. Fish.*, **3**, 85–102.

Salisbury, E. J. (1961) *Weeds and Aliens*, Collins, London.

Schmidt, W. (1981) Ungestörte und gelenkte Suksession auf Brachackern. *Scripta Geobotanica* XV, E. Goltze, Gottingen.

Shipley, B. (1987) The relationship between dynamic game theory and the Lotka-Volterra competition equations. *J. Theor. Biol.*, **125**, 121–3.

Shipley, B. and Peters, R. H. (1990) A test of the Tilman model of plant strategies: relative growth rate and biomass partitioning. *Amer. Nat.*, **136**, 139–53.

Shugart, H. H. (1984) *A Theory of Forest Dynamics: The Ecological Implications of Forest Succession Models*, Springer-Verlag, Berlin.

Thom, R. (1975) *Structural Stability and Morphogenesis* (trans. D. H. Fowler). Benjamin, Reading, Mass.

Tilman, D. (1982) *Resource Competition and Community Structure*, Princeton Univ. Press, Princeton, N.J.

Tilman, D. (1985) The resource ratio hypothesis of succession. *Amer. Nat.*, **125**, 827–52.

Tilman, D. (1987) Secondary succession and the pattern of plant dominance along experimental nitrogen gradients. *Ecol. Monogr.*, **57**, 189–214.

Tilman, D. (1988) *Dynamics and Structure of Plant Communities*, Princeton Univ. Press, Princeton, N.J.

Tilman, D. (1990) Mechanisms of plant competition for nutrients: the elements of a predictive theory of competition, in *Perspectives on Plant Competition* (eds J. B. Grace and D. Tilman), Academic Press, New York, pp. 117–141.

Tilman, D. and Wedin, D. (1991) Dynamics of nitrogen competition between successional grasses, *Ecology*, **72**, 1038–49.

Trenbath, B. R. (1974) Biomass productivity of mixtures. *Adv. Agron.*, **26**,

177–210.

Tuljapurkar, S. (1989) An uncertain life: Demography in random environments. *Theor. Pop. Biol.*, **35**, 227–94.

Turkington, R. and Aarssen, L. W. (1984) Local-scale differentiation as a result of competitive interactions, in *Perspectives on Plant Population Ecology* (eds. R. Dirzo and J. Sarukhan), Sinauer Ass., Sunderland, Mass, pp. 107–127.

Turelli, M. (1980) Niche overlap and invasion of competitors in random environments. II. The effects of demographic stochasticity, in *Biological Growth and Spread* (eds W. Jaeger *et al.*), Springer, Berlin, pp. 119–129.

Turelli, M. (1981) Niche overlap and invasion of competitors in random environments. I. Models without demographic stochasticity. *Theor. Pop. Biol.*, **20**, 1–56.

van Hulst, R. (1979) On the dynamics of vegetation: Succession in model communities. *Vegetatio*, **39**, 85–96.

van Hulst, R. (1980) Vegetation dynamics or ecosystem dynamics: Dynamic sufficiency in succession theory. *Vegetatio*, **43**, 147–51.

van Hulst, R., Shipley, W. and Theriault, A. (1984) Why is *Rhinanthus minor* L. (Scrophulariaceae) such a good invader? *Canad. J. Bot.*, **65**, 2373–9.

Walker, D. (1970) Direction and rate in some British post-glacial hydroseres, in *The Vegetational History of the British Isles* (eds D. Walker and R. West), Cambridge Univ. Press, Cambridge, pp. 117–39.

Werner, P. A. (1976) Ecology of plant populations in successional environments. *Syst. Bot.*, **1**, 246–68.

West, D. C., Shugart, H. H. and Botkin, D. B. (1981) *Forest Succession: Concepts and Application*, Springer-Verlag, New York.

Zeeman, E. C. (1980) Population dynamics from game theory, in *Global Theory of Dynamical Systems* (eds A. Dold and B. Eckman), Springer-Verlag, Berlin, pp. 471–497.

6 Statistical models of succession

Michael B. Usher

6.1 INTRODUCTION

There are four approaches to modelling the ecological processes of succession. Although these four processes appear discrete, in practice there is virtually a continuum of approaches from the purely descriptive to the highly theoretical, and indeed some authors purposely combine two or more approaches.

Historically the oldest approach is the *verbal model*, in which a series of words, a description, provides the model of the process. In the introductory chapters of this book the verbal models associated with names such as Clements or Gleason have been described. During the 75 or so years since these verbal models were originally formulated, other descriptive models have been put forward, either as series of words or as diagrams, such as Odum's (1969) diagram of the accumulation of biomass, a process that gradually slows as it approaches an asymptote.

The shift from words and diagrams to mathematical expressions and from these to computer code is more recent, having occurred during the last 30 years or so. The advent of computers led to the concepts and possibilities of *compartment models*, in which either single organisms or single parcels of land (each quite small) are modelled over a period of time. Summation of all compartments, be they individual organisms or small areas of land, then equates to the ecosystem.

Using a mathematical, as opposed to a computer-based, framework, there are two classes of models, depending on whether the emphasis is on the organisms themselves or on probabilities that various things will happen. The former leads to the *population dynamic models* where each species is modelled, and the integration of the species' models leads to a model of the community. The latter leads to *statistical models*, whereby every subprocess within the overall successional process is assigned a probability. The statistical approach could, therefore, be argued to have moved furthest from reality since the models are no longer directly concerned with the organisms, or with the areas of space that the organisms are occupying, but are overwhelmingly concerned with the probabilities that rather abstract events will happen.

Plant Succession: Theory and prediction Edited by David C. Glenn-Lewin, Robert K. Peet and Thomas T. Veblen © 1992 Chapman & Hall, London ISBN 0 412 26900 7

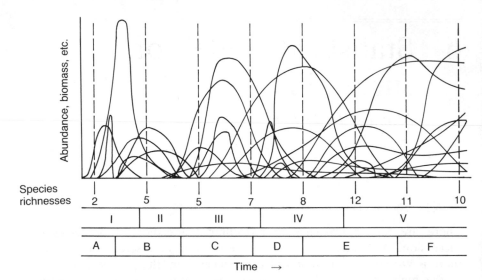

Figure 6.1 A hypothetical sequence of 30 species during a successional process (horizontal axis). The dashed vertical lines represent eight equally spaced 'snap-shots' of the sequence, and at each one of these times the species richness is given by the number of species lines crossing the vertical line. Below the species richnesses are two horizontal bars which could be used to divide the continuum into five or six separate states based either on dominance or collections of species: these divisions are discussed in the text.

To use a statistical model implies that the modeller can formulate the abstract nature of the event to be modelled. If one considers the process of gradual change from, say, bare ground to a forest community, then one must consider what changes are taking place, and how they are defined. The example in Fig. 6.1 illustrates some of the concepts in developing models. This example does not relate to a verbal model except in so far as the diversity, or species richness, is considered. Taking, arbitrarily, eight equally spaced divisions of the horizontal axis yields the species richness indicated in Fig. 6.1. The richness increases from 2 (having started at 0, it can only increase or stay the same) to 5, then increases to a maximum of 12, after which there is a slight decrease in species richness. A compartment model could be concerned with what was happening to any individual organism or to any particular piece of land, and, like the verbal model, is not clearly demonstrated in Fig. 6.1. The population dynamics model will be concerned with each of the 30 species represented by Fig. 6.1, with the rates of increase, with the carrying capacities of the environment, etc. For a statistical model one will be concerned with the probability that any parcel of land will change from one state to another state, but a fundamental problem is how to define these states.

State definition is an extremely arbitrary process, and it is essentially concerned with the age-old conundrum in biology of how to divide a continuum into a finite set of apparently discrete classes. One possible solution in Fig. 6.1 is to base the division on dominance, and hence one would probably divide the continuum into five classes, numbered I to V. Stage I has a single dominant species, and stage II also has one dominant species with one or two subdominants. However one views the continuum, a break is likely to be drawn between stages II and III: before this break there is a collection of species, which could be termed 'early successional', all of which have either died out or are about to die out. After this break a whole suite of new species is recorded, and stage III essentially has two dominant species, as does stage IV. The final stage, stage V, has dominance by one or more of a collection of three species. Another, and again arbitrary, way to divide the continuum is to look for collections of species (i.e. a classification) rather than dominance by one or two species. In attempting this, six stages, numbered A to F, are outlined in Fig. 6.1. The only place where the two methods coincide is in the division between stages II and III and between stages B and C. At this point there is a reasonably clear change in both the species complement and the pattern of dominance, and intuitively it is the only reasonable division of the continuum, though even here a fuzzy vertical line should probably be drawn.

The example shows that, for statistical modelling, it is essential *a priori* to define the states that are of interest to the modeller. Both dominance and classification have been used in Fig. 6.1. It will generally be helpful either to define the states in terms of something practical, such as dominance, or in terms of a numerical analysis, either using classification or ordination techniques. However, it is important to realize that whatever method is used for deciding states, even with the most objective methods of classification, both the choice of states and the number of states to be modelled are essentially subjective. The output from a statistical model is bound to be dependent on the choice of states to be modelled.

6.2   MATHEMATICAL BACKGROUND

6.2.1   Model formulation

A set of artificial, but nevertheless typical, data is shown in Table 6.1. This set of data will be used for the examples throughout section 6.2. The data can be viewed as being the observations on 50 quadrats in three consecutive years. Whatever the state of the vegetation, it is classified into one of three classes, A, B or C. Although there are, presumably, 50 randomly located quadrats, the fact that they were observed in a third season implies that observations are not statistically independent.

Table 6.1 Observations on three separate occasions of 50 sample points (for plants these would be quadrats, for animals they may be traps, water samples, etc.). Each observation has been classified into one of three states, designated A, B or C. The table gives, for each sample point, the time sequence of these states. The time step can be considered to be one year

Sample point	States			Sample point	States		
	Time 1	*Time 2*	*Time 3*		*Time 1*	*Time 2*	*Time 3*
1	A	A·	A	26	A	A	B
2	A	B	A	27	C	C	C
3	B	B	C	28	A	A	C
4	A	A	A	29	A	A	A
5	B	C	B	30	C	C	B
6	A	C	A	31	A	B	A
7	A	A	A	32	B	C	B
8	C	C	C	33	B	C	A
9	B	A	C	34	C	C	C
10	A	A	B	35	B	B	A
11	B	B	B	36	B	B	A
12	A	B	C	37	C	B	A
13	A	A	A	38	C	C	B
14	B	B	C	39	A	A	A
15	A	A	A	40	A	A	A
16	C	C	C	41	A	A	B
17	C	C	C	42	C	C	C
18	B	C	C	43	C	C	C
19	B	B	B	44	A	A	B
20	A	C	C	45	C	C	B
21	C	A	B	46	C	C	C
22	A	A	B	47	B	B	C
23	B	B	B	48	A	A	B
24	B	B	B	49	A	A	B
25	A	B	B	50	B	B	C

However, this lack of independence will be overlooked in sections 6.2.1 and 6.2.2, but it is of importance for the analyses in sections 6.2.3 and 6.2.4.

With 50 quadrats, each observed twice after the initial recording, there are a total of 100 annual transitions in the data. The first quadrat had a record

$$A \rightarrow A \rightarrow A$$

and hence scores two transitions from A to A. The second quadrat, with record

$$A \rightarrow B \rightarrow A$$

Table 6.2 The tally matrix derived from the 100 transitions that occur in the 50 sequences in Table 6.1

From state	To state			Row totals
	A	B	C	
A	24	12	4	40
B	6	15	9	30
C	3	6	21	30
Column totals	33	33	34	100

has a transition from A to B and another from B to A. A summary of the 100 transitions is given in Table 6.2: this can be put into matrix form, known as a *tally matrix*, thus

$$\begin{bmatrix} 24 & 12 & 4 \\ 6 & 15 & 9 \\ 3 & 6 & 21 \end{bmatrix}.$$

Dividing each row by the row total yields the *matrix of transition probabilities*. Thus, dividing the first row by 40, and the second and third rows each by 30, yields

$$\mathbf{P} = \begin{bmatrix} 0.6 & 0.3 & 0.1 \\ 0.2 & 0.5 & 0.3 \\ 0.1 & 0.2 & 0.7 \end{bmatrix}.$$

Each entry in $\mathbf{P}$, denoted p_{ij} for the element in the ith row and jth column, is the probability of observing a transition from the ith to the jth state. Thus, for a quadrat that is in state A, there is a probability of 0.6 (60%) that it will be in state A after a year has elapsed, a probability of 0.3 (30%) that it will be in state B, and a probability of 0.1 (10%) that it will be in state C.

$\mathbf{P}$ has two important mathematical uses, and one practical use. First, it can be used to predict what will happen to the system in the future. At the third inspection in Table 6.1 there were 15 quadrats in state A, 18 in state B and 17 in state C. These can be used as the elements of a row vector, known as the *state vector*, thus

$$\mathbf{p}_t = [15 \quad 18 \quad 17].$$

Post-multiplication of the state vector at time t by the matrix of transition probabilities yields the state vector at time $t + 1$, thus

$$\mathbf{p}_{t+1} = \mathbf{p}_t \mathbf{P} = [14.3 \quad 16.9 \quad 18.8] \tag{6.1}$$

or, to round to integer numbers of quadrats for biological reality,

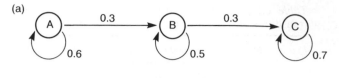

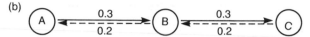

Figure 6.2 Diagrammatic representations of the matrix of transition probabilities based on Table 6.2. (a) All probabilities exceeding some arbitrary value, here 0.3, have been plotted. Note the conservative nature of the system since the probabilities of being in the same state at the end of a period of time are the largest. (b) The largest probabilities are indicated by continuous lines, and the next largest (0.2 to 0.3) by dashed lines (the probabilities of being in the same state have been omitted). In both diagrams the predominant flow from A to B to C is clearly seen.

$$\mathbf{p}_{t+1} = [14 \quad 17 \quad 19].$$

The process can be repeated, thus

$$\mathbf{p}_{t+2} = \mathbf{p}_{t+1}\mathbf{P} = [14 \quad 16 \quad 20] \tag{6.2}$$

where again the elements of $\mathbf{P}_{t+2}$ have been rounded to integers. Note that if equation (6.1) is inserted in equation (6.2), the result is

$$\mathbf{P}_{t+2} = \mathbf{p}_{t+1}\mathbf{P} = \mathbf{p}_t\mathbf{P}\mathbf{P} = \mathbf{p}_t\mathbf{P}^2$$

and this generalizes to

$$\mathbf{p}_{t+k} = \mathbf{p}_t\mathbf{P}^k \tag{6.3}$$

where k is any integer. One can, by the use of equation (6.3), predict the structure of the system any integer number of periods of time into the future.

The second use of $\mathbf{P}$ is to investigate when the process of change will stop. This is similar to saying that a succession has reached an endpoint or climax; the concept is illustrated in Worked Example 6.1.

The practical use of $\mathbf{P}$ is to portray, diagrammatically, what is occurring in the system of states that is being studied. One can arbitrarily choose what can be termed major transitions (i.e. probabilities larger than some level: 0.3 would be appropriate for matrix $\mathbf{P}$) and draw a diagram with these. Alternatively, one might also include a middle category, say between 0.2 and 0.3, to include in the diagram, as in Fig. 6.2. Such diagrams, used by Usher (1981), Greig-Smith (1983) and Digby and Kempton (1987), tend to highlight the important features of a system of states: for example, Fig. 6.2 clearly shows that there tends to be a sequence from A to B to C. The only rule for drawing such diagrams is to plot only sufficient probabilities so that the essential nature of the system becomes visually obvious. Usher (1981) found that about 15% of probabilities in the large class, 15% in the intermediate class, and 70% unplotted, was a reasonable compromise for clear visual presentation of a matrix of transition probabilities with eight states. These percentages would need to be decreased for matrices with more states, and increased for matrices with fewer states.

Worked Example 6.1: Convergence of a Markovian process to a stable endpoint

P is the matrix of transition probabilities that is defined in equations (6.1) and (6.3). Since all of the rows of **P** sum to 1, and since none of the elements of **P** can be negative, the Perron–Frobenius theorems, extended by Brauer (1957, 1962) to non-negative matrices (see Usher, 1966), indicate that the dominant eigenvalue of **P** will be 1. The theorems also indicate that this is associated with an eigenvector, all of whose elements can be chosen as positive numbers. The eigenvector is

$$\mathbf{p} = [9 \quad 11 \quad 14]$$

and it can be verified that

$$\mathbf{p} = \mathbf{pP}$$

i.e. that there is no change from one time to the next.

The eigenvector can be found by simple iteration. Guess that it is

$$\mathbf{p}_1 = [1 \quad 1 \quad 1],$$

and then calculate

$$\mathbf{p}_1 \mathbf{P} = [1 \quad 1 \quad 1]\begin{bmatrix} 0.6 & 0.3 & 0.1 \\ 0.2 & 0.5 & 0.3 \\ 0.1 & 0.2 & 0.7 \end{bmatrix} = [0.9 \quad 1.0 \quad 1.1].$$

Choosing the first element as an approximation to the eigenvalue, divide all the elements by it to give the first approximation to the eigenvector, thus

$$\mathbf{p}_2 = [1 \quad 1.0/0.9 = 1.11 \quad 1.1/0.9 = 1.22].$$

Then repeat the process, rounding to two decimal places, to give

$$\mathbf{p}_2 \mathbf{P} = [0.94 \quad 1.10 \quad 1.29],$$

with further approximations of 0.94 (the first element in the vector) for the eigenvalue and of the eigenvector as

$$\mathbf{p}_3 = [1 \quad 1.17 \quad 1.37].$$

Continuing the process gives

$\mathbf{p}_4 = [1 \quad 1.20 \quad 1.45]$
$\mathbf{p}_5 = [1 \quad 1.21 \quad 1.50]$
$\mathbf{p}_6 = [1 \quad 1.21 \quad 1.53]$
$\mathbf{p}_7 = [1 \quad 1.22 \quad 1.54]$
$\mathbf{p}_8 = [1 \quad 1.22 \quad 1.55]$
$\mathbf{p}_9 = [1 \quad 1.22 \quad 1.56]$.

All subsequent iterations are the same as $\mathbf{p}_9$. Multiplication by 9 turns all of the elements into integers, giving the eigenvector $\mathbf{p}$ above.

As the elements of $\mathbf{p}$ sum to 34, multiplication of the elements by 50/34, and rounding to the nearest integers, yields

$$\mathbf{p} = [13 \quad 16 \quad 21]$$

which is the eigenvector, whose elements sum to 50, that indicates the endpoint, i.e. when the system of 50 quadrats will change no more given the transition probabilities in matrix $\mathbf{P}$. The dominant eigenvector of $\mathbf{P}$ can thus be taken to represent the abundance of each of the states at the end of a successional sequence.

6.2.2   Tests of significance

In many processes that are essentially random in nature it can be observed that previous events influence, but do not rigidly control, subsequent events. Such processes are named *Markov processes*. Most simply, a Markov process can be defined as one in which the probability of being in a given state at some particular time can be deduced from a knowledge of the immediately preceding state. One particular form of Markov process is the *Markov chain*, which is a sequence, in time, of discrete states in which the probability of transition to any particular state at the next step in the chain depends only on the previous state. General introductions to Markov chains include those of Harbaugh and Bonham-Carter (1970) for applications in geology and Collins (1975) for applications in geography. A general introduction to matrix models in population biology is given by Caswell (1989). Despite the simplicity of these definitions, there is disagreement in the statistical literature about the definition of a Markov process (Usher, 1979).

As will be recognized, the model development in section 6.2.1 fits easily into the framework of Markov chains. This framework provides a basis for the first test of significance that may be appropriate when using successional data. Suppose that the matrix of transition probabilities in section 6.2.1 had had the form

$$\mathbf{P} = \begin{bmatrix} 0.6 & 0.3 & 0.1 \\ 0.6 & 0.3 & 0.1 \\ 0.6 & 0.3 & 0.1 \end{bmatrix},$$

where each row is the same. This indicates that the probability of a transition to any state does not depend on the present state, i.e. that there is always a probability of 0.6 that the system will be in state A irrespective of whether it was in state A, B or C in the previous time period. Thus, the matrix **P** above does not have a Markov property, and indeed it demonstrates a totally random process such as a child drawing coloured sweets out of a box (Usher, 1979).

A test of independence in a matrix of transition probabilities was proposed by Anderson and Goodman (1957). The null hypothesis, that successive steps in the sequence are statistically independent (i.e. as in the random process above), is tested against the alternative hypothesis that the steps are not independent (and therefore that they could form, but not necessarily are, a first-order Markov chain). Using the data in Tables 6.1 and 6.2, the tally matrix is

$$\mathbf{N} = \begin{bmatrix} 24 & 12 & 4 \\ 6 & 15 & 9 \\ 3 & 6 & 21 \end{bmatrix} \tag{6.4}$$

and the transition probability matrix is

$$\mathbf{P} = \begin{bmatrix} 0.6 & 0.3 & 0.1 \\ 0.2 & 0.5 & 0.3 \\ 0.1 & 0.2 & 0.7 \end{bmatrix}. \tag{6.5}$$

The appropriate test statistic is

$$-2(\ln \lambda) = 2 \sum_{i=1}^{m} \sum_{j=1}^{m} n_{ij} \ln(p_{ij}/p_j) \tag{6.6}$$

where 'ln' indicates the natural logarithm, m is the number of states in the system, n_{ij} is the element in the ith row and jth column of **N**, p_{ij} is the corresponding element in **P**, and p_j is the marginal probability of the jth column of **N**, given by

$$p_j = \sum_{i=1}^{m} n_{ij} \bigg/ \left(\sum_{i=1}^{m} \sum_{j=1}^{m} n_{ij} \right). \tag{6.7}$$

To apply this test to the example, $m = 3$ and both n_{ij} and p_{ij} are given in equations (6.4) and (6.5) respectively. Summing the columns of **N** in equation (6.4) gives values of $p_1 = 33/100$ (=0.33), $p_2 = 33/100$ (=0.33) and $p_3 = 34/100$ (=0.34), using equation (6.7). Thus, equation (6.6) is evaluated as

$$\begin{aligned}
-2(\ln \lambda) = 2 \, \{ &24 \ln(0.6/0.33) + 12 \ln(0.3/0.33) + 4 \ln(0.1/0.34) \\
&+ 6 \ln(0.2/0.33) + 15 \ln(0.5/0.33) + 9 \ln(0.3/0.34) \\
&+ 3 \ln(0.1/0.33) + 6 \ln(0.2/0.33) + 21 \ln(0.7/0.34) \} \\
= \, &37.978.
\end{aligned}$$

Anderson and Goodman (1957) show that $-2(\ln \lambda)$ is distributed asymptotically as χ^2 with $(m - 1)^2$ degrees of freedom. Reference to tables of χ^2 with four degrees of freedom indicates that the probability of observing $\chi^2 = 37.978$ is very much less than one in a thousand ($p \ll 0.001$), and hence the null hypothesis of statistical independence can be rejected.

It is important to remember that this test does *not* indicate that the process, shown in Tables 6.1 and 6.2, is a first-order Markov chain. All it does show is that the process is not completely random. If the probability of χ^2 was, say, greater than 5% ($p > 0.05$), then it would be possible to accept the null hypothesis that the process was random.

6.2.3   Order of process

Returning to the data in Table 6.1, one can derive three different transition probability matrices on the basis of the state of the system when it was first observed. Thus, if the state of the system was A at the first observation, the appropriate tally matrix and transition probability matrices between the second and third observations are:

$$\mathbf{N}_A = \begin{bmatrix} 8 & 7 & 1 \\ 2 & 1 & 1 \\ 1 & 0 & 1 \end{bmatrix}, \quad \mathbf{P}_A = \begin{bmatrix} 0.50 & 0.44 & 0.06 \\ 0.50 & 0.25 & 0.25 \\ 0.50 & 0 & 0.25 \end{bmatrix}. \tag{6.8}$$

The probabilities in $\mathbf{P}_A$ now represent the transition probabilities from state i to state j given that the system was in state A in the previous period of time. Thus, given that the quadrat was in state A at time $(t - 1)$, there is a probability of 0.50 that it will still be in state A after the time interval t to $(t + 1)$, and a probability that it will change from state A to state B of 0.44, etc. Similar matrices can be written for state B at time $(t - 1)$, thus

$$\mathbf{N}_B = \begin{bmatrix} 0 & 0 & 1 \\ 2 & 4 & 4 \\ 1 & 2 & 1 \end{bmatrix}, \quad \mathbf{P}_B = \begin{bmatrix} 0 & 0 & 1 \\ 0.20 & 0.40 & 0.40 \\ 0.25 & 0.50 & 0.25 \end{bmatrix}. \tag{6.9}$$

Similarly, for state C at time $(t - 1)$,

$$\mathbf{N}_C = \begin{bmatrix} 0 & 1 & 0 \\ 1 & 0 & 0 \\ 0 & 3 & 8 \end{bmatrix}, \quad \mathbf{P}_C = \begin{bmatrix} 0 & 1 & 0 \\ 1 & 0 & 0 \\ 0 & 0.27 & 0.73 \end{bmatrix}. \tag{6.10}$$

If the Markov chain is a first-order process – in other words, if the transition probabilities are dependent only on the current state of the system and not on any previous states of the system – then the three matrices $\mathbf{P}_A$, $\mathbf{P}_B$ and $\mathbf{P}_C$ should be identical to each other and to matrix $\mathbf{P}$ in equation (6.5).

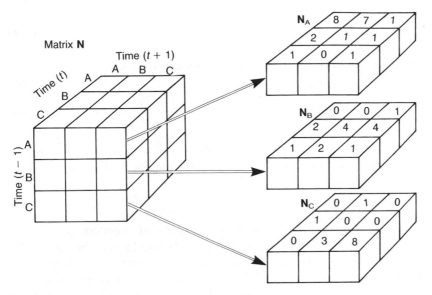

Figure 6.3 A diagrammatic representation of a three-dimensional matrix. The matrix at the left is denoted **N** in the text, and consists of three submatrices N_A, N_B and N_C (listed in equations (6.8) to (6.10)). In calculating p_{jk} in equation (6.12), the numerator terms are sums of columns through the **N** matrix (for example when $j = k = 1$, the column sum is 8 1 0 1 0 = 8), and the denominator terms are sums of vertical slices going back through this matrix (the sums are 15 for the left-hand slice, 18 for the central slice, and 17 for the right-hand slice).

Clearly, in formulating the three matrices N_A, N_B and N_C in equations (6.8) to (6.10) only 50 of the 100 transitions shown in Table 6.1 could be used since the first observation indicated to which of the three tally matrices any transition between the second and third observations should belong. There are, hence, far too few data available for accurate estimation of the transition probabilities in matrices P_A, P_B and P_C. However, the example does show that if models of ecological succession, which are of a second- or higher order process, are required, then large amounts of data are needed so as to obtain reasonable estimates of transition probabilities.

A test of significance is required to determine whether or not the **P** matrices are identical. Anderson and Goodman (1957) provide an appropriate test, but first it is essential to generalize the matrices in equations (6.4) and (6.5) from two dimensions to three dimensions. This is done for matrices N_A, N_B and N_C in Fig. 6.3. An element in the three-dimensional matrix **N** is now of the form n_{ijk}, indicating the number of transitions observed from state i to state j to state k. Notice that, as there are 50 entries in Table 6.1, the numbers in the cells in Fig. 6.3 add up to 50. Anderson and Goodman's test is

$$-2(\ln \lambda) = 2 \sum_{i=1}^{m} \sum_{j=1}^{m} \sum_{k=1}^{m} n_{ijk} \ln(p_{ijk}/p_{jk}) \tag{6.11}$$

where the symbols are as defined previously except that p_{jk} is the marginal probability given by

$$p_{jk} = \sum_{i=1}^{m} n_{ijk} \Big/ \sum_{i=1}^{m} \sum_{k=1}^{m} n_{ijk}, \tag{6.12}$$

and $-2(\ln \lambda)$ is asymptotically distributed as χ^2 with $m(m-1)^2$ degrees of freedom. The calculations involved in equation (6.11) are shown in Worked Example 6.2, from which

$$-2(\ln \lambda) = 17.502. \tag{6.13}$$

This is approximately a χ^2 with 12 degrees of freedom and has a probability of greater than 1 in 20 ($p > 0.05$). Hence, based on the extremely small sample size, the null hypothesis is not rejected, implying that the matrices $\mathbf{P_A}$, $\mathbf{P_B}$ and $\mathbf{P_C}$ in equations (6.8) to (6.10) can be considered as identical with matrix $\mathbf{P}$ in equation (6.5) (allowing, of course, for sampling variation). This implies that the process in Table 6.1 can be viewed as a first-order Markov chain.

In this example the probability of being in any state at time $(t + 1)$ is dependent only on the state at time t, and hence the chain has *single-dependence* and transitions involve a *single step* of unit length (one interval of time). Dependence, order and step can all vary. Thus, equation (6.13) tested the data in Table 6.1 for a double-dependence chain on the two immediately preceding states: if the value of χ^2 had been significant then the chain could be considered to have had double-dependence, it would have been of second order, and would have been of single step. Suppose that the state of a chain at time $(t + 1)$ was dependent on the state at time t and at time $(t - 2)$, then such a chain would have double-dependence, it would be of third order, and would have mixed step lengths. Tests for such a chain could be formulated as in equations (6.11) and (6.12): however, with the data given in Table 6.1, the test in equation (6.13) is the only alternative that can be formulated to a chain of single-dependence, first order and single step length.

Worked Example 6.2: The calculation involved in testing for the order of the Markov process in Table 6.1, using equations (6.11) and (6.12)

The notation is explained in the text. Since some probabilities, p_{ijk}, are zero, their expressions are indicated by an asterisk and they contribute nothing to the total.

i	j	k	$n_{ijk} \ln (p_{ijk}/p_{jk})$	
1	1	1	8 ln (0.50/(8/15))	= −0.516
1	1	2	7 ln (0.44/(8/18))	= −0.070
1	1	3	1 ln (0.06/(2/17))	= −0.673
1	2	1	2 ln (0.50/(5/15))	= 0.811
1	2	2	1 ln (0.25/(5/18))	= −0.105
1	2	3	1 ln (0.25/(5/17))	= −0.163
1	3	1	1 ln (0.50/(2/15))	= 1.322
1	3	2	*	0
1	3	3	1 ln (0.50/(10/17))	= −0.163
2	1	1	*	0
2	1	2	*	0
2	1	3	1 ln (1.00/(2/17))	= 2.140
2	2	1	2 ln (0.20/(5/15))	= −1.022
2	2	2	4 ln (0.40/(5/18))	= 1.459
2	2	3	4 ln (0.40/(5/17))	= 1.230
2	3	1	1 ln (0.25/(2/15))	= 0.629
2	3	2	2 ln (0.50/(5/18))	= 1.176
2	3	3	1 ln (0.25/(10/17))	= −0.856
3	1	1	*	0
3	1	2	1 ln (1.00/(8/18))	= 0.811
3	1	3	*	0
3	2	1	1 ln (1.00/(5/15))	= 1.099
3	2	2	*	0
3	2	3	*	0
3	3	1	*	0
3	3	2	3 ln (0.27/(5/18))	= −0.085
3	3	3	8 ln (0.73/(10/17))	= 1.727
Total				8.751

6.2.4 Stationarity

The data in Table 6.1 were collected at three instances in time, and hence over two time periods. This then means that the question can be posed: 'Are the transition probabilities the same for each of the two time periods?' Using the 50 transitions between the first and second observations, the tally and transition probability matrices are

$$\mathbf{N}_1 = \begin{bmatrix} 16 & 4 & 2 \\ 1 & 10 & 4 \\ 1 & 1 & 11 \end{bmatrix}, \quad \mathbf{P}_1 = \begin{bmatrix} 0.73 & 0.18 & 0.09 \\ 0.07 & 0.67 & 0.27 \\ 0.08 & 0.08 & 0.85 \end{bmatrix}, \tag{6.14}$$

where, due to rounding errors to two decimal places, it will be seen that not all of the rows of $\mathbf{P}_1$ sum exactly to unity. A similar pair of matrices can be formulated for the transitions between the second and third observations: these are

$$\mathbf{N_2} = \begin{bmatrix} 8 & 8 & 2 \\ 5 & 5 & 5 \\ 2 & 5 & 10 \end{bmatrix}, \quad \mathbf{P_2} = \begin{bmatrix} 0.44 & 0.44 & 0.11 \\ 0.33 & 0.33 & 0.33 \\ 0.12 & 0.29 & 0.59 \end{bmatrix}, \tag{6.15}$$

where again the rows of $\mathbf{P_2}$ do not all sum to unity due to rounding errors. A Markov chain is said to be *stationary* if the elements of the transition probability matrix do not change over time. In this example, this would imply that $\mathbf{P_1}$ and $\mathbf{P_2}$ were identical, and were also identical to $\mathbf{P}$ in equation (6.5).

In a test of significance, the null hypothesis is that a series of T transition probability matrices $\mathbf{P}(t)$, $t = 1, 2, \ldots, T$, are the same. The appropriate test, derived from Anderson and Goodman (1957), is

$$-2(\ln \lambda) = 2 \sum_{i=1}^{m} \sum_{j=1}^{m} \sum_{t=1}^{T} n_{ij}(t) \ln(p_{ij}(t)/p_{ij}) \tag{6.16}$$

where the symbols are as previously defined except that p_{ij} is the element in the ith row and jth column of $\mathbf{P}$ in equation (6.5). Again $-2(\ln \lambda)$ is distributed asymptotically as a χ^2 with $m(m - 1)(T - 1)$ degrees of freedom. The calculations involved in applying equation (6.16) to the example in Table 6.1 are shown in Worked Example 6.3. From this it will be seen that

$$-2(\ln \lambda) = 11.152$$

which, when treated as a χ^2 with six degrees of freedom, has a probability slightly in excess of 0.05. Hence, the null hypothesis that the elements $p_{ij}(t)$ are constant, with value p_{ij}, would not be rejected, and the Markov chain in Table 6.1 would be treated as stationary.

Stationarity, as described above, relates to a temporal process, and hence the Markov chain can be described as being stationary in time. It is also possible to think of spatial processes and investigate stationarity in space. Suppose that quadrats 1 to 25 in Table 6.1 had been collected in one field and quadrats 26 to 50 in another field. Treating these as the left- and right-hand sides of Table 6.1, one could derive two $\mathbf{P}$ matrices with elements of the form p_{ij} (left) and p_{ij} (right) and use equation (6.16) to test for stationarity in space. Undertaking such an analysis yields the following two tally matrices:

$$\mathbf{N}(\text{left}) = \begin{bmatrix} 12 & 6 & 3 \\ 2 & 11 & 5 \\ 2 & 1 & 8 \end{bmatrix}, \quad \mathbf{N}(\text{right}) = \begin{bmatrix} 12 & 6 & 1 \\ 4 & 4 & 4 \\ 1 & 5 & 13 \end{bmatrix},$$

and a χ^2 with six degrees of freedom of 6.217. A null hypothesis that the left- and right-hand sides are the same could not, therefore, be rejected. Whereas in ecological succession many studies have concentrated on the temporal aspects of the process, very few studies have yet investigated spatial aspects though, as the test above demonstrates, spatial aspects can be investigated simply. To do so would contribute to the resolution of several problems in succession theory (e.g. whether patterns of change are repeated in space).

It may also be appropriate to investigate stationarity in space and time jointly. Again using the data in Table 6.1, and separating the first and second time periods and the left- and right-hand sides as a spatial effect, four separate tally matrices can be formulated, thus:

$$\mathbf{N}(\text{time 1, left}) = \begin{bmatrix} 7 & 3 & 2 \\ 1 & 6 & 2 \\ 1 & 0 & 3 \end{bmatrix}, \quad \mathbf{N}(\text{time 1, right}) = \begin{bmatrix} 9 & 1 & 0 \\ 0 & 4 & 2 \\ 0 & 1 & 8 \end{bmatrix}$$

$$\mathbf{N}(\text{time 2, left}) = \begin{bmatrix} 5 & 3 & 1 \\ 1 & 5 & 3 \\ 1 & 1 & 5 \end{bmatrix}, \quad \mathbf{N}(\text{time 2, right}) = \begin{bmatrix} 3 & 5 & 1 \\ 4 & 0 & 2 \\ 1 & 4 & 5 \end{bmatrix}.$$

Use of equation (6.16) yields $\chi^2 = 30.255$ with 18 degrees of freedom. From the earlier analyses this can be apportioned into a time component ($\chi^2_{[6]} = 11.152$), a spatial component ($\chi^2_{[6]} = 6.217$) and a space–time interaction ($\chi^2_{[6]} = 12.886$). The latter χ^2 is significant with $p < 0.05$, indicating that there is some heterogeneity in these space–time matrices.

Worked Example 6.3: A test of the stationarity in time of the Markov process shown in Table 6.1

The symbols used are all defined in the text. Note that values of $p_{ij}(t)$ are given in equations (6.14) and (6.15), and below, with two decimal places, whereas in calculating the expression in the final column below four decimal places have been used.

t	i	j	$n_{ij}(t) \ln (p_{ij}(t)/p_{ij})$
1	1	1	16 ln (0.73/0.6) = 3.079
1	1	2	4 ln (0.18/0.3) = −2.004
1	1	3	2 ln (0.09/0.1) = −0.191
1	2	1	1 ln (0.07/0.2) = −1.098
1	2	2	10 ln (0.67/0.5) = 2.877
1	2	3	4 ln (0.27/0.3) = −0.471
1	3	1	1 ln (0.08/0.1) = −0.263
1	3	2	1 ln (0.08/0.2) = −0.956
1	3	3	11 ln (0.85/0.7) = 2.086
2	1	1	8 ln (0.44/0.6) = −2.402
2	1	2	8 ln (0.44/0.4) = 3.144
2	1	3	2 ln (0.11/0.1) = 0.211
2	2	1	5 ln (0.33/0.2) = 2.554
2	2	2	5 ln (0.33/0.5) = −2.028
2	2	3	5 ln (0.33/0.3) = 0.526
2	3	1	2 ln (0.12/0.1) = 0.324
2	3	2	5 ln (0.29/0.2) = 1.928
2	3	3	10 ln (0.59/0.7) = −1.740
Total			5.576

6.2.5 A review of applications

Markovian models have found their widest use in studies of change in forests (also see Chapter 5). Stephens and Waggoner (1970) and Waggoner and Stephens (1970) used data derived on a 10-year time interval for an unmanaged mixed hardwood stand in Connecticut. They used five states, defined on the basis of dominance by maple, oak or birch, with two states relating to 'minor' species and 'other' species. They used a transition probability matrix, which Usher (1979) tested for independence ($X^2_{[16]} = 468$). The results for this woodland were brought up to date with observations over a further 10-year period (Stephens and Waggoner, 1980). Horn (1975a,b, 1976) used a similar model for mixed hardwood stands in New Jersey, except that his observations were not at two time intervals. He observed canopy trees and then the seedlings and saplings beneath them, and estimated transition probabilities on the basis of the relative abundance of small trees beneath large trees. He defined his states on the basis of dominance, using categories such as gray birch, red maple, beech, etc. Usher (1979) also showed that Horn's matrices were not independent ($\chi^2_{[100]} = 2428$). A modified method of estimating transition probabilities was used by Tucker and Fitter (1981) when they employed a Markovian model to predict the course of succession in a woodland nature reserve. Again states were defined in relation to dominance by a single species, but they were able to show that birch was likely to decline in abundance dramatically, compensated for by a large increase in oak abundance, a small increase in willow, and virtually no change in abundance of the relatively scarce alder.

Heathland vegetational changes, especially after burning for management purposes, have also been modelled. In the Netherlands, Lippe *et al.* (1985) attempted to use Markovian models for studying the succession on *Empetrum nigrum* heathlands, using a matrix with nine states. However, temporal changes in the transition probability matrix over a 19-year period clearly indicated that the process was not stationary. In Scotland, the number of states used was large since dominance of small quadrats by different species, including heathers (*Erica cinerea* and *Calluna vulgaris*), other dwarf shrubs (*Vaccinium vitis-idaea* and *Arctostaphylos uva-ursi*) and herbs, mosses and lichens, as well as combinations of these species or groups, were used to categorize the states. The resulting transition probability matrices were unusual due to the large number of zero elements (see, for example, Gimingham *et al.* (1981) and Hobbs and Legg (1983)). These authors tested for independence using the $-2(\ln \lambda)$ test, but the validity of such tests when a large number of matrix elements is zero is unknown. (Hobbs and Legg (1983) quote $\chi^2_{[169]} = 1112$ for their matrix which contained only 48% of non-zero elements.) An interesting extension of the ideas of modelling systems subjected to burning is described by Isagi and Nakagoshi (1990). Their Markovian matrix contained two series of probabilities, an s-series relating to the successional change between states and a b-series indicating the probability of a fire and hence rejuvenation of the successional sequence. This therefore presented a more holistic model for an ecosystem subject to periodic fires rather than a model just of the recovery process after fire.

A move away from dominance to classifications as a means of describing the states to be included in Markov models can be seen in the studies of lawns by Austin (1980) and Austin and Belbin (1981) and of grasslands by Usher (1981). Although these studies used either ordination or cluster analysis techniques, there is still a marked use of dominance in describing the states that were modelled. Austin (1980) was able to show that drought had an impact on the transition probability matrix, and hence that the process was not stationary, while Austin and Belbin (1981) concluded that spatial heterogeneity on a lawn made the predictions of Markov models less reliable.

Moving away from studies of plant communities, van Hulst (1979) applied Markov models to information gathered from pollen diagrams. Slatyer (1977) postulated that transition matrix models would be useful for formulating hypotheses about the mechanism of succession (discussed further in section 6.5) and that they would have a role in making forecasts that would be useful in the management of ecosystems. Applications in animal ecology are much scarcer. Usher (1975) and Usher and Parr (1977) used these models to investigate the wood-feeding termite community of a cleared forest site in West Africa. Fortunately, there was nearly always only one species of termite on a baitwood block at any one time, and hence states could be defined in terms of species, or genera, except for two states which related to 'other species' (18 other species were too infrequently found to merit states in their own right) and 'unknown species' (where a termite was no longer present but presence was inferred from damage to the baitwood). The dominant eigenvector of the transition probability matrix was

[354   29   62   72   30   23   312],

a reasonably close approximation to the state vector after a long period of use of the test site. Since application of equation (6.6) gives $\chi^2_{[36]} = 610$, the termite process can be viewed as Markovian rather than purely random.

Grahame's (1976) data on the zooplankton in Kingston Harbour, Jamaica, are particularly interesting since they are truly random data. Usher (1979) analysed both the complete data set, and the data for each sampling station, and in no case was the value of χ^2, based on equation (6.6), significant.

Markov models have also been used to model the interactions between predators and prey. Usher (1979) attempted such an analysis using the classical data published by Huffaker (1958) on the orange-feeding six-spotted mite, *Eutetranychus sexmaculatus*, and its predator, *Typhlodromus occidentalis*. More recently, Woolhouse (1983) applied this modelling approach to a pest of apple orchards in Canada. The pest was the European red mite, *Panonychus ulmi*, which has two predators, *Zetzellia mali* and *Amblyseius fallacis*. States for the transition probability matrix were defined in terms of the density of the prey population and the presence or absence of predators. The largest probabilities show a cycle between prey abundance, then the presence of predators, then a crash in prey density, followed by absence of predators, and then a repeat of the cycle.

In order to illustrate the application of Markov models more comprehensively, one particular example is demonstrated. The example is based on a grassland data set used by Reed (1980) and Usher (1981).

6.3   AN EXAMPLE: THE BRECKLAND GRASSLANDS

A large number of small quadrats was repeatedly surveyed, on a year-to-year basis, in the Breckland grassland of southeast England (see, for example, Watt, 1960a,b, 1962, 1971, 1981a,b). The example is drawn from two permanent transects, one of which was grazed (mainly by rabbits) and one of which was fenced to exclude grazing. Each transect contained 128 contiguous quadrats, each of which measured 10 cm across the transect by 1.25 cm along the transect. Each quadrat was divided into eight 1.25 cm square subquadrats, and species' presence or absence in each subquadrat was noted. The published data concern frequencies of species at the quadrat level, and hence all frequencies are in the range 0 (the species was absent from the quadrat) to 8 (the species was present in each of the subquadrats). The published data are for the years 1936 to 1957 inclusive, except for 1950 when no observations were made.

The data consist of records of six taxa. Three of these are grasses, *Festuca ovina*, *Aira praecox* and *Agrostis* spp., one is the woodrush *Luzula campestris*, and the other two taxa are herbs, *Rumex tenuifolius* and *Galium saxatile*. The data matrices are extensive, consisting of 2688 observations (128 quadrats each measured in 21 years), for each of which there are frequencies for the six taxa of higher plants. Using a relocation method in the CLUSTAN classification package, Reed (1980) analysed each data set, the grazed and the ungrazed, separately. The dendrogram for the grazed grassland is shown in Fig. 6.4, where, arbitrarily, two divisions into 3 and 7 groups have been selected. Details of the mean frequencies of the taxa in these groups are given in Table 6.3.

Group G of the grazed plot is dominated by *Galium*, more strongly so in group G1 than in group G2, which has a slightly more diverse flora. Group F is dominated by *Festuca*, all the other taxa being relatively infrequent. Group M is the most diverse, and is relatively heterogeneous:

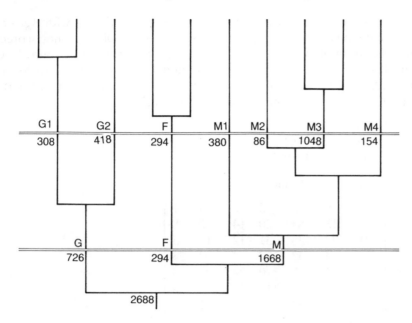

Figure 6.4 A dendrogram showing the final agglomeration of 2688 quadrats of grazed Breckland grassland data from ten groups to one group. The (unmarked) vertical axis is a measure of similarity. The two horizontal breaks indicate the 3- and 7-state systems that are discussed in the text. The letters above each break indicate the naming of the groups in the text, and the numbers below the break indicates the number of quadrats assigned to each group. The dendrogram is taken from Reed (1980).

Table 6.3 Using Watt's (1960a,b) data on the Breckland grasslands, Reed (1980) undertook a classification of the community into either 3 or 7 states (cf. Fig. 6.4). In this table the 3-state system is indicated by letters in brackets, the 7-state system by letters and numbers without brackets. The table lists the mean frequencies of the six taxa of higher plants in all quadrats included within a state (range of frequencies is 0 to a maximum of 8)

Taxa	States								
	F & (F)	G1	G2	(G)	M1	M2	M3	M4	(M)
Agrostis spp.	0.5	0.4	0.7	0.6	3.9	0.5	0.6	1.1	1.4
Aira praecox	0.1	0	0.1	0	0.3	0	0.2	0.3	0.2
Festuca ovina	3.5	0.1	0.2	0.1	0.1	0.1	0.1	0.2	0.1
Galium saxatile	0.4	6.5	3.1	4.5	0.3	0.7	0.2	0.3	0.2
Luzula campestris	0.4	0.2	0.3	0.3	0.2	2.6	0.2	0.3	0.3
Rumex tenuifolius	0.2	0.1	0.2	0.2	0.3	0.2	0.2	2.8	0.4

M1 is dominated by *Agrostis*, M2 by *Luzula*, M4 by *Rumex* with *Agrostis* as a subdominant, and M3 possibly by lichens (which were not scored) since the mean abundance of all higher plant taxa is small. Groups F, G and M are used for a 3-state Markov model and groups G and M are subdivided, as in Table 6.3, for a 7-state model. Tally matrices for the states in the order that they appear in Table 6.3 are

$$\mathbf{N}_3 = \begin{bmatrix} 179 & 11 & 24 \\ 32 & 404 & 264 \\ 61 & 198 & 1259 \end{bmatrix}$$

for the 3-state model, and

$$\mathbf{N}_7 = \begin{bmatrix} 179 & 6 & 5 & 2 & 2 & 19 & 1 \\ 11 & 113 & 83 & 20 & 14 & 58 & 8 \\ 21 & 72 & 136 & 31 & 12 & 116 & 5 \\ 14 & 4 & 32 & 177 & 4 & 94 & 23 \\ 4 & 1 & 7 & 0 & 24 & 28 & 3 \\ 37 & 18 & 87 & 119 & 16 & 624 & 48 \\ 6 & 12 & 37 & 14 & 7 & 41 & 37 \end{bmatrix}$$

for the 7-state model. It can be seen that $\mathbf{N}_3$ can be derived from $\mathbf{N}_7$ by adding together the second and third rows and columns as well as adding together the fourth to seventh rows and columns. The matrices of transition probabilities are

$$\mathbf{P}_3 = \begin{bmatrix} 0.84 & 0.05 & 0.11 \\ 0.05 & 0.58 & 0.38 \\ 0.04 & 0.13 & 0.83 \end{bmatrix}$$

and

$$\mathbf{P}_7 = \begin{bmatrix} 0.84 & 0.03 & 0.02 & 0.01 & 0.01 & 0.09 & 0 \\ 0.04 & 0.37 & 0.27 & 0.07 & 0.05 & 0.19 & 0.03 \\ 0.05 & 0.18 & 0.35 & 0.08 & 0.03 & 0.30 & 0.01 \\ 0.04 & 0.01 & 0.09 & 0.51 & 0.01 & 0.27 & 0.07 \\ 0.06 & 0.01 & 0.10 & 0 & 0.36 & 0.42 & 0.04 \\ 0.04 & 0.02 & 0.09 & 0.13 & 0.02 & 0.66 & 0.05 \\ 0.04 & 0.08 & 0.24 & 0.09 & 0.05 & 0.27 & 0.24 \end{bmatrix}$$

respectively. Using the largest probabilities, the diagram in Fig. 6.5 indicates three important aspects of the sequences observed in this vegetation. First, the system is conservative; in general, the largest probabilities lie along the main diagonal of $\mathbf{P}_7$. Second, state F is extremely isolated, being linked in Fig. 6.5 to no other state. Third, there is a tendency to 'home in' on state M3 as many arrows lead to it and only one arrow leads away from it.

Use of equation (6.6) yields values of $\chi^2 = 1795$ for the 7-state matrix (36 degrees of freedom) and $\chi^2 = 1220$ for the 3-state matrix (4 degrees of

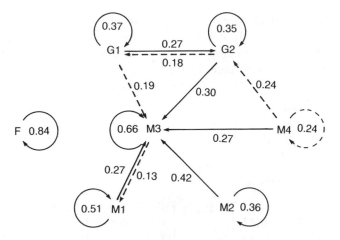

Figure 6.5   A diagrammatic representation of the main transition probabilities in the first-order Markov model for the 7-state system indicated in Fig. 6.4. Continuous lines indicate $p_{ij} > 0.25$, and dashed lines $0.125 \leqslant p_{ij} \leqslant 0.25$. Probabilities smaller than 0.125 have been omitted.

freedom). The null hypothesis of independence would be rejected in both cases. Assuming to begin with that it is a first-order process, the dominant eigenvectors of $\mathbf{P}_3$ and $\mathbf{P}_7$ are

[20   20   60]   and   [20   7   13   14   3   38   5]

respectively. These eigenvectors have been adjusted so that the elements represent percentages of the various states. In both cases the ratio of the largest to the second largest eigenvalue is 1.26, which indicates that there would be a relatively fast approach to a state vector similar in com-position to the eigenvectors. (The ratio of the first two eigenvectors indicates the speed of approach to the stable structure in Markov models.) Mean recurrence times can also be calculated; these are the mean times taken for a quadrat in a particular state to be back in that state again. These are

[4.9   4.8   1.7]

for the 3-state system (the numbers represent years, since the time step of the model is 1 year). For the 7-state system they are

[4.9   14.9   7.7   7.1   30.5   2.6   22.0].

Thus, for state M2 which is shown in Fig. 6.4 to be the least common, it is only likely to occupy a particular quadrat once every 30.5 years, on average. Interestingly, the predictions are that all states will remain in the community, even though two of the states (M2 and M4) will be uncommon.

In order to investigate the order of the process, only the 3-state system will be analysed. The only system tested in these data is a second-order process with double-dependence and a single time step, using transitions in observations over three consecutive years. If F was the state in the first of these three years, the transition probability matrix is

$$\mathbf{P}_F = \begin{bmatrix} 0.88 & 0.04 & 0.08 \\ 0.20 & 0.70 & 0.10 \\ 0.39 & 0.06 & 0.56 \end{bmatrix}$$

The other two matrices, with quadrats starting in states G and M in the first of the three years, are

$$\mathbf{P}_G = \begin{bmatrix} 0.90 & 0.03 & 0.06 \\ 0.05 & 0.48 & 0.47 \\ 0.03 & 0.13 & 0.84 \end{bmatrix}, \quad \mathbf{P}_M = \begin{bmatrix} 0.77 & 0.02 & 0.21 \\ 0.07 & 0.61 & 0.31 \\ 0.03 & 0.11 & 0.85 \end{bmatrix}.$$

Using the test in equation (6.11) gives $\chi^2 = 83.9$ with 12 degrees of freedom, and hence the null hypothesis that these three matrices are identical would be rejected. Inspection of the matrices shows that $\mathbf{P}_G$ and $\mathbf{P}_M$ are relatively similar but that $\mathbf{P}_F$ is different. If the system has once been in state F and changed to state G or M, then the probability of it

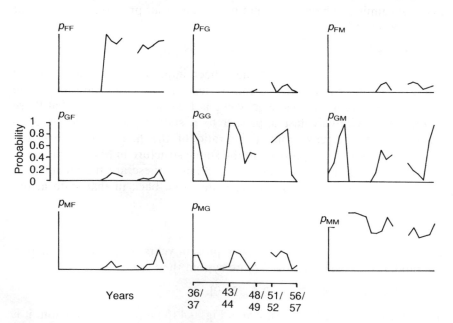

Figure 6.6 Transition probabilities for the 3-state system indicated in Fig. 6.4 as estimated on a year-to-year basis. The first probabilities plotted represent the transition probabilities between 1936 and 1937; the last plotted those between 1956 and 1957. As no observations were made in 1950, the breaks in the lines indicate missing values for the 1949/50 and 1950/51 transitions.

returning to F is very much greater than if it had not been in F (p_{GF} = 0.20 as opposed to 0.05 and 0.07, and p_{MF} = 0.39 as opposed to 0.03 and 0.03 in the matrices above). Accepting a second-order process with a single time step, and running a complete simulation until a stable structure was obtained, yielded the stable vector as

[28   16   56].

The prediction is that 28% of the quadrats will be in the *Festuca* (F) state as opposed to 20% with the simple first-order model. This increase is a reflection of the relative attraction of state F, as discussed for matrix $\mathbf{P_F}$ above.

It is also appropriate to test the 3-state system for stationarity in time. Using the year-to-year transitions for the period 1936–57 (1950 missing) yields 19 yearly matrices of transition probabilities. The separate matrices are not shown here, but the time trajectories of the transition probabilities are shown in Fig. 6.6. With the 3-state model, and using equation (6.16), Reed (1980) obtained χ^2 = 752 with 108 degrees of freedom. The null hypothesis that the year-to-year matrices are identical would therefore be rejected.

These Breckland data demonstrate three features of the use of Markov models in succession; Usher (1979) considered these almost to be rules.

1. Successional processes are not random processes [χ^2 in equation (6.6) is significant].
2. Successional processes are not first-order but higher order processes [χ^2 in equation (6.11) is significant].
3. Successional processes are not stationary in time [χ^2 in equation (6.16) is significant].

Apart from the Kingston Harbour data on zooplankton collected by Grahame (1975), no studies appear inconsistent with these three suggestions; the study of Dutch heathlands by Lippe *et al.* (1985) also supports them. However, if they are all true, they do imply that very complicated models are required if the model is to reflect accurately the processes of succession, and to provide reasonable forecasts on which to base the management of ecosystems or renewable resources.

6.4   THE USE OF SERIES OF FREQUENCY DATA

6.4.1   The mathematical development

In the Breckland example in section 6.3 it will have become apparent that very large amounts of data are required for even simple models. In order to investigate double-dependence in the 3-state system, a total of 27 (3^3) transition probabilities had to be estimated from the data: to investigate this with the more realistic 7-state system would have required that 343

Table 6.4 Frequently in the ecological literature one has the frequency with which species occur or the frequency of different states. If such estimates are available for a series of equally spaced time intervals, usually yearly, they can be used in a table such as this, where $n_i(t)$ represents the proportion of samples, quadrats, etc., in state i at time t. The table contains $(T + 1)$ rows: the first T rows are used in matrix **X**, and the last T rows in matrix **Y** (see text)

Time	Observed proportions of state				
	1	2	3	...	m
0	$n_1(0)$	$n_2(0)$	$n_3(0)$	...	$n_m(0)$
1	$n_1(1)$	$n_2(1)$	$n_3(1)$		$n_m(1)$
2	$n_1(2)$	$n_2(2)$	$n_3(2)$		$n_m(2)$
.	.				
.	.				
.	.				
$T - 1$	$n_1(T - 1)$	$n_2(T - 1)$	$n_3(T - 1)$		$n_m(T - 1)$
T	$n_1(T)$	$n_2(T)$	$n_3(T)$		$n_m(T)$

transition probabilities be estimated. Similarly, investigation of stationarity involved estimates of 9 transition probabilities in each of 19 matrices for the 3-state system. Watt's data are extremely unusual in the ecological literature because of both their detail and the length of time over which they were collected. Such data sets will never be widely available in ecology, and are extremely 'expensive' to collect in terms of both time and resources.

It therefore becomes important that 'cheaper' data sets be used in modelling succession. Cooke (1981) discussed one approach whereby more typical data sets could be applied to Markov models. States need to be decided *a priori*, but, rather than following the time trajectory of an individual quadrat, one needs to know the proportional abundance of each of the states during a sequence of time intervals. Symbolically this is shown in Table 6.4. Each entry in the table is of the form $n_i(t)$, where n is a proportion, i represents one of the m states ($1 \leq i \leq m$), and t is the time; note that

$$\sum_{i=1}^{m} n_i(t) = 1 \tag{6.17}$$

for all values of t (i.e. each row must sum to 1, or 100%). As previously, the probability that a quadrat (or unit of observation) will make a transition from state i to state j in one time interval is p_{ij}. Using a Markov model to predict proportions at time t from a knowledge of the proportions at time $(t - 1)$ gives the following estimate for the first state:

$$n_1(t) = p_{11}n_1(t - 1) + p_{21}n_2(t - 1) + \ldots + p_{m1}n_m(t - 1), \tag{6.18}$$

with similar expressions for $n_2(t), \ldots, n_m(t)$; t takes values $1, 2, \ldots, T$. Since values of all expressions of the form $n_i(t)$ are known, the problem is essentially one of estimating the expressions p_{ij}; the methods are discussed by Cooke (1981), He applied the method to observations of grassland in Snowdonia, with a series of twelve observations spanning the years 1960 to 1971 inclusive.

There are, however, some difficulties with Cooke's method. Two constraints on the estimates are that

$$0 \leqslant p_{ij} \leqslant 1 \quad \text{for all } i \text{ and } j$$

and

$$\sum_{j=1}^{m} p_{ij} = 1 \quad \text{for all } i.$$

(6.19)

If $\mathbf{P}$ is the transition probability matrix with elements p_{ij}, it is estimated as $\hat{\mathbf{P}}$ from the equation

$$(\mathbf{X}^T\mathbf{X} + c\mathbf{I})\hat{\mathbf{P}} = \mathbf{X}^T\mathbf{Y} + (c/m)\mathbf{ll}^T$$

(6.20)

where $\mathbf{X}$ is the matrix of proportions, as in Table 6.4, from time 0 to $(T - 1)$, $\mathbf{Y}$ is a similar matrix from time 1 to T, $\mathbf{I}$ is the identity matrix, $\mathbf{l}$ is a vector all of whose elements are unity, m is the number of states, c is an unknown constant, and the superscript T implies a transposed matrix. It is the constant c that provides the difficulty, and Cooke (1981) advises

initially c is set equal to zero ... Usually some of the estimates are out of range [i.e. constraints (6.19) are not satisfied] so c is increased by steps of 0.01 until the estimates of all the probabilities are within range. When this happens the estimation procedure is stopped unless any estimates are changing substantially at each iteration, in which case c is stepped on until the estimates change only slightly from one iteration to the next.

6.4.2   An application in the Breckland grasslands

The Breckland grassland states can be used to see how efficient Cooke's method is in recreating a probability transition matrix. As there are no data for 1950, it is appropriate to use only the data for the period 1936 to 1949, and, for simplicity, only to use the 3-state system.

The data for the grazed grassland are given in Table 6.5. Matrices $\mathbf{X}$ and $\mathbf{Y}$, in equation (6.20), are simply derived from this table by dividing each row by 128 to change it from a frequency to a proportion. Due to the matrix algebra involved in equation (6.20) all calculations have been undertaken by computer. Starting the process with $c = 0$, as recommended by Cooke (1981), three of the nine estimates of p_{ij} were outside the permitted range, as can be seen in Fig. 6.7. Stepping c

Table 6.5   Data from the grazed Breckland grassland classified into three states, F, G and M (see section 6.3 for details). The table gives the number of quadrats, along a transect of 128 permanent quadrats, in each state each year (note that no data are available for 1950)

Year	No. of quadrats in state		
	F	G	M
1936	0	106	22
1937	0	96	32
1938	0	73	55
1939	0	19	109
1940	0	0	128
1941	0	0	128
1942	0	0	128
1943	0	4	124
1944	0	11	117
1945	1	50	77
1946	8	62	58
1947	24	28	76
1948	27	15	86
1949	33	20	75

by 0.01, again as recommended by Cooke (1981), yielded a transition probability matrix within range when $c = 0.04$. Matrix $\hat{\mathbf{P}}$ was then

$$\hat{\mathbf{P}} = \begin{bmatrix} 0.89 & 0.03 & 0.08 \\ 0.04 & 0.69 & 0.27 \\ 0.02 & 0.05 & 0.93 \end{bmatrix} \tag{6.21}$$

where, as previously, the order of the states is F, G and M. The dominant eigenvector of this matrix is

$$\hat{\mathbf{p}} = [16 \quad 13 \quad 71],$$

where the elements have been adjusted so that their sum is 100, and hence they are percentages. These estimates were derived only from the data in Table 6.5, and can therefore be compared with the 'true' matrices derived from all of the observed transitions, $\mathbf{P}_3$, in section 6.3 and its dominant eigenvector

$$\mathbf{p} = [20 \quad 20 \quad 60].$$

The value $c = 0.04$ was the first to satisfy condition (6.19), though it might be argued that some of the estimates of transition probabilities were still changing substantially at each iteration: Cooke (1981) does not define precisely what a substantial change is. Hence in Fig. 6.7 the value

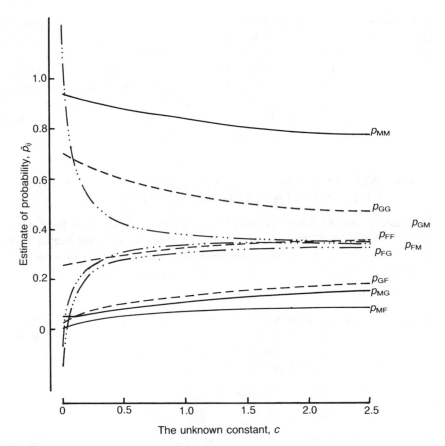

Figure 6.7 The dependence of estimates of p_{ij} using Cooke's method on the value of the unknown constant c. At least one estimate was out of range ($p_{ij} > 1$ or $p_{ij} < 0$) when c took values less than 0.04. As c tends to infinity, each estimate of p_{ij} tends to 0.333. The states F, G and M are as defined in Fig. 6.4.

of c has been allowed to increase substantially above 0.04. It is noticeable that three of the transition probabilities, which were out of range when $c < 0.04$, were changing considerably less rapidly when c was in the range 0.15–0.20. Taking $c = 0.15$,

$$\hat{\mathbf{P}} = \begin{bmatrix} 0.61 & 0.18 & 0.21 \\ 0.07 & 0.66 & 0.27 \\ 0.03 & 0.06 & 0.91 \end{bmatrix}, \tag{6.22}$$

which differs mainly from the estimate in equation (6.21) in the first row of the matrix. This matrix has a dominant eigenvector of

$$\hat{\mathbf{p}} = \begin{bmatrix} 8 & 17 & 75 \end{bmatrix}.$$

If Cooke's criterion of all probabilities within range is used, equation (6.21) defines a matrix that is similar to the 'true' matrix. If Cooke's criterion of lack of substantial change in the estimates of all p_{ij} is used, equation (6.22) is substantially different from the 'true' matrix. If c is continually increased in the hope that the iterative process will converge, convergence is only achieved when all of the matrix elements, p_{ij}, take the value $1/m$ (0.333 in the example here since $m = 3$ states). The problem remains; is there an objective way of estimating c?

6.5 ADVANTAGES, DISADVANTAGES AND THE ROLE OF STATISTICAL MODELS

It is interesting to note the differences in the models that have been applied to the Breckland grassland data. Using the simplest model, that of the stationary, first-order, single-dependence Markov process, the model predicted a stable structure of the grassland community of

[20   20   60]

where 20% of the quadrats would be dominated by *Festuca*, 20% by *Galium* and 60% with mixed communities in which *Agrostis*, *Luzula* and *Rumex* would all be represented. Use of annual frequency data, and analysis by Cooke's method, yielded

[16   13   71]   with $c = 0.04$, and

[8   17   75]   with $c = 0.15$.

It can be seen that these models predict that *Festuca* quadrats will be far less frequent, now only 16 or 8% of all quadrats, and that mixed communities would be more abundant. This is directly opposite to the predictions of the stationary, second-order, double-dependence Markov process where the eigenvector was

[28   16   56].

This more complex model predicts that *Festuca* quadrats would be much commoner, with reductions in both the *Galium* and mixed quadrats. *Festuca* certainly became dominant in the enclosed quadrats, and, as seen when grazing became less severe with the advent of myxomatosis (see Watt 1960b, 1981a,b), it is likely that the second-order process has provided the best of these predictions.

This leads to one of the disadvantages of Markovian models. For the simplest model for a system of m states, there are m^2 estimates of p_{ij}. To use a model that is a second-order process, then there are m^3 estimates of p_{ijk}. It is probably relatively simple to obtain estimates of the standard errors of any p_{ij}, but what are the effects of the error of estimation on the terms in the dominant eigenvector, on the ratio λ_1/λ_2, on the mean recurrence times, etc? So far these models have been used in a totally deterministic manner, and an application in a more stochastic manner remains an interesting possibility that is virtually a necessity.

A second problem depends on the purposes for which the model is required; it is linked to the question: 'How inaccurate a prediction is acceptable?' When Usher (1975) used one of these models for a termite community, his question related to whether a wood-feeding termite community would continue to exist after the changed conditions (clearance of farm scrub) of the test site. If one of the states of the Breckland grassland was important for conservation reasons, the question may be whether that state was likely to persist or to become extinct. Simple stationary, single-dependence models seem satisfactory when the question to answer is essentially so simple. More sophisticated models, with multiple dependence or without stationarity in time or space, will be required if more sophisticated questions are asked.

A third problem involves the effects of both time and space. The models developed for the Breckland grasslands are essentially models in time, and the spatial element has been ignored. If one takes into consideration Watt's (1947) thesis on pattern and process within communities, then the pattern of spread of species such as *Agrostis* and *Festuca* would be important in these models. In section 6.3 it was shown that the transition probabilities for the transitions $G \rightarrow F$ and $M \rightarrow F$ were much greater if the previous state of the system had been F rather than G or M. A question that has not been addressed relates to the effect of, say, F in an adjacent quadrat on the probabilities of transition from G or M to F. Since grasses such as *Festuca* spread by tillering, it is more than likely that there is no stationarity in space. Exploration of space/time interactions with these models is an exciting future development.

Other disadvantages of these models have been discussed by Usher (1981). However, one of the greatest disadvantages is the practical difficulty of collecting sufficient data on which to build anything but the crudest of models. The methods of analysis outlined in section 6.4 potentially provide a useful method of incorporating the kind of time-series data that have been collected by many ecologists. There are clearly many problems still to be overcome, especially in relation to estimation of the constant c in equation (6.20), but there may be some hope of finding a simpler method of data collection.

Markovian models of succession have an important role when the states of a system have been recognized and, particularly, when an answer is required for an unsophisticated question. The predictions of forest change in North America, of the change in a fen woodland in Yorkshire, of termites changing on baitwood in West Africa, or of the changes in predator and prey densities on oranges and in apple orchards, have all been reasonable. The basic Markov model has therefore found practical use.

The framework of the model may be important for distinguishing between mechanisms of succession. Both Slatyer (1977) and Usher (1981) have indicated the likely characteristics of matrices demonstrating the concepts of facilitation, tolerance or inhibition (Connell and Slatyer, 1977). Since the former tends to be a directed approach from one community to the next, the matrix of transition probabilities would take the form

$$
\begin{bmatrix}
XX & XX & X & \cdot & \cdot \\
\cdot & XX & XX & X & \cdot \\
\cdot & \cdot & XX & XX & X \\
\cdot & \cdot & \cdot & XX & XX \\
\cdot & \cdot & \cdot & \cdot & XX
\end{bmatrix},
$$

where a double 'X' indicates a large probability, a single 'X' a smaller probability, and a dot either a very small probability or a zero probability. Note that this system moves forward to the final state, and resembles none of the Breckland grassland matrices. The tolerance model will include many 'leaps' forward by more than one step, and hence the transition probability matrix may take the form

$$
\begin{bmatrix}
XX & X & X & X & X \\
X & XX & X & X & X \\
\cdot & X & XX & X & X \\
\cdot & \cdot & X & XX & X \\
\cdot & \cdot & \cdot & X & XX
\end{bmatrix}.
$$

The predominant movement is forward, from any state to any forward state, but there is a small amount of movement backwards. For the inhibition model, where there is an absorbing state, the matrix may take the following form

$$
\begin{bmatrix}
XX & XX & X & X & \cdot \\
X & XX & XX & X & X \\
0 & 0 & 1 & 0 & 0 \\
\cdot & X & \cdot & XX & XX \\
\cdot & \cdot & \cdot & X & XX
\end{bmatrix}.
$$

Here the third state is an absorbing state, which can be entered either from the first or second states. Once the system has got into the third state there is no escape (probability of remaining in the third state is 1) as all other transition probabilities in the third line of the matrix are 0. Finally, and more relevant to a discussion of Watt's Breckland data, there is the possibility of a cyclical climax. This is demonstrated in the following matrix

$$
\begin{bmatrix}
XX & XX & X & . & . \\
. & XX & XX & . & . \\
0 & 0 & . & XX & 0 \\
0 & 0 & 0 & . & XX \\
0 & 0 & XX & 0 & .
\end{bmatrix},
$$

where there is a cycle in the third, fourth and fifth states. Once the system has entered the third state, it will move to the fourth state (or possibly remain in the third state) at the next step. It will then move to the fifth, back to the third, and so on. Note that these three states, which form a cycle, are also an absorbing set of states since, once the system has entered one of these states, there is no way out of the set of them.

One feature of the Markovian models is very apparent: it is possible to list more disadvantages than advantages in their use. It is only when they are compared with the other models that it will be possible to assess whether the advantages outweigh the disadvantages for any particular modelling application.

SUMMARY

1. This chapter singles out one statistical model, the Markov process, for discussion. The model is described, and a worked example shows how to calculate the eigenvector. This is the endpoint of a system, after which no further change is expected.
2. Markovian models can be first order (i.e. the state of a system one time interval into the future can be predicted from the state now) or of higher order. Testing for the order of the process is described. A worked example is included.
3. Markovian models are often assumed to be stationary (i.e. they do not change over time or space); a test of stationarity is described and demonstrated with an example.
4. There is a brief review of applications of the Markovian model in plant and animal ecology, investigating forest succession, heathland succession, termite succession on decomposing wood, plankton changes in a tropical harbour and oscillations in predatory mites and their prey (phytophagous mites).
5. The techniques of Markov models are demonstrated on a set of grassland data collected over a 21-year period in the Brecklands, UK. The definition of the number of recognizable states of the grassland is important; the example compares models with three and seven states.
6. A method using annual frequency data is briefly described and applied to the Breckland grassland data. Although such methods are potentially useful, the example demonstrates that there are still many theoretical problems to be overcome.
7. The discussion of the advantages and disadvantages of Markovian

models in ecology contrasts the large amounts of data required for model formulation with the elegance of portraying theoretical concepts as pattern matrices.

ACKNOWLEDGEMENTS

Much of the work on the models of Breckland grasslands was undertaken by John Reed: I should like to thank him for permission to quote it. I should also like to thank Dr D. Cooke for making his computer program available to me, the late Dr A. S. Watt FRS for helpful comments about his data and Andrea Johnson for the final word processing.

REFERENCES

Anderson, T. W. and Goodman, L. A. (1957) Statistical inference about Markov chains. *Ann. Math. Stat.*, **28**, 89–110.

Austin, M. P. (1980) An exploratory analysis of grassland dynamics: an example of a lawn succession. *Vegetatio*, **43**, 87–94.

Austin, M. P. and Belbin, L. (1981) An analysis of succession along an environmental gradient using data from a lawn. *Vegetatio*, **46**, 19–30.

Brauer, A. (1957) A new proof of theorems of Perron and Frobenius on non-negative matrices: I. positive matrices. *Duke Math. J.*, **24**, 367–78.

Brauer, A. (1962) On the theorems of Perron and Frobenius on non-negative matrices, in *Studies in Mathematical Analysis and Related Topics* (eds S. Gilbarg *et al.*), Stanford Univ. Press, Stanford, pp. 48–55.

Caswell, H. (1989) *Matrix Population Models*, Sinauer, Sunderland, Mass.

Collins, L. (1975) *An Introduction to Markov Chain Analysis*, Geo. Abstracts, Norwich.

Connell, J. H. and Slatyer, R. O. (1977) Mechanisms of succession in natural communities and their role in community stability and organization. *Amer. Nat.*, **111**, 1119–44.

Cooke, D. (1981) A Markov chain model of plant succession, in *The Mathematical Theory of the Dynamics of Biological Populations, II* (eds R. W. Hiorns and D. Cooke), Academic Press, London, pp. 231–47.

Digby, P. G. N. and Kempton, R. A. (1987) *Multivariate Analysis of Ecological Communities*, Chapman and Hall, London and New York.

Gimingham, C. H., Hobbs, R. J. and Mallik, A. U. (1981) Community dynamics in relation to management of heathland vegetation in Scotland. *Vegetatio*, **46**, 149–55.

Grahame, J. (1976) Zooplankton in a tropical harbour: the numbers, composition, and response to physical factors of zooplankton in Kingston Harbour, Jamaica. *J. Exp. Marine Biol. Ecol.*, **25**, 219–37.

Greig-Smith, P. (1983) *Quantitative Plant Ecology*, 3rd edn, Blackwell, Oxford.

Harbaugh, J. W. and Bonham-Carter, G. (1970) *Computer Simulation in Geology*, Wiley, New York.

Hobbs, R. J. and Legg, C. J. (1983) Markov models and initial floristic composition in heathland vegetation dynamics. *Vegetatio*, **56**, 31–43.

Horn, H. S. (1975a) Forest succession. *Scient. Amer.*, **232**, 90–8.

Horn, H. S. (1975b) Markovian properties of forest succession, in *Ecology and Evolution of Communities* (eds M. L. Cody and J. M. Diamond), Belknap Press, Cambridge, Mass., pp. 196–211.

Horn, H. S. (1976) Succession, in *Theoretical Ecology: Principles and Applications* (ed. R. M. May), Blackwell, Oxford, pp. 187–204.

Huffaker, C. B. (1958) Experimental studies on predation: dispersion factors and predator-prey oscillations. *Hilgardia*, **27**, 343–83.

Isagi, Y. and Nakagoshi, N. (1990) A Markov approach for describing post-fire succession of vegetation. *Ecol. Res.*, **5**, 163–71.

Lippe, E., de Smidt, J. T. and Glenn-Lewin, D. C. (1985) Markov models and succession: a test from a heathland in the Netherlands. *J. Ecol.*, **73**, 775–91.

Odum, E. P. (1969) The strategy of ecosystem development. *Science*, **164**, 262–70.

Reed, J. (1980) *Markov Models of Succession with Reference to Data on the East Anglian Brecklands*, MSc Thesis, Univ. of York.

Slatyer, R. O. (1977) Dynamic changes in terrestrial ecosystems: patterns of change, techniques for study and applications to management. *UNESCO: MAB Tech. Notes*, **4**, 1–30.

Stephens, G. R. and Waggoner, P. E. (1970) The forests anticipated from 40 years of natural transitions in mixed hardwoods. *Bull. Connecticut Agric. Exp. Station*, 707.

Stephens, G. R. and Waggoner, P. E. (1980) A half century of natural transitions in mixed hardwood forests. *Bull. Connecticut Agric. Exp. Station*, 783.

Tucker, J. J. and Fitter, A. H. (1981) Ecological studies at Askham Bog Nature Reserve. 2. The tree population of Far Wood. *The Naturalist*, **106**, 3–14.

Usher, M. B. (1966) A matrix approach to the management of renewable resources, with special reference to selection forests. *J. Appl. Ecol.*, **3**, 355–67.

Usher, M. B. (1975) Studies on a wood-feeding termite community in Ghana, West Africa. *Biotropica*, **7**, 217–33.

Usher, M. B. (1979) Markovian approaches to ecological succession. *J. Animal Ecol.*, **48**, 413–26.

Usher, M. B. (1981) Modelling ecological succession, with particular reference to Markovian models. *Vegetatio*, **46**, 11–18.

Usher, M. B. and Parr, T. W. (1977) Are there successional changes in arthropod decomposer communities? *J. Environ. Mgmt*, **5**, 151–60.

van Hulst, R. (1979) On the dynamics of vegetation: Markov chains as models of succession. *Vegetatio*, **40**, 3–14.

Waggoner, P. E. and Stephens, G. R. (1970) Transition probabilities for a forest. *Nature*, **225**, 1160–1.

Watt, A. S. (1947) Pattern and process in the plant community. *J. Ecol.*, **35**, 1–22.

Watt, A. S. (1960a) The effect of excluding rabbits from acidiphilous grassland in Breckland. *J. Ecol.*, **48**, 601–4.

Watt, A. S. (1960b) Population changes in acidiphilous grass-heath in Breckland, 1936–57. *J. Ecol.*, **48**, 605–29.

Watt, A. S. (1962) The effect of excluding rabbits from grassland A (*Xerobrometum*) in Breckland, 1936–60. *J. Ecol.*, **50**, 181–98.

Watt, A. S. (1971) Factors controlling the floristic composition of some plant communities in Breckland, in *The Scientific Management of Animal and Plant*

Communities for Conservation (eds E. Duffey and A. S. Watt), Blackwell, Oxford, pp. 137–52.

Watt, A. S. (1981a) A comparison of grazed and ungrazed grassland A in East Anglian Breckland. *J. Ecol.*, **69**, 499–508.

Watt, A. S. (1981b) Further observations on the effects of excluding rabbits from grassland A in East Anglian Breckland: the pattern of change and factors affecting it (1936–73). *J. Ecol.*, **69**, 509–36.

Woolhouse, M. E. J. (1983) The application of a transition matrix model to a pest management problem, in *Mathematical Models of Renewable Resources* (ed. R. H. Lamberson), Vol. 2, Humboldt State Univ., pp. 90–5.

7 Individual-based models of forest succession

Dean L. Urban and Herman H. Shugart

7.1 INTRODUCTION

The basic tenets in ecology have given rise to a rich body of conceptual theory about succession and ecosystem dynamics (see Prologue and Chapter 1), yet it has been difficult to bring the rigour of mathematics to bear on this conceptual richness. The diversity and heterogeneity of natural systems have not yielded gracefully to mathematical formalization; we have guiding principles, but not much in the way of predictive theory in the form of models.

The development of ecological models in general, and succession models in particular, received a real boon as computers became increasingly available in the mid-1960s. Many extant models in ecology can be traced to this period, during which several independent lines of models evolved. Three important domains of model development were forestry and silviculture (Munro, 1974), forest ecology (Shugart and West, 1980) and ecosystems as embraced by the International Biological Programme (e.g. Reichle, 1981). These models witness an impressive range of focal concerns and applications, and a corresponding diversity of model structures and solution techniques.

Succession models have been classified according to a variety of criteria. Munro (1974) categorized forestry models according to whether the competition function was explicitly spatial or not. Shugart and West (1980) contrasted ecological succession models according to whether they simulated mixed or monospecies stands, the dimensionality of spatial representation (1D, 2D, 3D), and whether the original focus was on trees or the aggregate stand. Weinstein and Shugart (1983) sorted models based on how space and time were modelled (discrete or continuous), because these dictated model structure and the form of equations used. Shugart and Urban (1986) considered succession models in terms of their ability to deal with landscape attributes such as contagion effects, large extent, gradients, patch variety and contrast. In fact, there may be as many ways to classify models as there are reasons to classify (witness the schemes in Chapters 5 and 6), and particular applications tend to argue for certain modelling approaches. Yet, there is no one 'best' modelling

Plant Succession: Theory and prediction Edited by David C. Glenn-Lewin, Robert K. Peet and Thomas T. Veblen © 1992 Chapman & Hall, London ISBN 0 412 26900 7

approach, and models of different forms may often be brought to bear on the same application with only slight changes in detail or emphasis.

The focus of this chapter is the so-called gap model (Shugart and West, 1980), a class of models that has been applied to a broad spectrum of forests. Gap models simulate the establishment, annual diameter growth, and mortality of each tree on a small (c. $100–1000\,m^2$) model plot. Because demographics (growth, mortality and regeneration) vary among species and for trees of different sizes, these individual-based models are especially useful in simulating mixed-species, mixed-age forests, a capability not shared by some other models that average parameters over the entire canopy (e.g. Running and Coughlan, 1988) or population (e.g. Tilman, 1982).

In the following sections we trace the history and evolution of gap models, and detail the structure, algorithms and formulations in the basic model. We overview their diversification, pointing to selected variations on the general approach. We then present some applications that highlight their versatility and richness, focusing especially on successional studies. We speculate briefly on forests as complex systems with higher-level behaviour that is determined, in part, by the coupling of demographic processes of mortality and regeneration. Finally, we offer a prospectus on how gap models may contribute to the further development of succession theory.

7.2   THE BASIC GAP MODEL

The conceptual foundations of gap models can be traced to the classic work of A. S. Watt (1925, 1947) on patch dynamics and fine-scale pattern and process in forests. In this conceptual model, the death of an individual tree creates a canopy gap of varying suitability for subsequent establishment and/or growth of other trees. Gap-phase regeneration leads, in turn, to a conceptualization of a forest stand as a mosaic of fine-scale elements in gap, building (aggrading), and mature phases, which constitute a forest growth cycle (Bormann and Likens, 1979; Whitmore, 1982). This notion has spawned much empirical research on the nature of gap dynamics (e.g. Platt and Strong, 1989; see also Chapter 4), and directly underlies the development of gap models as described in this chapter.

The original gap model, JABOWA, was developed partly as an IBM research project in the early era of 'heavy computing' (Botkin *et al.*, 1972a,b). This model spawned the FORET model (Shugart and West, 1977), and a subsequent proliferation of variations on this theme has resulted in more than two dozen published versions for particular forests (Fig. 7.1). The basic approach of gap models has also been extended to grasslands (Coffin and Lauenroth, 1989, 1990).

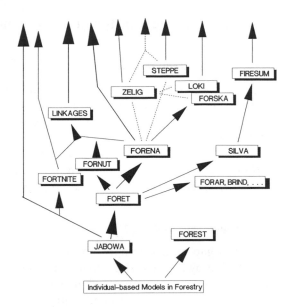

Model	Forest type (location)	Reference
JABOWA	Northern hardwood (New Hampshire)	Botkin et al., 1972
FORET	Appalachian hardwood (east Tennessee)	Shugart and West, 1977
FORMIS	Floodplains (Mississippi)	Tharp, 1978
FORAR	Pine, oak–pine (Arkansas)	Mielke et al., 1978
SWAMP	Floodplains (Arkansas)	Phipps, 1979
BRIND	Montane eucalypt (Australia)	Shugart et al., 1981
KIAMBRAM	Subtropical rainforest (Australia)	Shugart and Noble, 1981
FORICO	Rainforest (Puerto Rico)	Doyle, 1981
FORTNITE	Northern hardwood (New Hampshire)	Aber and Melillo, 1982
FORNUT	Appalachian hardwood (North Carolina)	Weinstein et al., 1982
CLIMACS	Pacific Northwestern conifer (Oregon)	Dale and Hemstrom, 1984
SILVA	Montane conifer (California)	Kercher and Axelrod, 1984
FORFLO	Floodplains (South Carolina)	Pearlstine et al., 1985
FORENA	Decidous forests (Eastern North America)	Solomon, 1986
LINKAGES	Deciduous forests (Eastern U.S.)	Pastor and Post, 1986
FORCAT	Deciduous (Cumberland Plateau)	Waldrop et al., 1986
FORECE	Southern central Europe	Kienast, 1987
FOREAL	Appalachian spruce–fir (North Carolina)	Busing and Clebsch, 1987
FORSKA	Boreal conifer (Sweden)	Leemans and Prentice, 1987
FORENZ	Mixed forests (New Zealand)	DeVelice, 1988
OUTENIQUA	Subtropical rainforest (South Africa)	van Daalen and Shugart, 1989
ZELIG	Appalachian hardwood (Tennessee)	Smith and Urban, 1988
LOKI	Boreal conifer (Alaska, Canada)	Bonan, 1989b
FIRESUM	Montane conifer (Montana)	Keane et al., 1989
STEPPE	Semi-arid grassland (Colorado)	Coffin and Lauenroth, 1989
OVALIS	Appalachian hardwood (Virginia)	Harrison and Shugart, 1990
JABOWA-II	(Revision of original model)	Botkin, 1992
SPACE	Appalachian cove hardwood (North Carolina)	Busing, 1991
BOS	Mixed forests (The Netherlands)	Mohren et al., 1991
FORBORS	Boreal forests (circumpolar)	Shugart et al., 1992

Figure 7.1 Diversification of gap models, keyed by location and primary citation. Not all models listed below are included in the figure, which emphasizes main lines of development.

7.2.1   Model structure and formulation

Each of the extant gap models varies slightly in emphasis, and consequently they vary in the level of detail at which various processes are simulated. Yet, although each model is unique in some respects, they all share a common set of assumptions and logic. The following overview applies to most versions of the basic gap model. Specific details cited here refer to the ZELIG model (Urban, 1990), a recent reformulation of a gap model which retains the essentials of the original JABOWA model.

(a)   Structure and state variables

The state variables of a gap model comprise a tally of all the trees on a plot. Each tree is labelled by species, size (diameter) and vigour (defined by its recent growth history). Through allometric relationships with diameter, each individual can be associated with a height, leaf area and woody biomass; these attributes are used to define the competitive environment of the plot. Competition among individuals is explicit but indirect in gap models. Trees do not interact with each other directly; rather, each tree contributes to the plot-level environment, and this aggregate environment affects each tree. Thus, competition works through a tree–plot–tree loop, rather than a tree–tree interaction.

Plot size is defined by the zone of influence of a single canopy-dominant tree, or equivalently, by the gap such a tree creates when it dies. In eastern deciduous forests, the median size of canopy gaps is on the order of $100\,m^2$ (Runkle, 1985), which corresponds well to the crown area of a large tree. The plot is considered homogeneous horizontally, but vertical heterogeneity is simulated in detail in terms of tree heights. Thus, each tree influences (shades) all shorter trees on the plot, and in turn is influenced by all taller trees on the plot. This assumption that a plot size can be defined such that spatial positions within it need not be accounted (i.e. the assumption of homogeneity) is central to gap models and represents a powerful simplification relative to spatially implemented competition models (Munro, 1974; Shugart and West, 1980). Shugart and West (1979) found the FORET model was sensitive to modelled plot size, with gap dynamics reproducible with plots 100–$1000\,m^2$, but degrading at smaller and larger sizes, a result which has been reproduced with ZELIG (Coffin and Urban, in press). Prentice and Leemans (1990) have speculated that gap dynamics are exhibited in forests only under a restricted set of conditions in which the horizontal zone of influence of a canopy-dominant tree is less than some small multiple (perhaps c. 5×) of the size of a single large tree. Because a tree's zone of influence is defined by its shadow length, hence by solar angle and tree height, the extent to which gap-phase regeneration is expressed should vary according to latitude (which affects solar angle) and tree stature. In accord with this,

Leemans and Prentice (1987) found that a comparatively large plot size was required to reproduce high-latitude forests, and Dale and Hemstrom (1984) used large plots to reproduce the dynamics of Pacific Northwestern conifer forests dominated by very large trees. For Appalachian hardwoods, Coffin and Urban (in press) found that between-plot variance was maximized with model plots of c. $250\,m^2$, which is larger than a tree crown but somewhat smaller than the maximum zone of influence as estimated by shadow lengths. This relationship has been explored further with spatially implemented gap models (see below).

(b)   Demographics

The models simulate the demographic processes of sapling establishment, annual diameter growth and mortality, and hence are appropriate for simulating multiple generations of tree replacement. Demographic processes are modelled by computing a potential behaviour expected under optimal (non-limiting) conditions, and then reducing this potential to reflect the constraining context of the simulated environment. The environment is defined in terms of available light, soil moisture and fertility, and temperature.

Gap models tend to use rather simple functions that can be parametrized readily (Table 7.1). The complexity of the models stems not so much from complicated processes or formulations, but from species diversity, size structure and feedback mechanisms in the models (Huston et al., 1988; Shugart and Urban, 1989).

The diameter growth equation is a logistic curve defined by a few species-specific parameters (Fig. 7.2); the derivative of this curve specifies the annual increment expected for a tree of a given size and species (Fig. 7.2, inset; see also Botkin et al., 1972b). Environmental constraints reduce this potential increment via simple scaling functions that are specified for species types or response categories (e.g. shade tolerance class) (Fig. 7.3). The annual increment, modified by the constraint multipliers, is added to the tree as an 'annual ring'.

Sapling establishment is largely stochastic in the models, with the probability of establishment conditioned by the plot environment. This amounts to environmental filtering (sensu Harper, 1977) of species available for establishment. Each simulation year, the eligibility of each species is determined by comparing its environmental tolerances (light compensation point, temperature limits, etc.) to the plot environment. A few individuals of eligible species are 'planted' each year. New individuals are established with initial diameters of c. $2\,cm$, with some random variation. The logistically daunting details of seedling germination and early survival are not simulated explicitly in most gap models (but see below).

Mortality is also stochastic, and may stem from two sources in the

Table 7.1 Formulations typically used in forest gap models[a]

Allometric relationships
(a) Height: $H = 137 + b_2D - b_3D^2$ Ker and Smith, 1955
      where $b_2 = 2(H_{max} - 137)/D_{max}$
      and $b_3 = (H_{max} - 137)/D^2_{max}$ Botkin *et al.*, 1972a
(b) Leaf area: $L = 0.16094\,D^{2.129}$ Sollins *et al.*, 1973
(c) Woody biomass: $B = 0.1193\,D^{2.393}$ Sollins *et al.*, 1973

Tree diameter growth
(d) $d[D^2H]/dt$ Botkin *et al.*, 1972a
                      $= rL(1 - DH/D_{max}H_{max})$
      Assume $L = cD^2$ and set $G = rc$; then on substitution and differentiation:
(e)
$$dD/dt = \frac{GD(1 - DH/D_{max}H_{max})}{(274 + 3b_2D - 4b_3D^2)}$$

Environmental response
(f) Light extinction: $Q_h = Q_0\exp[-kL(h')]$ Monsi and Saeki, 1953
(g) Light response: $r(Q_h) = c_1(1 - \exp[-c_2(Q_h - c_3)])$ Botkin *et al.*, 1972a
(h) Nutrients: $r(F) = c_1 + c_2F - c_3F^2$ Aber and Melillo, 1982
(i) Soil moisture: $r(M) = [(M^* - M)/M^*]^{1/2}$ Pastor and Post, 1986
(j) Temperature: $r(T) = \dfrac{4(T - T_{min})(T_{max} - T)}{(T_{max} - T_{min})^2}$ Botkin *et al.*, 1972a

Mortality
(k) Age-related: $m = 4.605/A_{max}$ Botkin *et al.*, 1972a
(l) Stress-related: $m = 0.369$ Botkin *et al.*, 1972a

[a] See text for further explanation. Symbols are: D, diameter at breast height (cm); H, height (cm); A, age (yr); L, leaf area (m²); Q, incident radiation (1 = full sun), (h, height; h', heights $> h$); G, growth rate scalar; F, relative soil fertility (dimensionless on 0–1); M, soil moisture index (on 0–1), M^* is maximum tolerable for a species; T, degree-day index (5.56°C base). Subscripts min, max are species minimum or maximum; k, b's and c's are fitted constants.

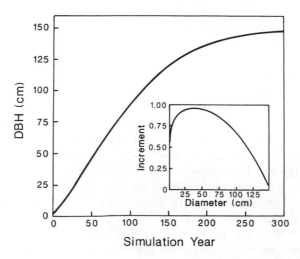

Figure 7.2 Diameter growth as simulated in JABOWA-derived gap models. *Inset*: diameter increment as a function of current diameter.

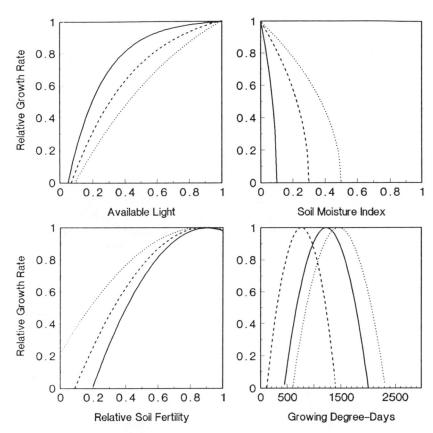

Figure 7.3 Environmental response functions typically used in gap models: available light, for three shade tolerance classes (solid line = tolerant); soil moisture, for three response classes (solid line = drought intolerant); soil fertility, for three nutrient response classes (solid line = responsive); and temperature, for three species (each species has its own curve).

models. A constant ambient probability of mortality depends on the expected longevity of a species. It is assumed that about 1% (JABOWA uses 2%; FORET derivatives, 1%) of individuals might survive to maximum age; the further assumption that this mortality is constant with age yields an annual probability of mortality on the order of 1–2% per year (Table 7.1). A second source of mortality is due to loss of vigour, and is invoked when a tree fails to meet a minimal-growth threshold. This rate of 'stress' mortality is an order of magnitude greater than the ambient rate: it is assumed that a tree might survive ten years under stress, which yields an annual stress mortality rate of 0.369. Slow-growth mortality is most pronounced in very small and very large trees. Small trees have very small diameter increments naturally, and because they are often suppressed by shading, experience a high mortality rate. Trees that approach maximum size eventually approach a potential increment of 0.0 (the growth curve is asymptotic), and so also enter the domain of stress mortality: they 'die of old age'. This implementation is somewhat arbitrary, but reproduces the U-shaped mortality schedule observed for trees (e.g. Harcombe, 1987). Importantly, because species differ in their inherent growth rates and tolerance to environmental stresses (suppression, drought), they may exhibit very different mortality schedules despite having the same basic implementation.

Of the demographic processes, the growth stage is deterministic and predictable while mortality and regeneration are stochastic and less predictable. Because trees may live for centuries, a stochastic event such as an episode of mortality induced by severe drought followed by a burst of wet-year regeneration, may be propagated into and preserved in forests for very long periods (Shugart and Urban, 1989).

(c)   Environmental feedbacks

Most gap models simulate the constraints of available light, soil moisture, soil fertility or nutrient availability, and temperature. All models emphasize the importance of light. Available light is computed in terms of percent of full sun, and falls off negative-exponentially within the canopy according to the Beer–Lambert law (Monsi and Saeki, 1953; Botkin et al., 1972b). The light available to each tree is thus determined by the aggregate leaf area of all taller trees on the plot. Competition for light is exploitative or asymmetric, with the intensity of asymmetry determined in part by the extinction coefficient in Beer's law (Table 7.1). For many forest canopies, this coefficient takes on values on the order of 0.40 (ranging c. 0.25–0.50: Jarvis and Levernez, 1983). A smaller extinction coefficient makes the canopy 'leaky' and reduces the asymmetry of light-competition, while a larger coefficient strengthens this asymmetry. Because this coefficient determines the extent to which small trees can subsist beneath taller trees and, in some models, the extent to which trees shade themselves, total stand-level productivity is also sensitive to this coefficient.

Soil moisture is typically modelled in terms of a drought index, based on a simple 'bucket' water balance computed on a monthly timestep (Pastor and Post, 1986; Botkin and Nisbet, 1991). Temperature affects trees directly via a growing degree-day heat sum (Botkin et al., 1972b; Solomon et al., 1984) and indirectly via its effect on potential evapotranspiration. Interannual variation in degree-days and soil moisture are simulated by generating pseudo-random variation in monthly temperatures and precipitation, based on long-term weather records for the study site.

Nutrient effects vary widely among models. A 'soil quality' constraint

Table 7.2   Level of resolution in coupling between plants and environment in gap models. Levels are STAND, whereby all trees have the same effect or response, regardless of species or size; TYPE, where responses are defined by functional category (e.g. shade-tolerance class), regardless of size; SPECIES, where each species has its own effect and response; and SIZE, where trees of different size (diameter or height) each have their own effect or response

	Plant effect	Plant response
Light	TYPE * SIZE: all models	TYPE * SIZE: all models[a]
Nutrients	SIZE: most models[b] TYPE * SIZE: FORTNITE *et seq.*[c]	STAND: JABOWA, FORET TYPE: FORTNITE *et seq.*
Moisture	STAND: LOKI[d]	SPECIES: most models after *c.* 1984
Temperature	STAND: LOKI	SPECIES: all models

[a] The FORSKA and ZELIG models compute light effects and responses for layers within each tree's crown.

[b] In models that do not simulate nutrient cycling, tree size affects plot-level biomass but all species have the same size-effect.

[c] FORTNITE *et seq.* are those nutrient-cycling versions that follow the FORTNITE approach; these include LINKAGES, JABOWA-II, and LOKI; FORNUT uses a similar approach.

[d] In LOKI, vegetation affects radiation incident to the ground and so affects PET and the soil thermal regime; still, trees do not 'transpire' water.

in the original JABOWA model was a simple boundary condition on plot-level biomass, i.e. a 'state-limiting' constraint. More recent versions of the model simulate details of litter decomposition and nutrient (nitrogen) effects on tree growth, which act in a qualitatively different mode as a 'rate-limiting' constraint.

The constraints vary in degree of coupling between plants and environment (Table 7.2). Light is fully coupled at the individual level, in that each tree affects the light profile of the plot via its leaf area; in turn, each tree reacts individually to the light available at its height. Nutrient availability (or its proxy, soil fertility) is variously coupled, depending on the model. The 'soil quality' boundary condition in some versions (e.g. Botkin *et al.*, 1972b; Shugart and West, 1977) is extrinsic in that trees do not affect its value, but the numbers and sizes of trees on the plot do affect how strongly this boundary condition is expressed (tree growth slows as the boundary is approached). Other models include feedbacks among soil nutrient status, forest stature and species composition (Aber and Melillo, 1982; Pastor and Post, 1986; Bonan, 1990a,b); potential productivity in these models is determined in part by the trees themselves, via the amount and chemistry of their litter. None of the current models strongly couples hydrology or temperature to vegetation (but see Bonan's model, below). In most gap models, trees are affected by

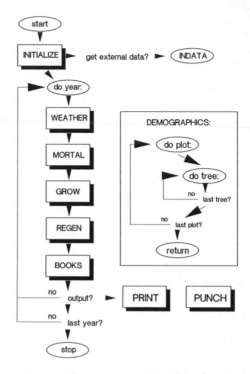

Figure 7.4 Flow diagram for a gap model. This flow is for ZELIG (Urban, 1990); in most other gap models, the outer loops – year and plot – are reversed. Specific routines vary among versions, but they all share the same general flow: an 'annual' loop and a 'plot' loop; demographic routines (*inset*) include a nested 'tree' loop in which individuals grow, die or are planted.

soil water status and by temperature, but they do not influence these parameters.

7.2.2   Implementation

A flow diagram for the ZELIG model is illustrated in Fig. 7.4. [Relative to previous gap models, the 'outer loops' in ZELIG are reversed: most models derived from JABOWA simulate each plot for as many years as required, then go on to the next plot; results from each plot are saved for output. The restructuring of ZELIG takes advantage of newer, more powerful computers and is geared towards spatial applications. Except for spatial applications, the two model structures are equivalent.] Because parts of the models (weather, seedling establishment and mortality) are stochastic, the models are solved with Monte-Carlo simulations. Stand-level phenomena are estimated by simulating a large (*c.* 20–100) number of model plots, and using the average to represent a forest in aggregate.

Table 7.3 Site and species parameters needed to implement a basic gap model[a]

Site parameters and units		
T, VT	°C	Mean monthly temperature (typically normals) and interannual standard deviation
R, VR	cm	Mean monthly precipitation and interannual standard deviation
FC, WP	cm	Soil water at field capacity and wilting point (estimated from soil texture and depth)
SF	Mg/ha/yr	Soil fertility, maximum above-ground woody production (approximated from other studies)
Tree species parameters		
A_{max}	yr	Maximum age typically achieved by species
D_{max}	cm	Maximum diameter at breast height (dbh)
H_{max}	cm	Maximum height of forest-grown trees
G	scalar	Growth rate constant (fitted from data)
DD_{min}	degree-days	Minimum and maximum growing degree-days
DD_{max}	(5.56°C base)	(estimated from elevation and latitudinal limits)
L	rank	Shade tolerance class (1–5, 1 = very tolerant)
M	rank	Soil moisture response (1–5, 1 = drought intolerant)
N	rank	Nutrient response class (1–3, 1 = very responsive)
SEED	rank	Establishment rate (relative to other species)

[a] Minimum data requirements to implement ZELIG version 1.0 (Urban, 1990). Other data may be required for model extensions or for other gap models.

(a)   Parametrization

One feature of gap models that has been fundamental to their widespread application and success is their rather simple formulations (Table 7.1) and, consequently, relative ease of parametrization (Table 7.3). Species parameters reflect maximum size and age, and environmental responses are typically specified as rank differences among species (e.g. relative shade tolerance). Often these parameters can be obtained from silvics manuals or forestry literature, with little site-specific fieldwork. Site parameters summarize soil characters and monthly normals for precipitation and temperature, which are also readily available for most sites.

This mode of model formulation and parametrization has occasionally been criticized due to its simplicity and lack of mechanistic (i.e. ecophysiological) detail. Yet one might also argue that this very aspect is a chief advantage of the JABOWA approach: one can implement a gap model with very limited site-specific demographic data. This is not to say the parameter values cannot be improved from initial estimates – they can be, and model refinement and calibration are typically an important phase of model implementation. Still, a preliminary model can be implemented quickly and easily from available information. This ease of implementation allows the modeller to proceed quickly to model testing and refinement, focusing on model improvements rather than initial development. This procedure for model implementation is probably responsible, in large part, for the proliferation of models evident in Figure 7.1.

(b) Model testing

A bonus of individual-based simulators is that model output can be compared directly to data collected in empirical studies. In its raw form, model output can have the same resolution and scope as field tally sheets on which trees are recorded by species and size, and a model plot corresponds nicely to a sample quadrat.

Testing procedures used with gap models have been detailed by Shugart (1984). Examples of 'target patterns' on which gap models have been verified include: matching forestry yield tables; matching stand structure (basal area, stem density, and/or diameter frequency distributions) or species composition (relative basal area, relative density) for stands of known age; reproducing elevation gradients in species composition; predicting forest response to disturbances such as clear-cuts, floods, fire and hurricanes; predicting average diameter increments for remeasured trees; and matching successional sequences inferred from chronosequences.

In many cases, simulations are tailored to be compatible with data available for model tests. Often, raw model output is post-processed or transformed before analysis (e.g. sorted by species or diameter classes, or aggregated to the stand level). The capability to post-process model output is important because model tests often use archived data or studies originally conducted for other purposes. For example, the FORET model was tested against field studies conducted several decades before the model was developed (Shugart and West, 1977). Often, valuable data are reserved during model development, specifically to be used in subsequent model tests. Partly as a result of this versatility in tailoring tests to available target data, gap models are rather well tested as a set of models.

It might be noted that gap models have the potential to generate an intimidating volume of simulated data (i.e. multiple plot tallies extended over hundreds of years). Moreover, the models can provide data not easily measured in field studies (e.g. growth histories for thousands of trees, time traces over multiple generations of tree replacement). Thus, the models have the potential to far outstrip data available from conventional field studies. In a sense, there are *never* sufficient data to completely validate such a model, and one's confidence in the model improves as additional tests are met (Mankin *et al.*, 1977). A continual challenge in model applications is to devise appropriate testing protocols.

7.3 DIVERSIFICATION AND MODEL EXTENSIONS

From the original JABOWA model, a large number of variations on the basic gap model have been developed. Many of these were modifications of the base model to simulate a particular forest; these geographic variants typically added or emphasized some site-specific detail, such as a local disturbance regime. In the following section, we highlight a few of these variations.

7.3.1   Model extensions

(a)   Geographical distribution

JABOWA was developed to simulate northern hardwood forests in the northeastern United States, and temperate deciduous forests are still the ecosystems to which gap models are most frequently applied. But gap models have been implemented in a wide variety of forests throughout the world, representing forests from the humid tropics to boreal taiga and from swamps to montane conifers (see legend to Fig. 7.1). In many cases, the successful implementation of a gap model has required the addition of phenomena specific to a particular system (flooding, permafrost, fires). Yet, the breadth of applications underscores the generality of the 'gap dynamics' paradigm and the power of individual-based simulators as forest models.

(b)   Environmental constraints

The original gap model emphasized the primary importance of light in controlling forest dynamics, and this emphasis has largely persisted in gap models. In most gap models, a tree is represented as a stem of a given diameter, with its leaf area (computed allometrically from diameter) concentrated at the top of the stem. JABOWA and most descendant gap models keep account of tree heights in 10-cm increments. As a result of this representation, a tree that is 10 cm taller than a second tree may shade the shorter tree completely because all of its leaf area is above the shorter tree. Leemans and Prentice (1987) revised this approach to distribute leaves along a tree's bole. They assumed that a tree in full sun would hang leaves to the ground. By dividing total leaf area by tree height, they compute a unit foliage density (leaf area per height increment); they then hang leaves at this foliage density, from the top of the tree downward, until the light compensation point for the tree (defined by its shade tolerance class) is reached. Leaf area below the compensation height is lost permanently, so trees 'prune' from below as the canopy develops. This approach has also been adopted in the ZELIG model (Urban, 1990; Urban et al., 1991). Distributing leaf area within tree crowns provides more realistic foliage profiles in simulated forests, which opens a new domain of model applications concerned with forest stratification (Urban et al., 1991; Weishampel et al., 1992). This approach also allows trees in the model to self-shade, and so improves the model's ability to simulate even-aged stands.

Soil moisture effects in the models were substantially revised in the 1980s (Pastor and Post, 1985, 1986; Botkin and Nisbet, 1991). Most gap models use a simple 'bucket' water balance routine, based on a single-layer of soil and the readily parametrized Thorthwaite–Mather approximation of potential evapotranspiration (PET) (Thorthwaite and Mather, 1957). Bonan (1989a,b) uses a Priestley–Taylor approximation of PET (Campbell, 1977), which is driven partly by solar radiation. Because insolation varies with slope and aspect, this version of the model can simulate topographic influences on soil moisture. Bonan's model also simulates the biophysical details of the soil thermal regime and feedbacks between vegetation and below-ground processes, which, in turn, determine the distribution of permafrost and so influence water availability in boreal soils (Bonan 1989a,b; Bonan et al., 1990). Because forest cover intercepts incident radiation, Bonan's model also couples vegetation to PET by shading the soil surface. No conventional gap model currently couples trees to site water balance to the extent that trees actually use ('transpire') water, but Friend's HYBRID model has this capability (Friend et al., in press; see below).

At least five gap models simulate nutrient cycling in some fashion, although details vary (Aber and Mellilo, 1982; Weinstein et al., 1982; Pastor and Post, 1986, 1988; Bonan, 1990a,b; Botkin and Nisbet, 1991). FORTNITE (Aber and Melillo, 1982) and models derived from it (Pastor and Post, 1986; Bonan, 1990a,b; Botkin, 1992) emphasize the feedbacks between vegetation and nutrient cycles. These models track the dynamics of annual cohorts of leaf litter and woody debris, thus propagating tree population dynamics into below-ground ecosystem processes. Bonan's model, again because of the detailed soil biophysics and insolation-driven PET routine, provides rich capabilities to couple vegetation to soil processes (Bonan, 1990a,b).

(c) Demographics

The FORSKA model of Leemans and Prentice (1987, 1989) further developed the original algorithm for tree growth by incorporating the relationship between leaf area and sapwood area (Waring et al., 1982; Waring and Schlesinger, 1985) and invoking elements of pipe-model theory (Shinozaki et al., 1964a,b; Makela, 1986). This model computes separate 'assimilation' and 'maintenance' components of growth, and uses the difference between these to estimate net production (diameter increment) for each tree, each year. This model was perhaps the first gap model to attempt to 'grow' trees via realistic physiological processes.

More recently Friend et al. (in press) have developed a very detailed, individual-based simulator that simulates the ecophysiology of photosynthesis, resource use and allocation within a gap-model (ZELIG) framework. This model interfaces gap models to ecophysiological 'big leaf' models (Running and Coughlan, 1988) and is appropriately named HYBRID. It should be noted that these models, while appealing in their mechanistic detail, require a larger number of parameters and are much more computer-intensive than a conventional gap model.

Other models have elaborated aspects of tree establishment and regeneration. Shugart *et al.* (1981), in developing a model for subtropical rainforests in Queensland, incorporated the details of seed pools and seedling establishment as well as unusual modes of regeneration used by 'strangler' figs. Busing and Clebsch (1987) implemented a spruce–fir simulator on a gridded model plot, and kept track of the spatial coordinates of each tree on the plot. They computed competition indices for the zone of influence of each large tree, and adjusted seedling establishment to account for these influences: seedlings were planted in positions between dominant trees (i.e. on safe-sites). In the SILVA model of Sierran mixed-conifer forests, Kercher and Axelrod (1984) explicitly modelled the dynamics of masting as a probabilistic event with a minimum lag time between good seed years. Finally, most gap models for eastern deciduous forests simulate regeneration via stump-sprouting, which reflects the importance of sprouting in these forests (Shugart and West, 1977). The inclusion of stump-sprouting also frees gap models somewhat from their original reliance on strictly gap-phase regeneration.

Still other versions have expanded the mortality routine in the basic gap model. In CLIMACS, a model of Pacific Northwestern conifer forests, Dale and Hemstrom (1984) used long-term empirical records from permanent plots to estimate mortality schedules for several classes of species; these schedules replaced the mortality rates used in the basic gap model. Leemans and Prentice (1987) replaced the implementation of stress mortality (originally a step function) with a continuous, probabilistic function based on the concept of growth efficiency (Waring, 1983). Growth efficiency E is a measure of biomass (or volume) gain per unit (m^2) leaf area. Leemans and Prentice defined relative growth efficiency E_{rel} as the ratio of realized efficiency to maximum efficiency (under optimal conditions); trees are subjected to an increasing probability of mortality as their relative efficiency decreases.

Many gap models have been used to simulate disturbances, both natural and management-related. Several versions simulate fire regimes (FORAR – for Arkansas pine stands: Mielke *et al.*, 1978; BRIND – Australian eucalypts: Shugart and Noble, 1981; SILVA – Sierran mixed conifers: Kercher and Axelrod, 1984; FIRESUM – Rocky Mountain conifers: Keane *et al.*, 1990). SILVA and FIRESUM in particular incorporate detailed feedback relationships linking stand condition, fire intensity, tree mortality and forest regeneration. A few models simulate the effects of flooding on bottomland forests (Tharp, 1978; Phipps, 1979; Pearlstine *et al.*, 1985). Doyle's (1981) FORICO model simulates the effects of hurricanes on Puerto Rican forests. Gap models have also been used to examine the effects of timber harvest on forest dynamics and recovery (Aber and Melillo, 1978, 1979). Smith *et al.* (1981) incorporated habitat classification functions into a gap model and used this simulator to examine the effects of timber management on habitat availability for forest songbirds (see also Smith, 1986).

(d)   Alternative formulations

Several versions have implemented alternative functional forms for various equations used in the basic model. These modifications have included alternative growth equations (Phipps, 1979; Smith and Huston, 1989), site-specific allometries (Dale and Hemstrom, 1984; Leemans and Prentice, 1987; Harrison and Shugart, 1990) and modified environmental response functions (Smith and Huston, 1989; Urban, 1990). These modifications reflect a concern borne of available data that argue against the default (original) formulations or simplifying assumptions. In particular, height–diameter allometries and diameter growth equations can often be improved substantially relative to the default estimations, if sufficient empirical records are available. Most current gap models are sufficiently modular that such model 'upgrades' are easily incorporated, typically by replacing a single equation in the computer program.

(e)   Spatial implementations

When Busing and Clebsch (1987) modified a version of FORET to consider fine-scale spatial heterogeneity within a model plot, they gridded each model plot and modified the regeneration routine so that seedlings were planted in 'safe sites' between dominant trees. This approach was later revised substantially, so that a new model (Busing, 1991) simulates a larger total area, and the zone effectively influencing each tree is computed with respect to all of the trees within a specified distance (10 m) of the tree. This approach 'blurs' the boundaries of a gap-scale plot, and thus departs somewhat from the general approach of gap models in a way reminiscent of spatially explicit, stand competition models (e.g. Ek and Monserud, 1974; see also Munro, 1974; Shugart and West, 1980).

Smith and Urban (1988) developed a spatial implementation of a gap model, ZELIG, which they used to examine the scaling implications of forest succession and the shifting-mosaic steady state. The original ZELIG model simulated a grid of crown-sized (100 m^2) plots which were linked together such that trees on adjacent plots influenced each other (Fig. 7.5(a)). Through this spatial coupling, a large tree (or the gap it creates when it dies) could propagate an effect beyond a single plot to the larger grid.

The spatial implementation of ZELIG was later extended to consider the effects of sun angle on tree shading within the forest canopy, and to

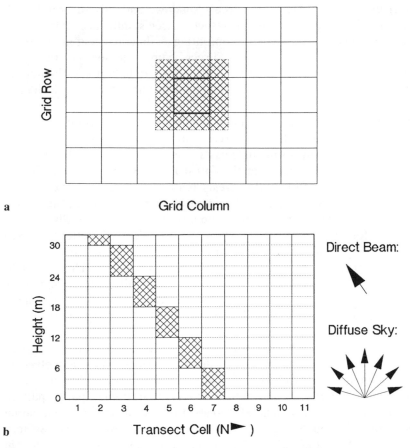

Figure 7.5 Spatial implementation of the gap model ZELIG. (a) The original gridded version, in which leaf area and biomass on a 0.04 ha plot aggregated around a 0.01 ha target cell (shaded) influence trees on the target cell (redrawn from Smith and Urban, 1988). (b) The transect version, which simulates the effects of solar angle and tree height to allow trees on adjacent cells to influence neighbouring cells. Direct-beam insolation impinges through a diagonal leaf-area profile (shaded) at solar incidence angle; diffuse-sky radiation is averaged over several diagonal profiles at various angles (redrawn from Urban *et al.*, 1991).

partition the light regime into direct-beam and diffuse sky radiation (Urban *et al.*, 1991). This extension incorporated the distributed-crown approach of the FORSKA model (Leemans and Prentice, 1987) and utilized general routines for estimating direct and diffuse radiation (Bonan, 1989a). This version of ZELIG is implemented on a transect of crown-sized plots; the transect is assumed to lie in a S–N orientation, so that shadows are cast from a given plot to the next (northward) plot. Sun angle (solar inclination angle as integrated diurnally over the growing season) defines the length of a tree's shadow, and this shadow length determines how far along the transect a tree's influence extends. In the model, this is estimated by computing a diagonal leaf-area profile oriented according to sun angle; direct-beam radiation is attenuated through this diagonal (Fig. 7.5(b)). Diffuse-sky radiation is estimated by 'sampling' the sky with a series of diagonal profiles at varying incidence angles.

In applications, this spatial version of the ZELIG model demonstrates a significant latitudinal effect in forest composition (Urban et al., 1991). At high latitudes a lower sun angle casts longer tree shadows; this reduces the amount of light available at the forest floor following the death of a single tree. Thus, high-latitude forests tend not to develop the pattern of light-mediated succession that is typical of temperate latitudes (see also Shugart, 1987; Prentice and Leemans, 1990). Reciprocally, forests simulated for tropical latitudes show an exaggerated pattern of gap dynamics; high sun angles maintain a greater proportion of shade-intolerant species in the forest, even in late succession. This is not to argue that other factors do not contribute substantially to forest dynamics at these latitudes, yet the simulation results suggest that sun angle alone can have a significant effect on forest pattern by modifying the scale and heterogeneity of the light regime.

This version of ZELIG has also been used to examine spatial aspects of the zone of influence of canopy-dominant trees. Weishampel et al. (1992) simulated a Pacific Northwestern conifer forest dominated by Douglas-fir (Pseudotsuga menziesii) and compared the spatial pattern (as semi-variance) in simulated forests to that measured from aerial videography of young, mature and old-growth forests in the western Cascades (Cohen et al., 1990). The model reproduced relative trends in spatial variance with forest age, although fine-scale details at less than 10 m resolution (i.e. plot width) could not be resolved with the model. This model output was also analysed with serial cross-correlation, relating the abundance of large (>50 cm diameter) to small trees (<20 cm) at increasing distances apart along the transect. The lag distance for maximum cross-correlations increased with forest age, and then decreased after the canopy broke up (Fig. 7.6). They interpreted this pattern to reflect an increase in the zone of influence – hence an increase in the spatial grain of the forest – as trees grew larger. The lesser grain of old-growth forests seems to be a result of averaging a variety of grain sizes.

The ZELIG model has also been extended, in a preliminary fashion, to landscape-scale applications. Urban and Smith (1989a) modified a transect version of ZELIG to simulate seed dispersal, and examined the effect of local seed availability on forest development across a topographic soil-moisture gradient. The model demonstrated strong positive-feedback effects that amplified species composition: soil-moisture patterns conferred an initial advantage to certain species on a given soil type, and

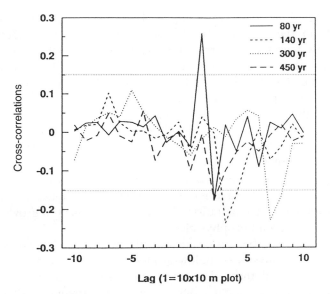

Figure 7.6 Spatial grain of forest, or zone of influence of canopy-dominant trees, as lagged cross-correlation between large and small trees along the transect version of ZELIG (from Weishampel *et al.*, 1992). With increasing forest age, the displacement of the maximum negative cross-correlation (as positive lag distance) indicates the distance along the transect at which large trees suppress small trees, i.e. their zone of influence. This zone reaches its maximum value in mature forests, before the canopy breaks up.

seed rain from these 'founders' reinforced the initial advantage until single species achieved nearly pure stands locally along the soil gradient. In fact, this result is dictated by model structure to some degree, in that the 'current composition–seed rain–future composition' loop is self-amplifying. The extent to which this feedback leads to monospecific dominance is very sensitive to the strength of coupling between local composition and local seedling establishment. In the model, this coupling is defined by a 'distance' parameter that determines how far along the transect seeds may disperse. We currently lack empirical records to quantify this parameter with precision and confidence. Still, this general approach of extending gap models to larger spatial scales holds much promise for landscape-scale applications.

7.3.2   Current themes in model extensions

In the proliferation of gap models in the 1980s, there was a seemingly undirected 'tinkering' with many aspects and details of the original model. This had the net benefit of providing a rather complete (if not exactly systematic) exploration of the basic model. An attendant result, less fortunate, is that very few if any of these model variants are directly comparable in their details. In the past few years, this tendency has been partially reversed: the trend now seems towards greater generality and comparability, often involving collaboration and direct sharing of computer code. Two specific trends seem especially noteworthy.

One trend in gap models is towards greater attention to mechanistic details rather than more phenomenological representations. The FORSKA (Leemans and Prentice, 1987, 1989; Leemans, 1991) approach to growing trees is a good example of this attention to ecophysiology. The HYBRID model of Friend et al. (in press) relies even more on ecophysiological process. While this latter approach is a bit parameter-intensive and computationally demanding for routine applications, it might be used to reformulate a computationally efficient approach to growing trees which is consistent with ecophysiology.

Similarly, there has been a trend towards greater mechanistic detail in simulating the biophysics of below-ground processes. Bonan's detailed treatment of the soil thermal regime and moisture balance is a good example of this trend (Bonan, 1989a, 1990a,b). Bonan's work suggests that the behaviour of gap models might change qualitatively according to the level of detail with which biophysical processes are simulated, particularly when models differ with respect to the inclusion of system-level feedbacks (Bonan 1990a,b; cf. 1991). This cautions that model output always should be interpreted at a level of resolution consistent with the level of detail simulated in the model, and that interpretations should always be made strictly relative to model assumptions (Urban et al., in press).

A second trend that seems noteworthy is a trend towards greater generality in model algorithms and parametrization. In part, this comes hand-in-hand with the simulation of mechanistic processes; biophysical processes in particular tend to be rather general. A recent project illustrates this goal to devise general models. An international collaboration was formed to attempt to develop a boreal forest simulator sufficiently general to apply to any boreal site in the northern hemisphere. Bonan's model represents a component of this project emphasizing below-ground processes, while the FORSKA model of Leemans and Prentice represents the above-ground component of the general model. These two models have since been merged into a unified boreal forest simulator, BORFORS, and the model has been implemented and tested in Alaska, Canada, Fenno-Scandinavia and Siberia (Shugart et al., 1992). The success of this project witnesses not only the value of general models, but also the benefits of a large network of collaborators with expertise in a broad array of systems.

The generality of individual-based simulators and gap models is underscored by their successful extension to non-forest vegetation. The STEPPE model (Coffin and Lauenroth, 1989, 1990) is an individual-based simulator for semi-arid grasslands which incorporates several life-forms (annual grass, perennial grass, forb, succulent, shrub). Compared to a forest gap model, STEPPE is more or less inverted: the relevant dynamics are below-ground, where rooting behaviour relative to soil-moisture availability determines plant success. Although the grassland and forest models have many similarities, some important differences accrue because of the contrasting life-history traits (plant size, growth rates and longevity) of the plant life-forms (Coffin and Urban, in press).

A new collaborative project further illustrates the trend towards generalizing gap models. In this project, generalized models are being used as a framework for comparisons of different ecosystems across broad (continent-scale) environmental gradients (Smith *et al.*, 1989; Coffin and Urban, in press; Lauenroth *et al.*, 1992). A particular goal is to implement the same individual-based vegetation model at each of several sites representing distinct climatic regimes. An important task of this project will be to generalize an individual-based simulator to such extent that the same model can be applied to vegetation dominated by various life-forms. The common model will provide a standard for comparison, to help identify site-specific processes and constraints by highlighting their relative importance at each site.

7.4   APPLICATIONS

In this section we present selected applications of gap models to the general issue of forest succession. The examples are selected to highlight the range of questions that the models may address, and their versatility in application.

7.4.1   The shifting-mosaic steady state

One of the original applications of a gap model was the use of JABOWA in formulating and illustrating the concept of succession as a shifting-mosaic steady state (Bormann and Likens, 1979). In this conceptual model, a forest is envisioned as a mosaic of small (gap-sized) patches, each of which is undergoing its own successional trajectory. Secondary forest succession can be divided into four stages: reorganization (which includes establishment), aggradation, a transitional stage of canopy breakup, and the steady state. Because the individual patches may be in different stages of succession at any given time, the forest as a whole is a dynamic mosaic, ever-changing at a small scale, yet remaining largely the same at a larger scale of reference (see Chapters 1, 3 and 4). Bormann and Likens also used JABOWA to estimate the proportions of a forest mosaic that might be in each of the four stages of succession at any given time, although they cautioned that other factors not in the model (especially disturbances) would alter these proportions.

(a)   Scaling implications of the shifting mosaic

The contrast between small-scale and aggregate dynamics of a forest stand has been a continued focus of model applications (Shugart, 1984). Smith and Urban (1988) used the original gridded ZELIG model to simulate a 9-ha stand of upland forest. They collated trees into 10-cm size classes and summarized forest structure by computing principal components from these size classes. The first two principal components reflected the abundance of small and large trees respectively, and corresponded to density of the understorey and overstorey. They then resampled the simulated forest at a range of spatial scales with square quadrats of 4, 9, 16, . . . , 100 grid cells, and recomputed principal component scores for the average of each aggregate sample. Qualitative differences in forest structural heterogeneity (as variance in principal component scores) were evident at different scales of reference (Fig. 7.7). Smith and Urban emphasized in particular the relationship between small-scale and large-scale successional dynamics: at the scale of the forest gap, a plot undergoes non-equilibrium dynamics driven by the demographics of large trees; at the aggregate-stand level, succession is a smooth trajectory that stabilizes at an equilibrium structure and composition determined by site conditions. Importantly, these two views are complementary: either is 'correct' but each is a scale-dependent description of succession.

7.4.2   Forest habitat dynamics and avian succession

Urban and Smith (1989b) used their framework of succession as a statistical construct (Smith and Urban, 1988) to examine the importance of forest dynamics for forest bird communities. They defined microhabitat variety in terms of statistical variance on principal component scores, and used these scores to illustrate the trend in microhabitat variety during succession. Their simulation produces a rapid increase in microhabitat diversity in early succession, which reaches a peak as the canopy breaks up and then equilibrates at a slightly lower steady-state diversity (Fig. 7.8, and recall the Odum–Margalef model: see Prologue; Odum, 1969; Margalef, 1968).

Urban and Smith defined bird species as ellipses in the 2-dimensional principal component space, analogous to conventional 'niche analyses' in avian ecology (Hutchinson, 1957; James, 1971). To develop a neutral model that focused on microhabitat effects, they generated hypothetical 'species' defined by random niche dimensions on the two axes. They then computed the abundance of available habitat for each bird species through succession, and used these habitat-abundance indices to illustrate patterns in bird community structure as dictated by microhabitat dynamics.

Their simulations reproduced many of the phenomenological patterns

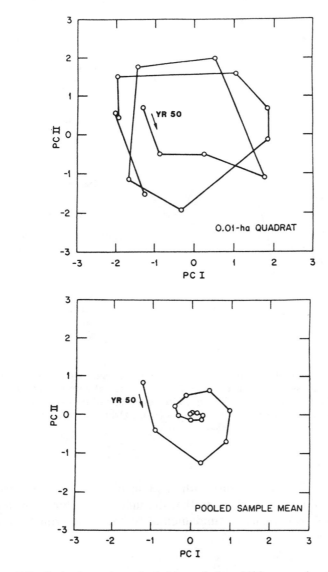

Figure 7.7 Scale-dependent depictions of the shifting-mosaic steady state, as simulated with ZELIG (Smith and Urban, 1988). Forest structure is summarized in terms of two principal components corresponding to understorey (PC I) and overstorey (PC II) trees. (a) Non-equilibrium dynamics at the plot level, driven by demographics of large trees; (b) succession to an equilibrium (steady-state) 'climax' at the aggregate stand level.

in bird community structure observed in empirical studies, including realistic trends in species diversity and turnover (Figs 7.9(a) and (b)). At steady-state, microhabitat dynamics dictated realistic patterns in the distribution of species abundance classes and a species-area effect (Figs

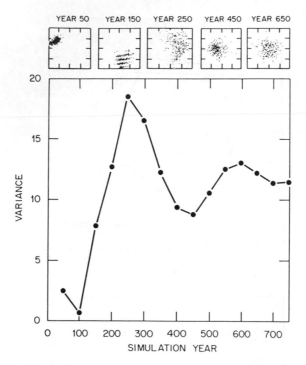

Figure 7.8 Microhabitat variety through succession as simulated for an upland forest in east Tennessee. Variance is variance on principal components of forest structure as identified in Fig. 7.7. *Insets*: scatter-diagrams of plot-scale variance in structural components, at selected forest ages (from Urban and Smith, 1989).

7.9(c) and (d)). The simulations with random bird 'species' conformed well to empirical results, suggesting that the influence of forest dynamics on microhabitat availability is fundamental to the structuring of forest bird communities.

7.4.3   Life-history traits and succession

Huston and Smith (1987) used a gap model to examine the implications of tree life-history strategies for forest successional pathways. To ensure some measure of control in their model experiments, they 'invented' hypothetical tree species that they defined in terms of five life-history traits: maximum size (height and diameter), longevity, maximum growth rate, shade tolerance, and sapling establishment rate. They argued that only certain combinations of these traits are biologically feasible, an argument reminiscent of the traditional *r*-K dichotomy and underpinning Tilman's resource-ratio hypothesis (Tilman, 1982, 1988) and the 'vital

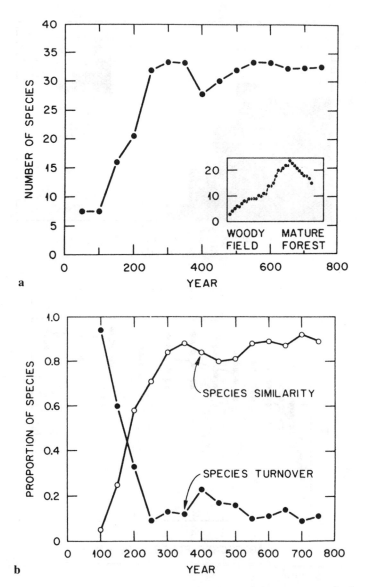

Figure 7.9 Patterns in forest bird communities as a consequence of microhabitat dynamics. Bird 'species' were defined as random niche ellipses in the microhabitat space illustrated in Figs 7.7 and 7.8, and habitat abundance within each ellipse at each time step was tallied as an index of potential species abundance. Predicted successional patterns in (a) bird species richness (average number of species) and (b) species turnover and similarity (as Sorenson's index). Steady-state patterns in (c) species abundance classes and (d) the species–area relationship. Insets are results from empirical studies (from Urban and Smith, 1989).

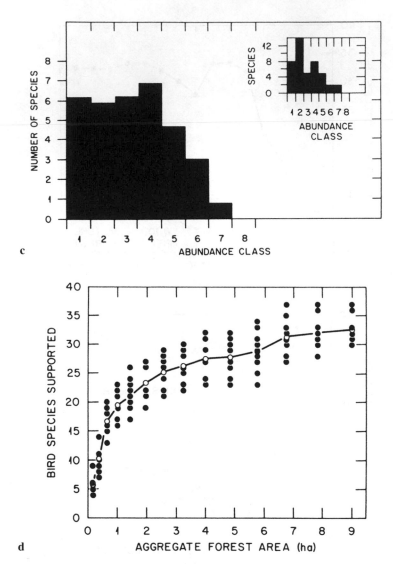

Figure 7.9   *Continued*

attributes' approach to modelling succession (Noble and Slatyer, 1980). Huston and Smith constructed a model experiment that simulated pairs of tree species which contrasted a single trait (e.g. shade-tolerant vs intolerant) while holding all other traits constant, analogous to a partial-correlation ANOVA design. They found that of the large number of possible life-history combinations, simulations could be grouped into five basic patterns. They summarized these patterns as classical successional replacement, divergence, convergence, total suppression, and pseudo-cyclic replacement (Fig. 7.10). An important result of this model experiment was that a single mechanism, competition for light, could result in a variety of successional pathways, depending on which life-history strategies were represented on a site.

Smith and Huston (1989) extended this approach to encompass two classes of resources, above- and below-ground. Their working hypothesis in this case invoked arguments about trade-offs in allocation strategies: that a tree cannot simultaneously maximize its efficiency above and below ground. Thus, they postulated a set of hypothetical species such that (a) a species that can grow rapidly under high light availability must also be intolerant of shade (conversely, a shade-tolerant tree also has a low maximum growth rate under full sun); (b) a species that can grow rapidly given abundant soil moisture must be intolerant of drought (conversely, a drought-tolerant tree has a low maximum growth rate); and (c) tolerance to low light and low moisture availability are inversely related, so that a shade-tolerant tree cannot also be drought-tolerant (although a tree could be intolerant to shade as well as drought). These premises are supported by physiological, anatomical and morphological evidence about plant strategies (Bazzaz, 1979; Bazzaz and Picket, 1980; Larcher, 1980; Orians and Solbrig, 1977; as discussed in Smith and Huston, 1989). They used the premises to generate an array of plant functional types on two resource-use axes, holding other life-history traits (maximum size, longevity) constant (Fig. 7.11(a)). An example of a shade-tolerant, drought-intolerant species might be American beech (*Fagus grandifolia*), corresponding to plant type 1 in the figure; scarlet oak (*Quercus coccinea*) might represent an opposite strategy of plant type 15, and tuliptree (*Liriodendron tulipifera*) is an example of a tree that grows very rapidly but only on very mesic sites (plant type 5).

Smith and Huston used these hypothetical species in a series of simulations representing a soil-moisture gradient. Their simulations illustrate the intuitive result that the species with the highest initial growth rate for a given soil-moisture regime had an early-seral advantage, while the most competitive shade-tolerant species on that site eventually came into late-successional dominance (Fig. 7.11(b)). Because of the trade-offs implicit in the plant types, species that compete well on mesic sites could not also persist on xeric sites. Likewise, species that did well early in succession on mesic sites could not persist into late succession on these sites. These same trade-offs dictated patterns in species diversity in time and over the soil-moisture gradient: on xeric sites diversity is low throughout succession because of the constraint that only drought-tolerant species can survive and persist; on mesic sites diversity is maximized in early succession but competition decreases diversity through time until the best-adapted species dominates (Fig. 7.11(b)).

The study of Smith and Huston (1989) also provides some insight into

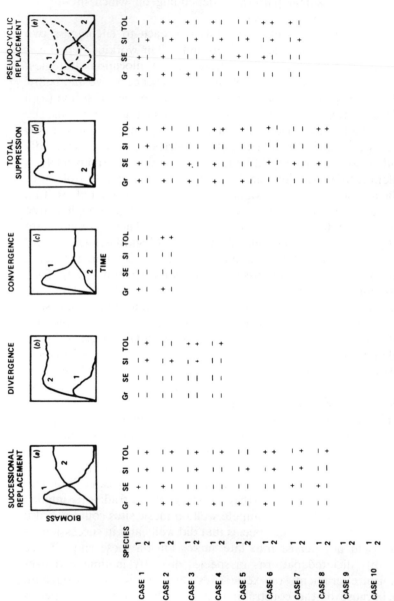

Figure 7.10 Typal successional patterns emerging from competition between tree species with various life-history strategies (from Huston and Smith, 1987). Hypothetical species were defined in terms of life-history traits representing an *r-K* dichotomy, and pairs of species were contrasted in each simulation; similar seres were then collated into five typal patterns. Life-history traits are growth rate (Gr), seedling establishment rate (SE), maximum size (SI), and shade tolerance (TOL); + and − denote comparatively high and low values.

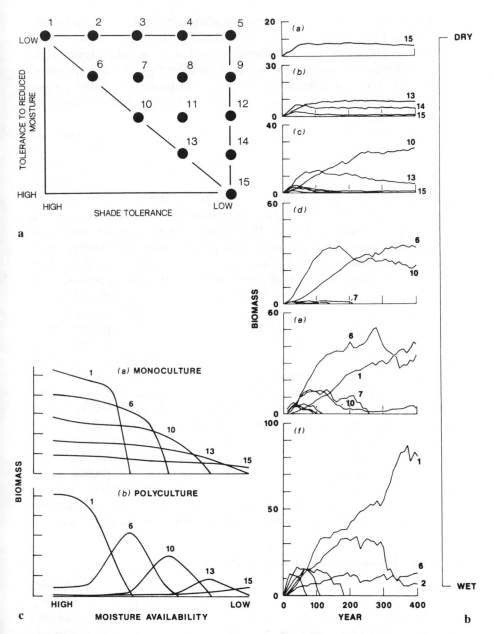

Figure 7.11 (a) Life-history strategies posed by Smith and Huston (1989), to represent trade-offs in adaptations toward above- and below-ground resources. (b) Seres along a soil-moisture gradient, involving species from (a). (c) Physiological modes (potential niches) vs ecological modes (realized niches) of tree species along the soil-moisture gradient.

the importance of competition in shaping plant communities. The premises they invoked suggest that most species would perform best under similar conditions of high availability of above- as well as below-ground resources. Yet, competition excludes most species from their physiological optimum, leaving only the best competitors to persist on these sites: most species realize an ecological mode somewhat removed from their physiological mode (Fig. 7.11(c)).

Austin and Smith (1989) used these results to formulate a revision of the continuum concept of vegetation. They defined direct physiological (e.g. temperature, pH) and resource (light, nutrients, water) gradients and postulated fundamental forms of species response to these gradients. They then invoked species life-history traits and the mechanism of competition to explain the wide variety of species-distributional patterns and successions observed in nature. They argue that the revised continuum concept subsumes many previous conceptual models (Ellenberg, 1953, 1954; Mueller-Dombois and Ellenberg, 1974; Gauch and Whittaker, 1972; Grime, 1979; Tilman, 1982, 1988; Bazzaz, 1979) in accounting for a greater variety of observations with the same conceptual framework. The salient features of this conceptual model are (a) fundamental species responses to environment, as dictated by trade-offs (physiological, anatomical, or morphological) in life-history strategy; (b) spatial and temporal variation in the environment; and (c) competition as a mechanism to sort individuals (hence also species) in a given environment. Relative to previous conceptual models of succession which are either inductive at the level of the observed case (sere), or deductive for all seres, the 'life history in context' conceptualization embodied in gap models invokes another level of explanation. To use the analogy of succession as a game, this model argues that only the 'rules' are fixed; which players are present and the arena in which they compete determine the outcome of the game, but any combination of players and any arena is possible. Importantly, given the players and arena, the outcome (succession) is generally predictable with an individual-based model.

7.4.4   Forest dynamics and ecosystem function

Pastor and Post (1985, 1986) developed a linked model of forest dynamics and nutrient cycling based on the FORENA version of the FORET model (Solomon, 1986) and an expansion of the nutrient-cycling scheme in FORTNITE (Aber and Melillo, 1982). The model tracks the inputs of litter and woody debris to the forest floor, and simulates the decomposition of this debris according to litter chemistry, temperature, moisture and canopy openings. Pastor and Post have been especially concerned with the feedbacks between tree population dynamics and ecosystem cycling of C and N. They postulate a scheme of feedback loops linking tree species composition, soil moisture and nutrient cycling.

The important links of these feedbacks are: (a) tree species respond differentially to soil moisture and to nutrient (N) availability; (b) species have variable litter quality; (c) litter quality and soil moisture determine decomposition and N mineralization rates.

Such a feedback loop may have profound implications over successional time spans. Pastor and Post (1986) examined their working hypotheses by simulating forests across a spectrum of soil types in central Wisconsin. Xeric, nutrient-poor soils favour species such as pines, which are drought-tolerant and have low N requirements. Because these species also have low-quality litter, they do not improve the soil substantially with their litter inputs, and thus their dominance on these sites persists. On more favourable mesic sites, species such as sugar maple can establish. These trees grow very well under conditions of high N availability and soil moisture, and can outcompete pines on these sites. Because maples return nutrients to the soil as high-quality litter, the soil improves with time and the advantage of maples is amplified. Over time, either of these two pathways might develop as a reasonable scenario. Over time, *only* these two endpoints are favoured, and any initial condition tends towards one or the other case, depending on initial soil conditions and species composition. Moderate sites support a mix of oaks with pine and maple, but build an increasingly favourable soil and so succeed to maples (a classical 'facilitation' model). Available light plays a role in this feedback system, in that shade-intolerant species tend also to have poor-quality litter (e.g. pine), while tolerant species tend to have better quality litter and higher N requirements. Thus, the same qualitative patterns are observed through time as occur across a gradient from xeric to mesic soils. The results of Pastor and Post (1986) are very much in accord with the arguments of Smith and Huston (1989) concerning correlations among life-history traits and their implications for succession and environmental gradients.

Similar 'divergent pathways' scenarios have been simulated for forests in other forest types (Pastor and Post, 1986). For northern Minnesota forests, the LINKAGES model generates divergent pathways characterized by either aspen or spruce (Pastor *et al.*, 1987), reflecting the very different tissue chemistry and N requirements for these two species. The extent to which this may occur generally in forests is not clear; not all forests may include such markedly different 'alternative' species with contrasting life-history traits.

Simulations with Bonan's boreal forest simulator suggest that similar divergent pathway scenarios may be mediated by feedbacks between forest cover and the soil thermal regime (Bonan, 1989a; Bonan *et al.*, 1990), a model result substantiated by empirical studies (Bonan and Shugart, 1989). It appears that as biophysical processes are incorporated into the models with greater detail and realism, the capacity for significant ecosystem-level feedbacks increases as well.

7.5 FORESTS AS COMPLEX SYSTEMS

Forests are difficult systems to study because the demographics that govern their long-term behaviours occur on disparate spatial and temporal scales, from square metres and weeks for germination and seedling establishment, to hectares and millennia for mortality events. The ability of individual-based simulators to explicity couple tree growth, mortality and regeneration has led to the frequent use of these models to explore forests as complex systems (Shugart and Urban, 1989).

Shugart (1984) defined forest systems in terms of functional roles of trees. Roles were defined according to simple dichotomies in the mode of coupling between tree mortality and regeneration: a tree may or may not create a gap when it dies, and may or may not require a gap to regenerate. This pair of dichotomies yields four roles (Fig. 7.12(a)). Shugart noted that some roles tend to be self-amplifying (Fig. 7.12(b)). For example, role-1 species, in dying, create the conditions they require for regeneration. Similarly, role-4 species do not create conditions that would favour other species, and so favour their own persistence. Other roles tend to give the advantage away: role-3 species create conditions that favour roles 2 and 4, and cannot help their own cause but must depend on extrinsic agents (disturbances) to create conditions for their regeneration. Given this simple conceptual model, Shugart noted that particular forest types contained various combinations of these roles (Shugart, 1984), and further, certain environments clearly favour some roles over others (Shugart, 1987). To be sure, functional roles might be defined according to finer distinctions or other criteria (e.g. Whitmore, 1982, 1989); yet this simple example suffices to illustrate the heuristic value of considering tree species in terms of their functional roles in systems.

The coupling between mortality and regeneration can lead to system-level behaviours that are self-amplifying. Shugart *et al.* (1980) simulated a forest system comprising *Liriodendron* and *Fagus*, which represent roles 1 and 2, respectively. The two species have very different temperature limits, and a model experiment was designed to gauge the nature of system response to a slowly changing climatic forcing. Instead of a smooth replacement of one species by the other, the model showed hysteretic behaviour: the transition was very rapid and occurred at a different temperature depending on whether the climate was warming or cooling. In fact, this behaviour was largely a consequence of the mortality–regeneration dynamics of the two species. *Liriodendron* requires light-gaps to regenerate, and it creates these when it dies; *Fagus* is very tolerant and can regenerate in its own shade. A forest dominated by either species tends to have stand dynamics that favour the dominant species, and so is self-sustaining. Here, climate was an extrinsic forcing that shifted the model between alternative stable states, which gave rise to hysteretic behaviour. This example concerns roles 1 and 2, but

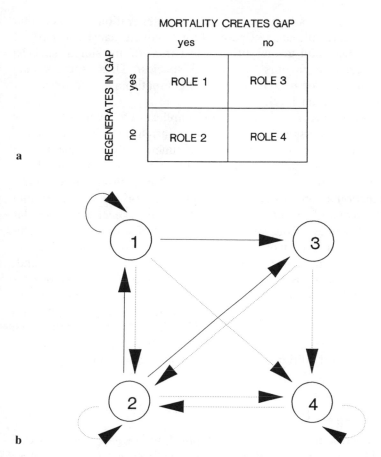

Figure 7.12 (a) Functional roles of tree species as defined by their mode of mortality and regeneration. (b) Roles as they might interact in long-term forest dynamics. Solid lines denote effect of a tree's dying at a large size; dotted lines, tree's effect when dying as a small tree (redrawn from Shugart, 1984).

analogous cases might be envisioned for other roles; role-3 species, in particular, are prone to stable cycles triggered by periodic disturbances such as fires, a system-level feedback originally posed as the 'flammability hypothesis' (Mutch, 1970).

The analogy of functional roles also implies that there may be two levels of competition in forest systems (Shugart, 1984; Shugart and Urban, 1989). At one level, trees within the same functional role compete in the conventional sense, and the 'winners' gain representation in the forest, as measured in terms of relative numbers or biomass. But, at another level, trees of contrasting roles also compete, with slightly more profound implications: winning roles also tend to favour the same roles in the next generation (recall the positive feedbacks in Fig. 7.12). This potential to pass the advantage to the next generation suggests that species of the same role are not competitors, but mutualists: any role-1 species that is successful favours other role-1 trees of any similar species; likewise, disturbance favours any role-3 species. Thus, species may compete *within* roles or *between* roles, and competitors at one level may be mutualists at another level.

This two-level model of competition implies two levels of species diversity, and this perspective offers some insight into the 'diversity/ complexity' issue. Species within roles are functionally redundant from a systems perspective: they are interchangeable cogs in the machine. Different roles are the diversity of importance in a system: roles determine the functional complexity of a system (van Voris *et al.*, 1980), and roles are the focus of examples of ecological convergence under similar environmental regimes. Role diversity is the stuff of systems, and system-level arguments must be invoked to explain its contribution to ecosystem function and its maintenance within a given system. On the other hand, within-role redundancy is a large component of the biodiversity of conservationists and evolutionary biologists who rely on taxonomic definitions of diversity. This diversity has its explanation in biogeographic and genetic mechanisms, which determine the origin and persistence of gene pools. While the two forms of diversity are related (every taxonomic species has its role), different mechanisms and constraints govern the two.

7.6 CONCLUSIONS

A general conclusion from the many applications of gap models is that forests tend to be characterized by positive-feedback mechanisms. These feedbacks are typically invoked at the level of the species by assigning each species a set of model parameters that define a life-history strategy. The specific action of these feedbacks is at the level of the individual tree: a tree's successful establishment, growth relative to other trees present, or eventual mortality sets the stage for the success of individuals on the plot. These feedbacks have implications at the level of the population, community, and ecosystem, although the specific details of the feedback loop may vary in each case.

At the population level, the self-amplifying tendency of certain roles can contribute to age-structure. Because gap-regenerating species tend to establish in discrete episodes (often following a disturbance), these populations exhibit cohort age-structures. Reciprocally, roles that readily establish under forest canopies tend to develop a mixed age-structure through continual (or at least frequent) recruitment. While this phenomenon is rather easy to verify in disturbance-driven systems (e.g. fire systems), less obvious cases of cohort age-structures have been observed in disturbance-free forest systems at high latitudes (Shugart, 1987).

At the community level, seed dispersal is an important mechanism by which species composition is amplified locally. This feedback is particularly sensitive if site conditions provide for initial compositional heterogeneity, but even in the absence of an underlying pattern, founder effects will be amplified over time.

From an ecosystem perspective, Pastor and Post (1986) have illustrated the feedback mechanisms that give rise to divergent successional pathways as a consequence of nutrient cycling. The fact that LINKAGES generates similar patterns for a variety of forest types suggests that this phenomenon may be rather general to forests, and Bonan's results (1989a,b; Bonan et al., 1990) suggest that such feedback scenarios may apply to other ecosystem processes. These system-level feedbacks also have direct couplings to population- and community-level patterns.

It is worth emphasizing that the model that generates these complex behaviours is actually rather simple: complex system behaviour does not require a complicated model. Yet it should also be emphasized that this behaviour is dictated to some extent by the model structure itself: the feedback loops specified in the model predispose it to simulate divergent pathways. In an analytic model, this sort of behaviour would be expressed in terms of 'attractors' in the parameter space and multiple stable states in the model solution. These sorts of feedback phenomena might be examined in more depth by interfacing individual-based simulators with more tractable analytic models, an exercise in cross-modelling that has been discussed but not yet accomplished (Horn et al., 1989; Cohen and Pastor, 1991).

A rather perplexing question that arises is, 'How real are these model results?' Although gap models are rather well tested as a group, it is difficult to verify model behaviours that develop over multiple generations of tree replacement. The simulations of Pastor and Post (1986) conform reasonably well to available data (which are rather sparse), but this in itself does not verify the longer term implications of the feedbacks postulated in their model. Indeed, it is difficult to identify specific field measurements that might definitively verify or disprove the model. The perplexing aspect of this is that the feedbacks implemented in the models are based on empirical results and field studies, and many conceptual or anecdotal models of forest succession invoke processes that are positive-feedback mechanisms. Thus, the goal in model testing is not so much to verify the models qualitatively (*do* these processes occur?), but rather to determine the *degree* to which these processes operate in forests; the latter is inherently a difficult task. Already, model versions exist which demonstrate qualitatively different behaviours depending on the level of detail with which specific processes and feedbacks are simulated. As individual-based simulators evolve towards increased mechanistic richness, they will outpace the capability of existing databases to verify their predictions. This should caution us that model results must be interpreted in terms of the assumptions implicit or explicit in the model. It should be emphasized that many of these are the same assumptions and working hypotheses that underly empirical research in forest gap dynamics. A real challenge in succession theory will be to devise appropriate procedures to verify and validate these models and the hypotheses they represent. This effort will proceed more efficiently by closely coordinating empirical studies and modelling.

7.7 PROSPECTUS

A number of recent trends in model development suggest that gap models may play a slightly different role in future research in forest ecology than they have in the past. Past versions of the models have tended to be site-specific and were developed for particular applications (i.e. to reproduce a local forest of interest). More recent versions of the models tend to place more emphasis on mechanistic details, and yet extrapolate these details to long time spans or large areas. As an upper bound on the range of these extensions, current gap models span time scales from hours (ecophysiology: Friend *et al.*, in press) to millennia (Holocene climate change: Solomon *et al.*, 1980; see also Chapter 8). Spatial implementations have extended the scale of applications from within the forest gap to landscapes and beyond.

As gap models are extended in these directions, they broach new sources of information from other disciplines. For example, gap models are currently being interfaced with ecophysiology processes, soil process and hydrological studies, radiative transfer models, and remote-sensing imagery. In this, gap models are serving a role more as integrators than as models in the traditional sense of predictive tools. As integrators, gap models incorporate new information as it becomes available, in the form of upgraded algorithms and new formulations. In applications and model exploration, critical areas of uncertainty can be identified which set priorities for further data needs and model refinements.

The role of integrator is natural for gap models in that individuals are an obvious focus for a wide variety of disciplines. Small-scale studies use individuals as a target towards which to integrate (e.g. to bring eco-physiology 'up to the plant level'), while larger scale studies often use individuals as convenient 'packages of data' to aggregate to larger scales. Individual-based models are especially powerful integrators for forest studies since field studies at most scales measure tree diameters, the state variables in the models.

SUMMARY

1. Forest gap models have been developed specifically to embody a conceptual model of a forest as a mosaic of fine-scale elements in

various stages of gap-phase regeneration. This conceptual model can be traced to classic works by Watt and others, and is fundamental to a large body of empirical and theoretical research in forest dynamics.

2. Gap models simulate the establishment, diameter growth and mortality of individual trees on a small model plot corresponding to the scale of a canopy gap. Because the inherent demographics and environmental responses of trees vary among species and according to tree size, gap models are especially useful in simulating the dynamics of mixed-species, mixed-age stands.

3. The models tend to rely on comparatively simple formulations and algorithms. Because gap models are individual based, they are easily reconciled to empirical studies that measure individual trees. The initial parametrization is rather straightforward, and additional data needed for model refinement and testing can be readily stated in terms of measurements made in field studies.

4. Demographic processes are simulated using a 'constrained potential' logic, which specifies a maximum behaviour (establishment rate, diameter growth or longevity) that might be achieved under optimal conditions, and then reduces this potential to account for the environmental constraints in effect on the plot. Most gap models simulate the constraints of available light, soil moisture and fertility, and temperature. Light is the primary constraint in most gap models and is simulated in the greatest detail.

5. Site-specific variations of the basic gap model have been implemented for a wide variety of forests, and the approach has also been extended to grasslands. These versions tend to emphasize biotic processes, environmental constraints or disturbance regimes which are important locally. Many of the various gap models have refined or extended demographic mechanisms, allometric relationships and abiotic processes. These modifications might be envisioned as a loosely consolidated 'toolbox' comprising an assortment of available functions, algorithms and submodels. The models seem to be evolving towards greater generality as well as greater fidelity to biophysical and ecophysiological mechanism.

6. Several simulation studies have used gap models to examine forest succession. Specific applications presented describe (a) scaling relationships of gap dynamics, which contrast the fine-scale non-equilibrium dynamics to a large-scale steady-state; (b) the role of gap dynamics in generating forest structural pattern, and the implications of this dynamic pattern to forest bird communities; (c) the importance of tree life-history traits and competition to forest succession and gradient responses; and (d) feedbacks between the details of forest pattern (species composition, forest age-structure) and ecosystem process.

7. Forests are characterized by feedbacks (a) among demographic pro-

cesses (especially mortality and regeneration) and (b) between abiotic and biotic processes (such as nutrient cycling and species composition). As complex systems, forests tend to develop emergent behaviours such as multiple stable states. These are seen as divergent successional pathways and hysteretic response to changing environmental conditions. These patterns are qualitatively reasonable but difficult to verify quantitatively because they develop over multiple tree generations.

8. Individual-based simulators increasingly are playing a role as research integrators, reconciling conceptual or empirical information on disparate spatiotemporal scales (e.g. tree ecophysiology and remotely sensed imagery). As an integrative framework, gap models incorporate available information while identifying critical uncertainties and data requirements, to focus further research.

REFERENCES

Aber, J. D., Botkin, D. B. and Melillo, J. M. (1978) Predicting the effects of differing harvest regimes on forest floor dynamics in northern hardwoods. *Canad. J. For. Res.*, **8**, 306–15.

Aber, J. D., Botkin, D. B. and Melillo, J. M. (1979) Predicting the effects of differing harvest regimes on productivity and yield in northern hardwoods. *Canad. J. For. Res.*, **9**, 10–14.

Aber, J. D. and Melillo, J. M. (1982) *FORTNITE: A Computer Model of Organic Matter and Nitrogen Dynamics in Forest Ecosystems*, University of Wisconsin Research Bulletin #R3130.

Austin, M. P. and Smith, T. M. (1989) A new model for the continuum concept. *Vegetatio*, **83**, 35–47.

Bazzaz, F. A. (1979) The physiological ecology of plant succession. *Ann. Rev. Ecol. System.*, **10**, 351–71.

Bazzaz, F. A. and Pickett, S. T. A. (1980) Physiological ecology of tropical succession: a comparative review. *Ann. Rev. Ecol. System.*, **11**, 287–310.

Bonan, G. B. (1989a) A computer model of the solar radiation, soil moisture, and soil thermal regimes in boreal forests. *Ecol. Model.*, **45**, 275–306.

Bonan, G. B. (1989b) Environmental factors and ecological processes controlling vegetation patterns in boreal forests. *Landscape Ecol.*, **3**, 111–30.

Bonan, G. B. (1990a) Carbon and nitrogen cycling in North American boreal forests. I. Litter quality and soil thermal effects in interior Alaska. *Biogeochem.*, **10**, 1–28.

Bonan, G. B. (1990b) Carbon and nitrogen cycling in North American boreal forests. II. Biogeographic patterns. *Canad. J. For. Res.*, **20**, 1077–88.

Bonan, B. G. (1991) A biophysical surface energy budget analysis of soil temperature in the boreal forests of interior Alaska. *Water Resources Res.*, **27**, 767–81.

Bonan, G. B. and Shugart, H. H. (1989) Environmental factors and ecological processes in boreal forests. *Ann. Rev. Ecol. System.*, **20**, 1–28.

Bonan, G. B., Shugart, H. H. and Urban, D. L. (1990) The sensitivity of some high-latitude boreal forests to climatic parameters. *Climatic Change*, **16**, 9–29.

Bormann, F. H. and Likens, G. E. (1979) *Pattern and Process in a Forested Ecosystem*, Springer-Verlag, New York.

Botkin, D. B. (1992) *The Ecology of Forests: Theory and Evidence*, Oxford University Press, Oxford.

Botkin, D. B., Janak, J. F. and Wallis, J. R. (1972a) Rationale, limitations, and assumptions of a northeastern forest growth simulator. *IBM J. Res. Develop.*, **16**, 101–16.

Botkin, D. B., Janak, J. F. and Wallis, J. R. (1972b) Some ecological consequences of a computer model of forest growth. *J. Ecol.*, **60**, 849–73.

Botkin, D. B. and Nisbet, R. A. (1991) Forest response to climatic change: effects of parameter estimation and choice of weather patterns on the reliability of projections. *Climatic Change*.

Busing, R. T. and Clebsch, E. E. C. (1987) Application of a spruce–fir forest canopy gap model. *For. Ecol. Mgmt*, **20**, 151–69.

Busing, R. T. (1991) A spatial model of forest dynamics. *Vegetatio*, **92**, 167–91.

Campbell, G. S. (1977) *An Introduction to Environmental Biophysics*, Springer-Verlag, New York.

Coffin, D. P. and Lauenroth, W. K. (1989) Disturbances and gap dynamics in a semiarid grassland: a landscape-level approach. *Landscape Ecol.*, **3**, 19–27.

Coffin, D. P. and Lauenroth, W. K. (1990) A gap dynamics simulation model of succession in a semiarid grassland. *Ecol Model.*, **49**, 229–36.

Coffin, D. P. and Urban, D. L. (in press) Model-based comparisons of forests and grasslands: implications of life-history traits to ecosystem dynamics.

Cohen, Y. and Pastor, J. (1991) The responses of a forest model to serial correlations of global warming. *Ecology*, **72**, 1161–5.

Cohen, W. B., Spies, T. A. and Bradshaw, G. A. (1990) Semivariograms of digital imagery for analysis of conifer canopy structure. *Remote Sens. Env.*, **34**, 167–78.

Dale, V. H. and Hemstrom, M. A. (1984) *CLIMACS: A Computer Model of Forest Stand Development for Western Oregon and Washington*, USDA For. Serv. Research Paper PNW-327.

DeVelice, R. L. (1988) Test of a forest dynamics simulator in New Zealand. *New Zealand J. Bot.*, **26**, 387–92.

Doyle, T. W. (1981) The role of disturbance in the gap dynamics of a montane rain forest: an application of a tropical forest succession model, in *Forest Succession: Concepts and Applications* (eds D. C. West, H. H. Shugart and D. B. Botkin), Springer-Verlag, New York, pp. 56–73.

Ek, A. R. and Monserud, R. A. (1974) *FOREST: A Computer Model for the Growth and Reproduction of Mixed Species Forest Stands*, Res. Rep. A2635, College of Agric. and Life Sciences, Univ. Wisconsin, Madison.

Ellenberg, H. (1953) Physiologisches und ökologisches Verhalten derselben Pflanzenarten. *Ber. Deut. Bot. Gesell.*, **65**, 351–62.

Ellenberg, H. (1954) Über einige Fortschritte der Kausalen Vegetationskunde. *Vegetatio*, **5/6**, 199–211.

Friend, A. Shugart, H. H. and Running, S. W. (in press) *HYBRID: A Physiology-Based Gap Model of Forest Dynamics*.

Gauch, H. G. and Whittaker, R. H. (1972) Coenocline simulation. *Ecology*, **53**, 446–51.

Grime, J. P. (1979)*Plant Strategies and Vegetation Processes*, Wiley, Chichester.

Harcombe, P. A. (1987) Tree life tables. *BioScience*, **37**, 557–68.

Harrison, E. A. and Shugart, H. H. (1990) Evaluating performance of an Appalachian oak forest dynamics model. *Vegetatio*, **86**, 1–13.

Harper, J. L. (1977) *Population Biology of Plants*, Academic Press, London.

Horn, H. H., Shugart, H. H. and Urban, D. L. (1989) Simulators as models of forest dynamics, in *Perspectives in Ecological Theory* (eds J. Roughgarden, R. M. May and S. A. Levin), Princeton Univ. Press, Princeton, New Jersey, pp. 256–67.

Huston, M. A. and Smith, T. M. (1987) Plant succession: life history and competition. *Amer. Nat.*, **130**, 168–98.

Huston, M., DeAngelis, D. L. and Post, W. M. (1988) New computer models unify ecological theory. *BioScience*, **38**, 682–91.

Hutchinson, G. E. (1957) Concluding remarks. *Cold Spring Harbor Symp. Quant. Biol.*, **22**, 415–27.

James, F. C. (1971) Ordinations of habitat relationships among breeding birds. *Wilson Bull.*, **83**, 215–36.

Jarvis, P. G. and Leverenz, J. W. (1983) Productivity of temperate evergreen and deciduous forests, in *Physiological Plant Ecology* (eds O. L. Lange, P. S. Novel, C. B. Osmond and H. Ziegler), Vol. IV, Springer-Verlag, Berlin, pp. 234–261.

Keane, R. E., Arno, S. F. and Brown, J. K. (1990) Simulating cumulative fire effects in ponderosa pine/Douglas-fir forest. *Ecology*, **71**, 189–203.

Kercher, J. R. and Axelrod, M. C. X. (1984) A process model of fire ecology and succession in a mixed-conifer forest. *Ecology*, **65**, 1725–42.

Kienast, F. (1987) *FORECE – A Forest Succession Model for Southern Central Europe*, ORNL/TM-2989, Oak Ridge National Laboratory, Oak Ridge, TN.

Larcher, W. (1980) *Physiological Plant Ecology*, Springer-Verlag, Berlin.

Lauenroth, W. K., Urban, D. L. Coffin, D. P. Parton, W. J. Shugart, H. H. Kirchner, T. B. and Smith T. M. (1992) Modeling vegetation pattern-ecosystem process interactions across sites and ecosystems. *Ecol. Model.* (in press).

Leemans, R. (1991) Sensitivity analysis of a forest succession model. *Ecol. Model.*, **53**, 247–62.

Leemans, R. and Prentice, I. C. (1987) Description and simulation of tree-layer composition and size distributions in a primaeval *Picea-Pinus* forest. *Vegetatio*, **69**, 147–56.

Leemans, R. and Prentice, I. C. (1989) *FORSKA: A General Forest Succession Model*, Medd. Vaxbiol. Inst. Uppsala 89/2, Uppsala, Sweden.

Makela, A. (1986) Implications of the pipe model theory on dry matter partitioning and height growth in trees. *J. Theor. Biol.*, **123**, 103–20.

Mankin, J. B., O'Neill, R. V., Shugart, H. H. and Rust, B. W. (1977) The importance of validation in ecosystems analysis, in *New Directions in the Analysis of Ecological Systems*, Part I (ed. G. S. Innis), Simulation Council of America, La Jolla, CA, pp. 63–71.

Margalef, R. (1968) *Perspectives in Ecological Theory*, Univ. Chicago Press, Chicago.

Mielke, D. L., Shugart, H. H. and West, D. C. (1978) *A Stand Model for Upland Forests of Southern Arkansas*, ORNL/TM-6225, Oak Ridge National Lab, Oak Ridge, TN.

Mohren, G. M. J., van Hees, A. F. M. and Bartelink, H. H. (1991) Succession models as an aid for forest management in mixed stands in The Netherlands. *For. Ecol. Mgmt*, **42**, 111–27.

Monsi, M. and Saeki, T. (1953) Ber der Lichtfakto in den Pflanzengesellschaften und seine Bedeutung für die Stuffproduktion. *Japan. J. Bot.*, **14**, 22–52.

Mueller-Dombois, D. and Ellenberg, H. (1974) *Aims and Methods of Vegetation Ecology*, Wiley, New York.

Munro, D. D. (1974) Forest growth models: a prognosis, in *Growth Models for Tree and Stand Simulation* (ed. J. Fries), Res. Notes 30, Dept Forest Yield Res., Royal College of Forestry, Stockholm, pp. 7–21.

Mutch, R. W. (1970) Wildland fires and ecosystems – a hypothesis. *Ecology*, **51**, 1046–51.

Noble, I. R. and Slatyer, R. O. (1980) The use of vital attributes to predict successional changes in plant communities subject to recurrent disturbance. *Vegetatio*, **43**, 5–21.

Odum, E. P. (1969) The strategy of ecosystem development. *Science*, **164**, 262–70.

Orians, G. H. and Solbrig, O. T. (1977) A cost-income model of leaves and roots with special reference to arid and semi-arid areas. *Amer. Nat.*, **111**, 677–90.

Pastor, J. and Post, W. M. (1985) *Development of a Linked Forest Productivity-Soil Process Model*, ORNL/TM-9519, Environmental Sciences, Oak Ridge National Lab, Oak Ridge, TN.

Pastor, J. and Post, W. M. (1986) Influence of climate, soil moisture, and succession on forest carbon and nitrogen cycles. *Biogeochem.*, **2**, 3–27.

Pastor, J. and Post, W. M. (1988) Response of northern forests to CO_2-induced climate change. *Nature*, **334**, 55–8.

Pastor, J., Gardner, R. H., Dale, V. H. and Post, W. M. (1987) Successional changes in nitrogen availability as a potential factor contributing to spruce declines in boreal North America. *Canad. J. For. Res.*, **17**, 1394–1400.

Pearlstine, L., McKellar H. and Kitchens, W. (1985) Modeling the impacts of river diversion on bottomland forest communities in the Santee River Floodplain, South Carolina. *Ecol. Model.*, **29**, 283–302.

Phipps, R. L. (1979) Simulation of wetlands forest dynamics. *Ecol. Model.*, **7**, 257–88.

Platt, W. J. and Strong, D. R. (eds) (1989) Treefall gaps and forest dynamics. *Ecology*, **70**, 535–76.

Prentice, I. C. and Leemans, R. (1990) Pattern and process and the dynamics of forest structure: a simulation approach. *J. Ecol.*, **78**, 340–55.

Reichle, D. E. (ed.) (1981) *Dynamic Properties of Forest Ecosystems*, US/IBP No. 23, Cambridge Univ., Press, Cambridge.

Runkle, J. R. (1985) Disturbance regimes in temperate forests, in *The Ecology of Natural Disturbance and Patch Dynamics* (eds S. T. A. Pickett and P. S. White), Academic Press, Orlando, pp. 17–33.

Running, S. W. and Coughlan, J. C. (1988) A general model of forest ecosystem processes for regional applications. I. Hydrological balance, canopy gas exchange and primary production processes. *Ecol. Model.*, **42**, 125–54.

Shinozaki, K., Yoda, K., Hozumi, K. and Kira, T. (1964a) A quantitative analysis of plant form – the pipe model theory. I. Basic analyses. *Japan. J. Ecol.*, **14**, 97.

Shinozaki, K., Yoda, K., Hozumi, K. and Kira, T. (1964b) A quantitative analysis of plant form – the pipe model theory. II. Further evidence of the theory and its application in forest ecology. *Japan. J. Ecol.*, **14**, 133.

Shugart, H. H. (1984) *A Theory of Forest Dynamics*, Springer-Verlag, New York.

Shugart, H. H. (1987) The dynamic ecosystem consequences of coupling birth and death processes in trees. *BioScience*, **37**, 596–602.

Shugart, H. H., Hopkins, M. S., Burgess, I. P. and Mortlock, A. T. (1980) The development of a succession model for subtropical rain forest and its application to assess the effects of timber harvest at Wiangaree State Forest, New South Wales. *J. Environ. Mgmt*, **11**, 243–65.

Shugart, H. H., Leemans, R. and Bonan, G. B. (eds) (1992) *A Systems Analysis of the Global Boreal Forest*. Cambridge University Press, Cambridge.

Shugart, H. H. and Noble, I. R. (1981) A computer model of succession and fire response to the high altitude *Eucalyptus* forest of the Brindabella Range, Australian Capital Territory. *Austral. J. Ecol.*, **6**, 149–64.

Shugart, H. H. and Urban, D. L. (1986) Overall summary: a researcher's perspective, in *Modeling Habitat Relationships of Terrestrial Vertebrates* (eds, J. Verner, M. L. Morrison, and C. J. Ralph, Univ. of Wisconsin Press, Madison, pp. 425–29.

Shugart, H. H. and Urban, D. L. (1989) Factors affecting the relative abundances of forest tree species, in *Toward a More Exact Ecology* (eds P. J. Grubb and J. B. Whittaker), Blackwell, Oxford, pp. 249–73.

Shugart, H. H. and West, D. C. (1977) Development of an Appalachian deciduous forest succession model and its application to assessment of the impact of the chestnut blight. *J. Environ. Mgmt*, **5**, 161–70.

Shugart, H. H. and West, D. C. (1979) Size and pattern of simulated forest stands. *Forest Sci.*, **25**, 120–2.

Shugart, H. H. and West, D. C. (1980) Forest succession models. *BioScience*, **30**, 308–13.

Smith, T. M. (1986) Habitat simulation models: integrating habitat classification and forest simulation models, in *Modeling Habitat Relationships of Terrestrial Vertebrates* (eds J. Verner, M. L. Morrison, and C. J. Ralph), Univ. of Wisconsin Press, Madison, pp. 389–93.

Smith, T. M. and Huston, M. (1989) A theory of the spatial and temporal dynamics of plant communities. *Vegetatio*, **83**, 49–69.

Smith, T. M., Shugart, H. H. and West, D. C. (1981) The use of forest simulation models to integrate timber harvest and nongame bird habitat management. *Proc. North Amer. Wildl. and Nat. Resource Conf.*, **46**, 501–10.

Smith, T. M., Shugart, H. H., Urban, D. L., Lauenroth, W. K., Coffin, D. P. and Kirchner, T. B. (1989) Modeling vegetation across biomes: forest-grassland transition. *Studies Plant Ecol.*, **18**, 47–9.

Smith, T. M. and Urban, D. L. (1988) Scale and resolution of forest structural pattern. *Vegetatio*, **74**, 143–50.

Sollins, P., Reichle, D. E. and Olson, J. S. (1973) *Organic Matter Budget and Model for a Southern Appalachian Liriodendron Forest*, EDFB/IBP-73/2, Oak Ridge National Laboratory, Oak Ridge, TN.

Solomon, A. M. (1986) Transient response of forests to CO_2-induced climate change: simulation modeling experiments in eastern North America. *Oecologia*, **68**, 567–79.

Solomon, A. M., Delcourt, H. R., West, D. C. and Blasing, T. J. (1980) Testing a simulation model for reconstruction of prehistoric forest stand dynamics. *Quat. Res.*, **14**, 275–93.

Solomon, A. M., Tharp, M. L., West, D. C., Taylor, G. E., Webb, J. M. and Trimble, J. C. (1984) *Response of Unmanaged Forests to CO₂-Induced Climate Change: Available Information, Initial Tests, and Data Requirements*, US Dept of Energy, Washington, D.C.

Tharp, M. L. (1978) *Modeling Major Perturbations on a Forest Ecosystem*, Ms. Thesis, Univ. of Tennessee, Knoxville.

Thornthwaite, C. W. and Mather, J. R. (1957) Instructions and tables for computing potential evapotranspiration and the water balance. *Pub. Climatol.*, **10**, 183–311.

Tilman, D. (1982) *Resource Competition and Community Structure*, Princeton Univ. Press, Princeton.

Tilman, D. (1988) *Plant Strategies and the Dynamics and Structure of Plant Communities*, Princeton Univ. Press, Princeton.

Urban, D. L. (1990) *A Versatile Model to Simulate Forest Pattern: A User's Guide to ZELIG Version 1.0*, Environmental Sciences, Univ. of Virginia, Charlottesville.

Urban, D. L., Bonan, G. B., Smith, T. M. and Shugart, H. H. (1991) Spatial applications of gap models. *For. Ecol. Mgmt*, **42**, 95–110.

Urban, D. L., Harmon, M. E. and Halpern, C. B. (in press) Potential response of Pacific Northwestern forests to climatic change: Effects of Stand age and initial composition.

Urban, D. L. and Smith, T. M. (1989a) Extending individual-based forest models to simulate large-scale environmental patterns. *Bull. Ecol. Soc. Amer.*, **70**, 284.

Urban, D. L. and Smith, T. M. (1989b) Microhabitat pattern and the structure of forest bird communities. *Amer. Nat.*, **133**, 811–29.

van Daalen, J. C. and Shugart, H. H. (1989) OUTENIQUA – a computer model to simulate succession in the mixed evergreen forests of the southern Cape, South Africa. *Landscape Ecol.*, **2**, 255–67.

van Voris, P., O'Neill, R. V., Emanuel, W. R. and Shugart, H. H. (1980) Functional complexity and ecosystem stability. *Ecology*, **61**, 1352–60.

Waldrop, T. A., Buckner, E. R., Shugart, H. H. and McGee, C. E. (1986) FORCAT: A single-tree model of stand development on the Cumberland Plateau. *Forest Sci.*, **32**, 297–317.

Waring, R. H. (1983) Estimating forest growth and efficiency in relation to canopy leaf area. *Adv. Ecol. Res.*, **13**, 327–54.

Waring, R. H. and Schlesinger, W. H. (1985) *Forest Ecosystems: Concepts and Management*, Academic Press, Orlando.

Waring, R. H., Schroeder, P. E. and Oren, R. (1982) Application of the pipe model theory to predict canopy leaf area. *Canad. J. For. Res.*, **12**, 556–60.

Watt, A. S. (1925) On the ecology of British beechwoods with special reference to their regeneration. *J. Ecol.*, **13**, 27–73.

Watt, A. S. (1947) Pattern and process in the plant community. *J. Ecol.*, **35**, 1–22.

Weinstein, D. A. and Shugart, H. H. (1983) Ecological modeling of landscape dynamics, in *Disturbance and Ecosystems* (eds H. Mooney and M. Godron), Springer-Verlag, New York, pp. 29–45.

Weinstein, D. A., Shugart, H. H. and West, D. C. (1982) *The Long-Term*

Nutrient Retention Properties of Forest Ecosystems: *A Simulation Investigation*, ORNL/TM-8472, Oak Ridge National Lab, Oak Ridge, TN.

Weishampel, J. F., Urban, D. L., Smith, J. B. and Shugart, H. H. (1992) A comparison of semivariograms from a forest transect model and remotely sensed data. *J. Veg. Sci.* (in press).

Whitmore, T. C. (1982) On pattern and process in forests, in *The Plant Community as a Working Mechanism* (ed. E. I. Newman), Blackwell, Oxford, pp. 45–60.

Whitmore, T. C. (1989) Canopy gaps and the two major groups of forest trees. *Ecology*, **70**, 536–8.

8 Climate change and long-term vegetation dynamics

I. Colin Prentice

8.1 INTRODUCTION

This final chapter describes the chain of causation that leads from variations in the earth's orbit, through changes in incoming solar radiation (insolation) via atmospheric circulation to surface climate and the effects of the changing climate on vegetation as recorded in pollen samples from sediments. It focuses on the Quaternary ($\approx$1.6 Ma: 1 a = 1 year), a time scale on which there have been continual, major changes in global climate yet little macro-evolution in plants. The term Late Quaternary is used here to mean the period since the last glacial maximum, $\approx$18 ka; the Holocene is the past 10 ka. These are the time scales on which vegetation changes have been studied most extensively by pollen analysis.

Climate changes; vegetation responds

If climate were stationary, vegetation dynamics would consist of fluctuations due to year-to-year weather variations, succession triggered by disturbances and gap-phase regeneration (Chapters 1 and 3). These are stochastic processes, but their average behaviour can be predicted from physiological and life-history characteristics of the available species and information on climate and natural disturbance (Chapters 2, 4 and 7). Vegetation processes vary spatially in response to the broad-scale pattern of climate, and change through time in response to long-term variations in climate.

Climate acts on vegetation directly and indirectly

Growing-season warmth and winter cold directly affect which species can grow and how fast. Changes in growth rates affect species' relative competitive abilities, which affect the patterns of succession and, indirectly, mortality and gap formation (Shugart, 1984; see also Chapter 7). Precipitation and potential evapotranspiration affect establishment and growth indirectly, through their effects on soil moisture. Climate also influences the frequency, magnitude, type and extent of disturbances.

Plant Succession: Theory and prediction Edited by David C. Glenn-Lewin. Robert K. Peet and Thomas T. Veblen © 1992 Chapman & Hall, London ISBN 0 412 26900 7

Each species has a unique response to variation in these different aspects of climate. Virtually any sustained climate change will produce some response in the vegetation, and an enormously wide variety of vegetation responses is documented in the Quaternary record.

The major causes of vegetation change on a Quaternary time scale are different from the causes of vegetation change on the more familiar time scale of succession, and different species' characteristics may be important in explaining vegetation dynamics on the different time scales. The control of present vegetation by climate is most obvious at the broad spatial scale of world vegetation maps. At this scale the effects of variations in slope, aspect, soil and time-since-disturbance, which are paramount at the scale of field studies, are reduced to noise in comparison with the effects of geographic variation in climate. Analogously, the effects of climate change are most obvious on a time scale of thousands of years, long enough for individual cycles of succession and disturbance to become noise relative to the long-term trend. To understand succession requires a knowledge of species' life-history characteristics such as longevity, regeneration requirements and tolerance; to understand long-term vegetation dynamics requires a knowledge of species' responses to environment and an understanding of the nature and causes of climate change.

8.2   THE STUDY OF THE QUATERNARY

8.2.1   Reconstructing past environments

Quaternary research is concerned with the reconstruction and explanation of climate changes and their effects on earth surface phenomena, including life. Quaternary scientists draw on a wide range of techniques to reconstruct past climates (Lowe and Walker, 1984; Bradley, 1985; Hecht, 1985; Berglund, 1986).

On the continents, geomorphic and sedimentological evidence are used to reconstruct changes in lake levels, which record changes in water balance (Street and Grove, 1979; Street-Perrott and Harrison, 1985); river regimes, which record changes in runoff (Knox, 1983; Schumm and Brakenridge, 1987); and glaciers, which record changes in their mass balance (Grove, 1979; Barry, 1985; Karlén, 1988). Geochemical evidence for climate changes comes from changes in the chemical composition of ombrotrophic peat (Barber, 1981) and lake sediments (Engstrom and Wright, 1984), and isotopic ratios (primarily $^{18}O/^{16}O$, D/H and $^{13}C/^{12}C$) in tree-rings (Stuiver and Burk, 1985; Stuiver and Braziunas, 1987), lake sediments (Siegenthaler and Eicher, 1986), fossil wood (Edwards and Fritz, 1986) and polar and mountain ice caps (Lorius et al., 1979; Robin, 1983; Oeschger et al., 1984; Paterson and Hammer, 1987; Thompson

and Mosley-Thompson, 1987; Oeschger and Langway, 1989). Palaeoecological evidence for climate change comes from assemblages of various
organisms preserved in sediments, including diatoms (Mannion, 1982),
ostracodes (Forrester, 1987) and beetles (Coope, 1970, 1979; Atkinson
et al., 1986), and from tree-ring series (Brubaker and Cook, 1983;
Stockton et al., 1985). Terrestrial palaeoclimates can also be recon-
structed from pollen and plant macrofossil assemblages in so far as
regional vegetation patterns can be assumed to be in equilibrium with
climate (section 8.4.4; Webb and Bryson, 1972; Bartlein et al., 1984;
Webb, 1985). Such data have contributed substantially to present knowledge of Quaternary climate change.

Studies of ocean sediments complement terrestrial palaeoclimate
studies by providing long, continuous records of changes in glaciation,
sea-surface temperatures and atmospheric CO_2, which together with
insolation are key factors in the global system that determines the nature
of terrestrial as well as marine climates. The record of $^{18}O/^{16}O$ ratios in
benthic foraminifera preserved in ocean sediments is a proxy for changes
in global ice volume (Shackleton and Opdyke, 1973; Mix and Ruddiman,
1984; Mix, 1987). Ice is isotopically light, so the growth of large continental ice sheets leaves the global ocean enriched in ^{18}O; this is believed
to be the main factor controlling the isotopic composition of deep ocean
water. The species composition of preserved planktonic foraminiferal
and radiolarian assemblages provides data on sea-surface temperatures
(Imbrie and Kipp, 1971; CLIMAP Project Members, 1976, 1981;
Ruddiman, 1985). Changes in atmospheric composition can be estimated
most directly from ice cores (Neftel et al., 1982; Barnola et al., 1987;
Paterson and Hammer, 1987) but changes in CO_2 content can also be
inferred from foraminiferal $^{13}C/^{12}C$ ratios in deep-sea cores (Shackleton
et al., 1983; Shackleton and Pisias, 1985). Isotopically-light carbon fixed
by marine phytoplankton is ingested by planktonic foraminifera, some of
which sink to the bottom leaving the surface waters enriched in ^{13}C. The
difference between $^{13}C/^{12}C$ ratios in planktonic and benthic foraminifera
preserved in ocean sediments is thus an index of primary production in
surface waters, which increases with atmospheric CO_2.

The oxygen isotope record can be precisely correlated with known
variations in the earth's orbital geometry (section 8.3.2; Imbrie et al.,
1984; Martinson et al., 1987), providing a global Quaternary chronology.
A higher resolution chronology for the past 40 ka is provided by radio-
carbon (^{14}C) dating (Olsson, 1986; Saarnisto, 1988), which is the main
technique used to estimate the age of terrestrial organic materials. Radio-
carbon dates are not exactly equivalent to calendar dates because of
natural variations in atmospheric ^{14}C content (Stuiver, 1989). They are
expressed in conventional ^{14}C years BP (before 1950). Direct dating
techniques with a longer time range exist but are much less accurate
(Saarnisto, 1988). Long, continuous terrestrial records can, however,

be dated by orbital correlation in the same way as oceanic records (Hooghiemstra, 1988).

8.2.2 Reconstructing past vegetation and its dynamics

Quaternary palaeoecology is the branch of Quaternary research concerned with the responses of organisms, communities and ecosystems to environmental change. Evidence for Quaternary vegetation changes comes primarily from pollen and plant macrofossils preserved in lake sediments and peat (Jacobson and Bradshaw, 1981; Jacobson, 1988; Prentice, 1988a), mor humus in humid regions (Iversen, 1969; Jacobson and Bradshaw, 1981) and *Neotoma* (packrat) middens in arid regions (Wells, 1976; Thompson, 1985).

Pollen can typically be identified to genus or family and only occasionally to species. Some taxa are better represented than others, with strong biases towards the tallest plant life-forms (e.g. trees in forested regions) and towards wind-pollinated taxa with light pollen grains. Only a proportion of the flora of any region is quantitatively represented. This proportion is typically lowest in tropical forests, where pioneer species are represented strongly but many climax species are palynologically 'silent' (Bush and Colinvaux, 1988). Despite these limitations, the pollen record contains abundant information about vegetation changes, in environments ranging from Arctic tundra and boreal forests to tropical rainforests, grasslands and semi-deserts (Huntley and Webb, 1988).

Plant macrofossils are dispersed more erratically than pollen, and being more prone to oxidation they are not as consistently preserved in most sediments. Macrofossils in lake sediments can, however, prove the local presence of taxa, and can often be identified to species (Watts, 1978). Macrofossils are also an important data source in the arid regions of western North America, where packrat middens are found (Spaulding *et al.*, 1983; Spaulding, 1984; van Devender *et al.*, 1987; Thompson, 1988).

Vegetation releases large quantities of pollen annually into the turbulent layers of the atmosphere. Most of this pollen is deposited on the canopy nearby, but the fraction that remains airborne is dispersed over a much wider area (Janssen, 1970, 1973; Prentice, 1985). The production, dispersal and deposition of pollen grains across vegetated surfaces can be approximated by models based on Sutton's equation (Tauber, 1965; Chamberlain, 1975; Prentice, 1985, 1988a). These models show that variations in size and shape cause significant differences in the dispersal range of different pollen types. The models also predict large differences between the pollen source areas of 'open' sampling sites, such as lakes, and sites in closed vegetation, such as mor humus samples from the forest floor. Comparisons of surface pollen deposition with modern vegetation have confirmed that samples from moderate-sized lakes ($\approx 1-100\,\text{ha}$)

essentially record the regional vegetation, averaged over some tens of kilometres, plus contributions from the fringing vegetation and from local aquatic plants (Janssen, 1973; Bradshaw and Webb, 1985), while forest-floor moss and mor-humus samples typically reflect canopy composition within tens of metres of the sampling site (Andersen, 1970; Bradshaw, 1981, 1988). Samples from mires reflect the regional abundances of upland taxa (e.g. trees) and the local abundances of mire plants (Jacobson and Bradshaw, 1981).

The temporal resolution obtainable in pollen records from lake sediments is often limited by the vertical mixing effect of benthic organisms (Davis, 1974). One sample per 100 years is a typical useful sampling density. Higher temporal resolution (<10 years) can, however, be obtained in the undisturbed, annually-laminated sediments of meromictic lakes (Craig, 1972; Saarnisto, 1986; Turner and Peglar, 1988). The laminations provide a precise, calendar-year chronology. Temporal resolution down to ≈10 years can also sometimes be achieved in mor humus (Bradshaw, 1988; Bradshaw and Zackrisson, 1990) and in peats (Barber, 1985; Turner and Peglar, 1988).

Pollen analyses may be complemented by analyses of windblown charcoal as an indicator of past fires (Patterson and Backman, 1988). Total charcoal can be tallied on the same microscope slides as pollen grains, or assayed independently (Winkler, 1985a). Petrographic thin sections can yield a separate record of the larger charcoal particles, which more accurately reflect the occurrence of fires in the catchment (Clark, 1988a,b). High-resolution pollen and charcoal analyses have contributed to an understanding of the relationship between long-term vegetation dynamics and fire.

Pollen analysis thus gives considerable freedom of choice over the spatial and temporal scales at which vegetation changes are observed (Fig. 8.1; Webb *et al.*, 1978; Jacobson and Bradshaw, 1981; Solomon and Webb, 1985; Prentice, 1986a,b, 1988a). Mor-humus pollen analysis extends the time frame of observations at the spatial scale of the field ecologist's sample plot and is a suitable (though under-used) tool for the study of forest succession. Pollen analysis based on cores from moderate-sized lakes or mires can provide information on changing landscape composition and regional vegetation patterns. Annually laminated sediments provide the opportunity to resolve the detailed dynamics of these changes at the landscape scale; studies with high time resolution, including charcoal analyses, can give information about disturbance regimes and their relationship with long-term vegetation change.

8.2.3   Quantitative information from the pollen record

Pollen data are most commonly presented as percentages of total land pollen (or for some purposes, total tree pollen), based on counts of

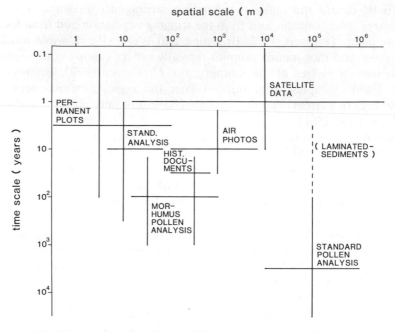

Figure 8.1   Space and time scales of different observational techniques in vegetation dynamics. The ranges shown run from the finest resolution or grain (Allen *et al.*, 1984) to the broadest scope or extent of data obtainable with a particular technique. The points where the lines cross indicate the approximate scales at which the techniques are most often used.

200–1000 grains. Pollen percentages can be quantitatively correlated with tree percentages within 10–100 km (Webb, 1974; Prentice, 1978; Delcourt *et al.*, 1984).

Alternatively, the input of pollen grains to sediment can be estimated as a flux density (grains $cm^{-2}a^{-1}$) (Thompson, 1980). Pollen flux densities are estimated from counts of $10^2–10^3$ grains of all taxa in samples of sediment, 'spiked' with a known number of marker particles:

$$Y_i = (n_i/n_x)(N/V)S \qquad (8.1)$$

where Y_i is the (estimated) flux density of pollen type i, n_i and n_x are the counts of pollen type i and the marker, N is the number of marker particles added to volume V of sediment, and S is the sediment accumulation rate (estimated from the profile of estimated age against depth). When averaged over the length of time represented by the pollen sample, the pollen flux density of each taxon is assumed to be approximately proportional to the abundance of that taxon in the surroundings.

Pollen flux-density measurements have been used successfully to estimate changes in population density over short intervals in homogeneous sediments (section 8.4.3). They have also permitted the reconstruction of shifts in the Arctic tree-line, and in the boundaries between subarctic forest types (Hyvärinen, 1976; Ritchie, 1984; Lamb and Edwards, 1988). Large changes in total pollen flux density in these environments provide important evidence to supplement changes in the relative abundance of the rather small number of pollen types involved. Some problems limit the more general use of pollen flux-density measurements. Sediment focusing, the bulk movement of seston from shallower to deeper water in response to changing lake bathymetry, can produce large and spurious changes in the pollen flux density to deep-water sediments (Lehman, 1975; Davis et al., 1984) and may be responsible for relatively weak correlations observed between tree abundances and pollen flux densities in surface lake sediments (Davis et al., 1973). Measured flux densities in peat also seem to be highly variable (Jacobson and Bradshaw, 1981).

The concept of pollen flux density is nevertheless important in analysing the quantitative information content of pollen percentage data. Taxa differ in pollen productivity and dispersal. These two factors together determine a taxon-specific *representation factor*, represented by a_i in:

$$Y_i = a_i X_i \tag{8.2}$$

where X_i is the average abundance of taxon i, inversely weighted by distance from the site (e.g. by equation (4) of Prentice, 1988a). Representation factors can be estimated empirically from pollen percentages in surface sediments and relative abundances in the surrounding vegetation (Prentice, 1986c). From equation (8.2), pollen percentages (p_i) and percentage relative abundances (v_i) are related non-linearly by

$$p_i = a_i v_i / \sum_j (a_j v_j) \tag{8.3}$$

where the denominator is a function of all the taxa present. The non-linearity is slight provided the vegetation is fairly diverse (Webb et al., 1981; Prentice and Webb, 1986), so variation in pollen percentages are approximately linearly related to variations in relative abundance.

The pollen source area can be shown to be different for different taxa (Prentice et al., 1987). This complication is, however, avoided in studies of regional patterns. Mapping across a region smoothes the data to an extent that depends on the spacing of the sampling sites rather than on the dispersal properties of the pollen types (Prentice, 1988a). In practice 'isopoll maps' usually give a good approximation to the spatial patterns of relative plant abundance at regional to continental scales.

8.2.4 The palaeoecological perspective

The coverage of data on past vegetation is extremely uneven at a global scale. Recent syntheses of Late Quaternary pollen data for the continents include: for North America, Jacobson *et al.*, (1987), Webb *et al.*, (1987), Webb (1988), Delcourt and Delcourt (1987), Barnosky *et al.*, (1987), van Devender *et al.*, (1987), Thompson (1988), Ritchie (1987); for Central and South America, Markgraf (1989); for Europe, Huntley (1988); for the former USSR, Grichuk (1984), Khotinskiy (1984); for Africa, Rognon (1987), Deacon and Lancaster (1988), Lézine (1989); for Japan, Tsukada (1988); for Australasia, McGlone (1988), Dodson (1989); for the Arctic, Lamb and Edwards (1988); for the Antarctic, Heusser (1989). There are still major gaps. Most of the Amazon basin, most of Siberia, large parts of central and southern Asia and most of northern and western Australia, for example, still have no pollen records. In contrast, the record from Europe and eastern North America has allowed synoptic pollen mapping with 0.5–1 ka resolution (Davis, 1976, 1981a; Bernabo and Webb, 1977; Huntley and Birks, 1983; Delcourt and Delcourt, 1987; Jacobson *et al.*, 1987; Webb, 1987, 1988; Huntley, 1988). Eastern North America is especially favourable for pollen mapping because of its relatively simple topography and vegetation gradients. These well-studied regions offer excellent possibilities for testing hypotheses about the mechanisms of regional climate change and vegetation response. The less-studied parts of the world offer opportunities for new insights into floristic and vegetation history, and for the addition of key information to help in reconstructing the patterns of global climate change.

Information on older (>18 ka) vegetation changes in glaciated regions, or regions peripheral to the ice sheets, comes from pollen analysis of organic sediments buried under till or loess. An impressive number of interglacial records has been accumulated in northern Europe (West, 1980; Watts, 1988), the northwestern portion of the former USSR (Grichuk *et al.*, 1984; Velichko, 1984) and North America (Heusser and King, 1988). Continuous pollen profiles back to >100 ka have been obtained from unglaciated regions of Europe (Woillard, 1978; de Beaulieu and Reille, 1984; Follieri *et al.*, 1988; Behre, 1989) and elsewhere. A few records extend to >1 Ma (Fuji, 1984; Hooghiemstra, 1984, 1988; van der Hammen, 1988).

Palaeoecological studies on these various time scales have given rise to a distinct perspective on vegetation and its dynamics that in some respects contradicts the conventional wisdom in phytogeography and evolutionary biology. A few generalizations have emerged repeatedly from studies in different parts of the world, irrespective of the particular region or particular mechanisms and patterns of climate change:

1. Climate change is ubiquitous on a Quaternary time scale. Climate changes in the wet and dry tropics, and in the south temperate zone, have been as large as those of the north with its waxing and waning ice sheets (McGlone, 1988; Markgraf, 1989; Deacon and Lancaster, 1990). These climate changes have universally affected vegetation to

the extent of radically changing composition, structure and disturbance regimes and changing the geographic distributions of plant taxa and vegetation formations through time.

2. Climate change is spatially differentiated and complex in structure. Changes in any one region typically include independent variations in summer and winter temperatures and moisture, forming a non-repeating sequence. Comparisons between regions show a variety of temporal relationships. Superimposed on a general (though by no means uniform) temperature trend with ≈100 ka periodicity, the glacial–interglacial cycle of the high latitudes, are changes in climatic variables that may show diachronous or antiphase relationships, high-frequency events and cycles with global effects, and other events restricted to particular sectors or regions (COHMAP Members, 1988).

3. Taxa have migrated individualistically in response to these climate changes, adjusting their ranges and abundances according to their unique response characteristics (Watts, 1973, 1983; Davis, 1976, 1981a; Walker and Flenley, 1979; Huntley and Birks, 1983; Walker, 1982a; Colinvaux, 1987; Walker and Chen, 1987; Webb, 1987; and several articles in Huntley and Webb, 1988). Taxa have not been tied to a particular community or formation. Some formations, such as the modern boreal forest, can be shown to have existed as extensive formations for less than 10 ka (West, 1964; Davis, 1976; Webb, 1987, 1988). The ephemeral nature of plant communities and formations unequivocally supports the individualistic hypothesis on plant community dynamics (Gleason, 1926; see also Chapter 1).

8.3 CLIMATE CHANGE

8.3.1 The spectrum of climatic variation

Climatic variations at different frequencies have different causes (Mitchell, 1976; Berger, 1981; Imbrie, 1985; Bartlein, 1988), and different physical and biological effects (McDowell *et al.*, 1991). Imbrie (1985) distinguished six broad frequency bands: tectonic (>400 ka), orbital ('Milankovitch') (10–400 ka), millennial (0.4–10 ka), decadal (10–400 a), interannual (2.5–10 a) and annual (0.5–2.5 a). The entire variance spectrum of climate is broadly U-shaped (Kutzbach, 1976), with relatively little variance in the broad region. between the 'astronomical' periodicities in the orbital and annual bands.

Climate changes in the *tectonic band* include notably the effects on the climate system of changes in the distribution of land, sea and mountain ranges, and long-term changes in the earth's energy balance due to solar evolution and atmospheric composition (Kutzbach, 1985; Bartlein, 1988; Ruddiman *et al.*, 1989). The global cooling during the Cainozoic

($\approx$65 Ma), and the initiation of glaciations in the Late Cainozoic, belong to this band.

The major climate changes in the *orbital band* are caused by regular variations in the earth's orbital geometry, which affect the distribution of incoming solar radiation (insolation) by latitude and season (Berger, 1978a,b). Effects of orbital variations are dominant at a global scale even in the climate history of the Holocene (COHMAP Members 1988).

Climate changes in the *millennial band* include phenomena such as the historically documented Medieval Warm Period and Little Ice Age (Le Roy Ladurie, 1971; Schuurmans, 1981; Grove, 1988). These phenomena have been attributed to variations in volcanic activity (affecting atmospheric transparency: Bryson and Goodman, 1980; Porter, 1986), changes in solar output (Harvey, 1980; Stuiver, 1980), and surges in the Antarctic ice sheet (Flohn, 1979). Such changes are slight on a Late Quaternary time scale but typically become important over intervals of a few thousand years, for example in the climate records from tree-rings and small glaciers.

Climate changes in the *decadal band* have been attributed to various external causes, including volcanic and solar activity and lunar tidal cycles (Pittock, 1983; Gilliland and Schneider, 1984). These changes are beyond the limits of sampling resolution of most sedimentary records, but have been detected in tree-ring series (Guiot, 1987).

Apart from the dominant annual cycle, climatic variations in the *interannual* and *annual bands* are attributed primarily to free (unforced) oscillations in the climate system, especially interactions between the atmosphere and oceans such as those responsible for the El Nino/ Southern Oscillation (ENSO) phenomenon (Rasmusson, 1985). Individual ENSO cycles can rarely be resolved in the palaeoclimatic record, but long-term changes in their average strength or frequency may be an important component of climate change in the lower frequency bands (Thompson *et al.*, 1984).

8.3.2   Orbital variations and glaciation

The time of year when the earth is closest to the sun (perihelion) varies cyclically with periods at 19 and 23 ka. This variation is called precession. Precession alternately increases and reduces the insolation contrast between summer and winter; its effect is opposite in the northern and southern hemispheres. The obliquity (tilt) of the earth's axis also varies, with a period of 41 ka, alternately reducing and increasing the seasonal contrast simultaneously in both hemispheres. The eccentricity of the orbit determines the strength of the precession effect, and varies with periods near 100 ka. The superimposition of these components of variation causes a virtually non-repeating pattern of changes in the latitudinal and seasonal distribution of insolation (Fig. 8.2).

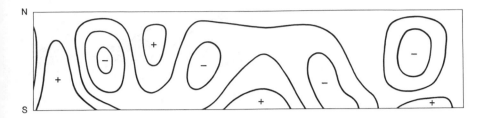

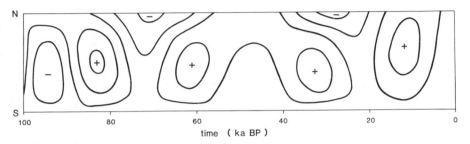

Figure 8.2 Variations in insolation in winter (top) and summer (bottom) as a function of latitude and time, from Berger (1978b). 'N' is 90°N, 'S' is 90°S.

Orbital variations are the 'pacemaker of the ice ages' (Hays *et al.*, 1976; Imbrie *et al.*, 1984; Imbrie, 1985), as proposed by Milankovitch (1930). Milankovitch's hypothesis – that low summer insolation at high northern latitudes builds ice sheets and high insolation melts them – has been confirmed by spectral analysis of the deep-sea oxygen-isotope record, which possesses spectral peaks at precisely the same frequencies as the orbital variations, and by modelling studies. Imbrie and Imbrie (1980), for example, related global ice volume (y) and insolation at a given latitude and season (x) by

$$dy/dt = (1/T_i)(x - y) \qquad (8.4)$$

where x and y are in standardized units and T_i has one value, T_c when $x < y$ (ice sheets growing) and another, T_w when $x > y$ (ice sheets decaying). Daily insolation is given by

$$x = \varepsilon + \alpha e \sin(\omega - \phi) \qquad (8.5)$$

where ε is the tilt of the earth's axis, e is the eccentricity of the orbit, ω indicates the timing of perihelion and α and ϕ indicate the latitude and season. The values $T_c = 42.5\,ka$, $T_w = 10.6\,ka$, and α and ϕ corresponding to July insolation at 65°N produced a fair simulation of the oxygen isotope record for the later part of the Quaternary.

The strongest periodicity in the ice-volume record is at 100 ka, corresponding to the eccentricity cycle. By modulating precession, the

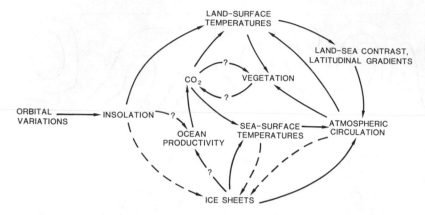

Figure 8.3   Causal connections and feedbacks in the climate system's response to orbital variations. Broken lines indicate responses with longer lags.

eccentricity cycle produces an exceptionally high northern-hemisphere summer insolation peak about once every 100 ka. Each time this occurs it triggers a 'termination' (a relatively rapid transition from a glacial to interglacial state in *both* hemispheres: Broecker, 1984). Imbrie and Imbrie's (1980) model underestimated the true amplitude of the 100 ka cycle. The physical mechanism that amplifies the 100 ka cycle is still unknown but is thought to involve atmospheric CO_2, which shows strong periodicity at 100 ka, increases substantially during terminations, and can be invoked as a cause of synchronous deglaciation in the northern and southern hemispheres (Pisias and Shackleton, 1984; Shackleton and Pisias, 1985; Genthon *et al.*, 1987). Possible factors in the observed CO_2 changes include changes in ocean circulation (Sarmiento and Toggweiler, 1984; Siegenthaler and Wenk, 1984), changes in primary production at the ocean surface (Broecker, 1982; Knox and McElroy, 1984; Oeschger *et al.*, 1984) and changes in the amount of carbon stored in vegetation and soils in response to climate change (Shackleton, 1977; Grove, 1984).

8.3.3   How the climate system responds to orbital variations

Climate changes in the orbital band depend on a combination of effects with different response times (Fig. 8.3; Wright, 1984). Immediate effects include differential heating or cooling of the land and ocean surface. Such effects are transmitted to the atmospheric circulation and bring about immediate changes in the terrestrial distribution of heat and moisture. Lagged effects include temperature and circulation changes due to the growth and decay of continental ice sheets.

The history of changes in insolation and ice sheets during the past 18 ka illustrates the hysteresis that results from this superimposition of

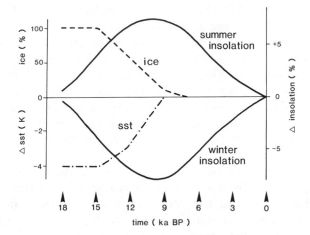

Figure 8.4 Late Quaternary changes in boundary conditions, as used by Kutzbach and Guetter (1986) for 'snapshot' simulations of global climate. Sea-surface temperature is indicated as 'SST'.

processes with different lags. For example, summer insolation in the northern hemisphere rose from a minimum around 18 ka to a maximum during the last termination (13–9 ka). Now it has returned to its 18 ka value, but the ice sheets have disappeared. The global climate went through different states during the insolation build-up and decline.

Complex changes in the climate system are susceptible to climate modelling (Kutzbach, 1985; Saltzman, 1985), including the use of atmospheric general circulation models (GCMs) to simulate the effects of boundary condition changes on circulation patterns (e.g. effects of the 18 ka ice sheets; Gates, 1976; Manabe and Hahn, 1977). GCMs are three-dimensional simulation models of the atmospheric circulation, with varying amounts of detail and realism in the way they represent exchanges of moisture and energy at the land and ocean surfaces.

To model the whole system portrayed in Fig. 8.3 would be a formidable challenge, but a simplification can be made by using the geological evidence (e.g. Denton and Hughes, 1981; CLIMAP Project Members, 1976, 1981; Barnola *et al.*, 1987) to define ice-sheet extent and height, sea-ice extent, sea-surface temperatures and atmospheric CO_2 as boundary conditions for the climate models (Fig. 8.4; Kutzbach and Guetter, 1986; Kutzbach, 1987). This approach has been used to predict changes in the behaviour of the atmosphere, and its effects on land climates. The simulated climates are remarkably consistent with broad features of the evidence from palaeoecological and lake-level records for changing spatial patterns in seasonal temperatures and annual water balance (COHMAP Members, 1988).

Refinements of the COHMAP approach include the use of GCMs that model the thermal response of the oceans and thereby predict, e.g., early Holocene sea-surface temperature anomalies (Mitchell *et al.*, 1988; Kutzbach and Gallimore, 1988), and sensitivity experiments using such models to indicate the relative importance of different boundary conditions (e.g. ice sheets versus CO_2 in the climate of the glacial maximum: Broccoli and Manabe, 1987). A combination of model studies and comparisons with Quaternary palaeoclimatic data has resulted in an understanding of many of the processes that have determined climate change during the Late Quaternary.

At the last glacial maximum, the high albedo of the ice sheets lowered annual temperatures through most of the northern hemisphere. Low CO_2 enhanced this cooling and also cooled the high latitudes of the southern hemisphere (Broccoli and Manabe, 1987). Global cooling may have been further enhanced by low atmospheric CH_4 and high aerosol (dust) concentrations (Raynaud *et al.*, 1988; Stauffer *et al.*, 1988; Legrand *et al.*, 1988; Harvey, 1988). The North American ice sheet acted as an enormous mountain range, splitting the westerly flow of air aloft and affecting the circulation over a large part of the hemisphere (Kutzbach and Wright, 1985; Webb *et al.*, 1987, COHMAP Members, 1988). The area south of this ice sheet may have experienced anomalously mild winters because of strong, adiabatically warmed winds off the ice sheet (Bryson and Wendland, 1967) and an unfamiliar climate with cool, dry summers and mild winters may have existed temporarily in the Great Lakes region around 12 ka (Wright, 1984), south of the still-large but rapidly retreating ice sheet.

At lower latitudes, the cool ocean surface produced a general reduction in the intensity of the hydrological cycle (Manabe and Hahn, 1977; Street and Grove, 1979; Peterson *et al.*, 1979; Velichko, 1987) and a reduction in the strength of the monsoons. The shift in the mean jet stream location caused by the North American ice sheet caused major changes in the storm tracks, and brought large amounts of moisture to the southwestern USA, maintaining the so-called pluvial lakes (Kutzbach and Wright, 1985; Benson and Thompson, 1987). Thus, the climatic anomalies associated with the glacial maximum were spatially very heterogeneous, with much wetter than present conditions in some areas even though global precipitation was reduced.

As the ice sheets decayed, the other effects of the insolation changes – not related to glaciation – began to predominate. Early Holocene summers were warmer than today in high northern latitudes (Ritchie *et al.*, 1983; Huntley and Prentice, 1988), and winters were colder than today in northern mid-latitudes (Prentice *et al.*, 1990), in direct response to the seasonal insolation pattern. Total annual insolation was greater than present at high latitudes, reducing the extent and duration of sea-ice and increasing winter temperatures in the polar regions (Kutzbach and

Gallimore, 1988; Mitchell *et al.*, 1988). The high land–sea contrast produced further, indirect effects: strengthened subpolar low-pressure cells, transporting warmth and moisture to high northern latitudes (Harrison *et al.*, 1992); strengthened subtropical high-pressure cells, causing dry summers in the northern mid-latitudes (Barnosky *et al.*, 1987; Harrison *et al.*, 1992); and at lower latitudes, major extensions of the African and Asian monsoons (Kutzbach, 1981; Kutzbach and Otto-Bliesner, 1982; Kutzbach and Street-Perrott, 1985). In North America, an extension of the southwest monsoon brought wet summers to the south-western deserts, simultaneous with dry summers in the Pacific Northwest (Thompson, 1988). Highest summer temperatures in eastern North America were delayed until the North American ice disappeared, ≈6 ka BP (Webb *et al.*, 1987). The broad climatic trends of the Holocene reflected the gradual decay of these effects.

Climate model studies and data-model comparisons have also thrown light on the longer term history of Quaternary climates. Changes in monsoon strength and extent are the dominant feature of the Quaternary climatic record in much of the present-day arid zone. These changes represent an immediate atmospheric response to orbital variations (Rossignol-Strick, 1983). Prell and Kutzbach (1987) used a combination of GCM simulations, representing various combinations of precession, tilt and glaciation, to derive a continuous time series of simulated African monsoon strength over the past 150 ka. The simulation closely paralleled the record of the windblown pollen of moisture-requiring plants in marine sediments off the east African coast (van Campo *et al.*, 1982). Similar approaches may make it possible to explain the time course and spatial patterns of climate in other parts of the world, including the similarities and differences noted by Andersen (1966), West (1980) and Watts (1988) between the pollen records of successive interglacial phases in Europe.

Higher frequency climatic changes cannot be generally attributed to orbital variations, but interactions between processes with different response times (Fig. 8.3) may produce short-lived events or reversals. The classic example of a climatic reversal is the Younger Dryas, a period between ≈11 and 10 ka when deglacial warming in the North Atlantic region was briefly interrupted by a phase of cold temperatures and Scandinavian glacial readvance. This interruption appears strongly in pollen records and glacial and periglacial geomorphic records from western Europe and the eastern seaboard of North America, and in foraminiferal sea-surface temperature records from the North Atlantic itself (Watts, 1980; Mangerud, 1980; Ruddiman and McIntyre, 1981; Rind *et al.*, 1986).

One hypothesis to explain the Younger Dryas suggests that meltwater flowing from the Laurentide ice sheet via the St Lawrence temporarily put a cold, low-salinity 'cap' on the surface of the North Atlantic (Ruddiman and McIntyre, 1981). A GCM study showed that a cold North

Atlantic could produce a temperature anomaly pattern resembling the Younger Dryas (Rind *et al.*, 1986; Overpeck *et al.*, 1989). Another hypothesis involves a change in ocean circulation, in which the production of North Atlantic Deep Water (NADW) was stopped (Broecker *et al.*, 1985; Boyle and Keigwin, 1987). The rapid beginning and end of the Younger Dryas suggest an abrupt transition between alternative stable states of the ocean–atmosphere system (Oeschger *et al.*, 1984; Broecker *et al.*, 1985). Experiments with a coupled three-dimensional model of the ocean–atmosphere system (Manabe and Stouffer 1988) suggest that the system does have two stable states, one with a warm North Atlantic surface and vigorous NADW production, the other with a colder, low-salinity North Atlantic surface and no NADW production. The second of these states may have prevailed during the Younger Dryas.

Cessation of NADW production during the Younger Dryas would also explain a contemporaneous phase of low lake levels that interrupted the general progression from dry conditions at the glacial maximum to wet conditions during the early part of the Holocene in northern Africa (Street-Perrott and Perrott, 1990; Gasse *et al.*, 1990). There is controversy about the wider global distribution of climate changes contemporary with the Younger Dryas (e.g. Heusser and Rabassa, 1987; Markgraf, 1989). More data, especially from the Southern Hemisphere, are required to discriminate among the various causal hypotheses (Schneider *et al.*, 1987; Harvey, 1989). The Younger Dryas is important because it provides a hint of how the complex interactions of land, sea and ice may have surprising consequences for global climates.

8.4   THE RESPONSE OF VEGETATION TO CLIMATE CHANGE

8.4.1   Scale and response time

Section 8.3 illustrated how physical models, ranging from simple caricatures of global ice-volume response to complex representations of the dynamics of the atmosphere and oceans, have contributed to the development of a theoretical framework for Quaternary palaeoclimatology. Climate is envisaged as an interacting system that responds to changes in its boundary conditions. The system's response to this 'forcing' can be predicted by climate models of varying complexity, and the predictions tested by reference to the Quaternary record of past climates.

Vegetation can similarly be considered as an interacting system (Roberts, 1987; Prentice, 1986a) that responds to external forcing. Climate provides the changing boundary conditions to which the vegetation adjusts through a combination of processes with different response times. The processes of adjustment can be explored with a combination of model studies and comparisons with the Quaternary record of past vegetation.

The nature of vegetation's response to climate change depends on the space and time scales on which it is observed (Fig. 8.1; White, 1979; Prentice, 1986a,b; Ritchie, 1986; McDowell et al., 1991). According to a general principle of systems theory, a system forced simultaneously at a wide range of frequencies will average out the high-frequency components of the forcing, while responding to – but staying close to equilibrium with – the low-frequency components. Intermediate frequencies, commensurate with the system's response time, produce a response that is damped or lagged (Levins, 1974; Levin, 1976; Prentice, 1983, 1986b). The mode of response (stationary, dynamic equilibrium or disequilibrium) depends on the product of the forcing frequency (μ) and the system's response time (T) (Webb, 1986). Stationary behaviour occurs if $\mu T \gg 1$, disequilibrium if $\mu T \approx 1$ and dynamic equilibrium if $\mu T \ll 1$. Vegetation can be observed on time scales ranging from years to millions of years and on spatial scales ranging from the sample plot up to the globe. Different frequencies and strengths of climate change are important on different time scales, and different response times apply to the different spatial scales. The various scales of data summarized in Fig. 8.1 therefore yield different and complementary perspectives on the processes of long-term vegetation dynamics.

8.4.2   The patch scale

A patch of vegetation can be defined functionally as an area just small enough that all the individual plants composing the vegetation interact strongly with one another (Prentice and Leemans, 1990). This 'natural patch size' should generally be about the same as the gap size required to allow regeneration of pioneer species (Prentice and Leemans, 1990; see also Chapters 2 and 4). The natural patch size is thus about $100–1000\,m^2$ in forests (Whitmore, 1982; Brokaw, 1985), but is much smaller in grass or shrub communities. Patches of this size are *non-equilibrating systems* that never reach a stable climax (McIntosh, 1981; Shugart, 1984). Eventually, either an individual grows so large that its death creates a gap about as large as the patch (Watt, 1947), or the patch suffers some external disturbance (White, 1979). Such disturbances happen in forests at a wide range of frequencies, depending on the nature of the climate and the susceptibility of the forest itself (Chapter 4). In many forests, natural disturbance occurs at intervals comparable with the life span of the trees (Runkle, 1985).

Pollen analyses from fast-aggrading mor humus can document the course of change in forest composition at the patch scale of $100–1000\,m^2$ in response to small-scale disturbances, such as individual treefalls, storm-felling and pathogen attacks (Bradshaw and Miller, 1988), or to larger scale disturbances of natural or human origin. Such investigations are comparable with observational studies based on permanent plots; they record succession after specific disturbance events. Vegetation that is composed of relatively short-lived plants may respond at the patch scale to short-term climatic changes in the annual to interannual bands, but forests are buffered against most changes on this time scale, both by the ability of trees to carry over resources between seasons, and by mechanisms such as facultative deciduousness and phenological plasticity. The main changes observed in forest composition at the patch scale over a century or so are usually related to the non-equilibrium dynamics of the forest patch, rather than to the effects of climate change. Modelling studies on this space and time scale commonly focus on the processes of forest succession, rather than responses to climate (e.g. Prentice and Leemans, 1990).

Climate change affects vegetation dynamics by changing the possibilities for establishment and potential growth rates of species, altering both the probabilities of gap formation and the direction of succession after disturbance (Neilson, 1987; Brubaker, 1986; Clark, 1988c, 1990; see also Chapter 7). Short-term climate fluctuations have little effect on long-term successions, but climate changes in the decadal to millennial bands alter the direction of succession as it proceeds (Finegan, 1984; Prentice, 1986b).

The same mechanisms (changes in which available species can grow and reproduce, and changes in their potential growth rates) are involved in the response to climate changes in the orbital band, but the effects are too slow to be observed during a single round of succession after disturbance. Instead, there is a change in the pattern of succession through successive gap-phase or disturbance cycles (Delcourt et al., 1983). Such changes can be observed at the patch scale in longer mor humus records (Bradshaw and Zackrisson, 1990). The course of succession is modified both by climate change and by changes in species availability caused by climate change (section 8.4.3).

8.4.3   The landscape scale

Individual patches of vegetation suffer disturbance. A vegetated landscape possesses a disturbance regime, that is each patch has a certain probability for disturbances of a given magnitude and type (Chapters 1, 3 and 4). Disturbance regimes may be natural (e.g. storm damage, wildfire) or artificial (e.g. logging, slash-and-burn cultivation). The area affected by a single disturbance can be very different under different climatic (White, 1979; Sousa, 1984) or management regimes. Some types of natural disturbance, such as minor windstorms, are selective and at any one time only affect areas comparable with a single patch of vegetation.

Others (notably major hurricanes and fires) are almost indiscriminate and can initiate succession over large areas.

The ratio of the landscape area to the disturbance area is crucial in determining the dynamics of the landscape as a whole (Romme and Knight, 1981; Shugart, 1984). Landscapes that are smaller than or comparable in area with the mean disturbance area are non-equilibrating systems that typically undergo quasi-cyclical sequences of succession and disturbance. Pollen diagrams with high time resolution from regions where fire is important (e.g. high-resolution pollen diagrams from laminated sediments in the boreal and northern hardwood zones) record this kind of dynamics. The pollen data show changes corresponding to the course of secondary succession after each successive fire (Swain, 1973; Cwynar, 1978). Such studies have also documented how major Holocene climate changes have affected forested landscapes by altering fire return times (Swain, 1978; Clark, 1988c, 1990), modifying the importance of different taxa in succession, and facilitating the invasion of new taxa (Green, 1981, 1982; Walker, 1982b).

The fire regime is not simply a product of the climate, but is influenced by the nature of the vegetation. Major changes in vegetation may be preceded by destructive fires (Green, 1982), or a change in fire frequency may follow a change in vegetation (Clark et al., 1989). High-resolution pollen diagrams can also provide detailed information on the effects of large-scale human impacts on the landscape, including human uses of fire as a management tool. Large-scale alterations of the disturbance regime have become an important feature of vegetation changes in Europe during the last few millennia (Iversen, 1973; Behre, 1988).

Landscapes that are many times larger than the mean disturbance area can be represented as mosaics of smaller areas in different stages of succession (Loucks, 1970; Heinselman, 1973; Whittaker and Levin, 1977; Bormann and Likens, 1979; Pickett, 1980; Shugart, 1984; see also Chapters 1 and 3) and show a different kind of dynamics in which the cycles of succession, gap formation and disturbance on individual patches are subsumed in the average dynamics of the whole. Landscapes that are all-aged mosaics can reach an equilibrium composition if given enough time, a constant disturbance regime, and a constant climate. If not in equilibrium, they change towards an equilibrium by adjustments both in the course of succession on individual patches (section 8.4.2) and in the age-distribution of patches. Climate change shifts the position of the equilibrium. The composition of the landscape then changes, through precisely the same mechanisms, in the direction of a new equilibrium.

Most pollen diagrams record areas larger than the mean disturbance area, or achieve a similar effect by averaging in time. Thus, *they do not register secondary succession*, only changes in the average species composition of the landscape. Such changes may occur in response to climate change, and other changes (e.g. in the disturbance regime and in the availability of species) brought about by climate change (Ritchie, 1986).

The response time of landscape composition is partly determined by the pace of secondary succession (Brubaker, 1986; Prentice, 1986b), but is also affected by the frequency of disturbance. Davis and Botkin (1985) used a gap model to simulate the response of a forested landscape to climate changes in the decadal to millennial bands. In one series of experiments, an array of patches was run for several hundred years without disturbance, before a climate change was imposed. In another series, the patches were subject to a stochastic disturbance regime. The response of the landscape without disturbance was lagged by 100–200 years, and became weaker as either the magnitude or the duration of the change was reduced. The effect of adding disturbance was to increase the rate of response and reduce the damping, but also to reduce the sensitivity (i.e. the magnitude of the response) to a given climate change. These results suggest that forested landscapes may be buffered against decadal climate changes but respond (with some lag) to millennial climate changes. Frequent disturbance may allow the landscape to respond to higher frequency changes, but only if they are of large enough amplitude.

Field investigations suggest that vegetation has responded quickly but not instantaneously (i.e. a disequilibrium response) to climate changes during the past 100–1000 years (Neilson and Wullstein, 1983; Davis, 1986; Kullman, 1986; Brubaker, 1986). There is also pollen evidence that vegetation responded to the Little Ice Age (Bernabo, 1981; Grimm, 1983; Gajewski, 1987), but because the spatial variation of this climate change is not well known, it is hard to know whether spatial variations in the response reflect variations in the sensitivity and response time of the landscape (Davis and Botkin, 1985) or variations in the intensity and duration of the climatic event. Pollen evidence also records differences in the response of landscapes under similar climates (Brubaker, 1975). The differential response of vegetation on soils with different water-holding capacities shown in Brubaker's study provided a test of a forest gap model (Solomon and Shugart, 1984). Bernabo's (1981) finding that sites on poor soils showed weaker vegetation effects of Little Ice Age-scale climate variations is consistent with further simulations by Davis and Botkin (1985), who predicted that vegetation on poorer soils would respond less strongly than vegetation on better soils. There is also evidence of different fire regimes on different soils and this, too, can result in different vegetation responses to climate change (Winkler, 1985b).

Hysteresis occurs when climate change temporarily allows the coexistence of two or more landscape equilibria (Chapter 3). Shugart et al. (1980) used a gap model to simulate hysteresis at the patch scale in a hypothetical transition between Fagus grandifolia and Liriodendron tulipifera as forest dominants in response to change in growing-season warmth. An abrupt transition took place at different points depending on the direction of change. Different mechanisms promoted the persistence of whichever species was originally present. Grimm (1983, 1984) gave an example of hysteresis at the landscape scale in the conversion of prairie (which burns easily and is well adapted to burning) via *Quercus* woodland to mesic deciduous 'Big Woods' vegetation (which burns less readily but is more damaged by burning) in response to a gradual increase in effective moisture from 5 ka to the present. The positive feedback between fire frequency and vegetation ensured that the prairie and woodland were separated by sharp (physiographically determined) spatial boundaries in pre-settlement times, and that transitions in time from prairie to woodland were relatively sharp and not directly related to the timing of the climate change. The later invasion of the woodland by the Big Woods taxa (*Ulmus* species, *Ostrya virginiana*, *Tilia americana*, *Acer saccharum*) was different in character because a woodland fire regime was already established. This transition took place gradually over 100–200 years in response to a further moisture increase at the beginning of the Little Ice Age.

Climate determines which species can grow in a region. However, when climate change first allows growth of a new species, then even if the species is already present nearby it will take some time to invade the landscape. Invasions are prominent features of most Holocene pollen diagrams; they are often fast (taking place over a few centuries) and separate periods of slower change. The dynamics of invasions are documented in pollen diagrams with high time resolution (Watts, 1973). The initial phase of an invasion can often be approximated by an exponential curve:

$$Y_t = Y_0 \, e^{t/T} \tag{8.6}$$

where Y_t is the pollen flux density at time t and $T/(\ln 2)$ is the 'population doubling time'. If flux density data are available, then doubling time can be estimated by regression of $\ln Y_t$ against t. Tsukada and Sugita (1982), Bennett (1983, 1986a) and Walker and Chen (1987) have used this method to estimate initial population doubling times, which ranged from 25–400 years with some tendency to be shorter for pioneers and longer for slow-growing taxa (Bennett, 1986b; Prentice, 1988). The full course of an invasion can be approximated by a logistic curve over the whole interval, typically 200–1000 years, between the start and completion of the invasion.

Changes in climate and species availability are causally related but not necessarily simultaneous. Although landscapes may take only a few hundred years to respond fully to *either* climate change *or* species immigration (Solomon et al., 1980, 1981), the arrival of a new species in a region could be delayed for an unknown period after the climate became suitable for it to grow (Solomon and Webb, 1985; Davis et al., 1986). There is still uncertainty over the length of time required for taxa to cross continents in response to changes in their potential ranges. This question has been approached through analyses of changes in regional vegetation patterns, as discussed in the following section.

8.4.4 The regional scale

Maps of changes in the large-scale geographic patterns of vegetation through time give a different perspective from records of vegetation change in particular landscapes. Invasions become migrations (Davis, 1976, 1981a); abrupt transitions (Grimm, 1983) become smooth boundary shifts (Webb *et al.*, 1983). Lags in the landscape's response to climate change, and related effects such as hysteresis, effectively disappear.

Regional vegetation patterns respond strongly to climate changes in the orbital band, whether caused directly by insolation and its effects on the atmospheric circulation, or mediated by changes in surface boundary conditions such as the occurrence of ice sheets (section 8.3.3). It was previously thought that the Holocene migration patterns of tree taxa in northern Europe (Iversen, 1960, 1973) and eastern North America (Davis, 1976, 1981a) reflected differential rates of spread from glacial-age refuges, implying long migrational lags (>3 ka; perhaps even >10 ka) for some taxa. However, the arguments for this hypothesis were largely predicated on an untenable model of climate change (Prentice, 1983, 1986b; Webb, 1986). More recent studies have begun to establish the outlines of Late Quaternary climate changes in each continent, and the specific ways in which these changes have continuously shifted each species' potential range.

Response surfaces form one of the tools used in these continental-scale studies. Response surfaces describe the present geographic variation in surface pollen percentages of taxa as non-linear functions of climate variables. Conventionally the variables used are mean July temperature, mean January temperature and annual precipitation; although these are not the variables to which plants directly respond, they often account for much of the continental-scale variation in seasonal temperature and moisture. Response surfaces have been constructed empirically for the major pollen taxa in eastern North America (Fig. 8.5; Bartlein *et al.*, 1986; Webb *et al.*, 1987; Prentice *et al.*, 1991) and Europe (Huntley *et al.*, 1989). Many of these pollen taxa include several species, but all show clear (most often unimodal) relationships to climate.

In some cases (e.g. *Fagus*), the response surfaces are strikingly similar on the two continents, although the species are different. This similarity supports a causal relationship between the pollen abundances and climate on both continents, and implies that this relationship has been evolutionarily conservative since the mid-Cainozoic separation of the North American and European floras (Huntley *et al.*, 1989). Not all the taxa show such similar responses, however, and some taxa present in one

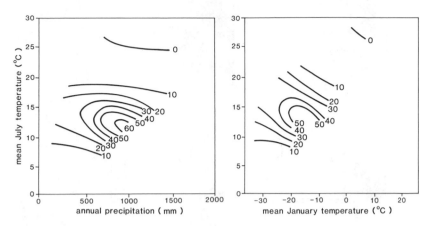

Figure 8.5 Response surfaces for *Picea* pollen percentages in eastern North America, based on >1000 surface pollen samples. These two 2D surfaces together give an impression of the form of the full 3D surface. From Webb *et al.* (1987).

continent (e.g. *Tsuga* and *Carya* in eastern North America) are now absent from the other despite an apparently suitable climate.

The forms of the response surfaces presumably reflect a combination of physiological controls and interspecific competition (Austin and Smith, 1989). In many cases the limits of each species' common occurrence (where pollen percentages fall to some low value) can be related to a known physiological mechanism, such as the growing degree-day requirements of trees, the different freezing tolerance mechanisms of broadleaved evergreens, deciduous trees and conifers, and differences in drought tolerance related to leaf morphology and habit (Larcher, 1983; Woodward, 1987). Some other limits can be explained by competition (e.g. high abundances of grasses and forbs are confined to climates that do not support trees).

Response surfaces allow hypothesized climate changes to be translated into simulations of the pollen record (Prentice *et al.*, 1991; Bartlein *et al.*, 1986). Webb *et al.* (1987) used Kutzbach and Guetter's (1986) GCM results for eastern North America (section 8.3.3), together with response surfaces, to simulate isopoll maps for six quantitatively important pollen taxa (northern and southern *Pinus*, *Picea*, *Betula*, *Quercus* and Prairie forbs) at 3 ka intervals from 18 ka to the present. Major distributional changes occurred over this period, and most were simulated accurately (e.g. *Picea*: Fig. 8.6). The absence of any major offset between the observed and simulated isopoll maps implies migrational lags <1.5 ka (Prentice *et al.*, 1991). Some discrepancies between observed and simulated maps pointed to possible inaccuracies in surface boundary conditions such as ice-sheet height (Webb *et al.*, 1987), which would affect surface temperatures south of the ice sheet.

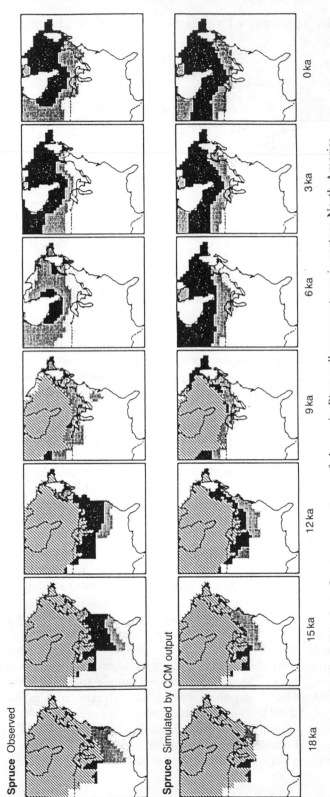

Figure 8.6 Late Quaternary patterns of change in *Picea* pollen percentages in eastern North America. Dark stippling, >20% *Picea* pollen; intermediate stippling, 5–20% *Picea* pollen; light stippling, 1–5% *Picea* pollen; no stippling, <1% *Picea* pollen. Diagonal hatching denotes the ice sheet. The upper series is based on pollen data; the lower series on the climate model simulations of Kutzbach and Guetter (1986) using known changes in boundary conditions (Fig. 8.4). From COHMAP Members (1988). (Copyright 1988 by the AAAS.)

Spruce Observed

Spruce Simulated by CCM output

0 ka 3 ka 6 ka 9 ka 12 ka 15 ka 18 ka

Prentice *et al.* (1991) obtained closer agreement between real and simulated isopoll maps, using a climatic scenario *inferred* from the pollen data via the response surfaces and tested against independent pollen data (Prentice, 1984; ter Braak and Prentice, 1988). This scenario agreed qualitatively with the GCM simulations (Webb *et al.*, 1987), but differed quantitatively, particularly during 18–12 ka. The inferred climate scenario was used to simulate patterns of change, first for the six taxa used to derive it, then for a further, independent set (*Carya*, *Fagus*, *Tsuga*, *Abies*, Cyperaceae, *Alnus* and *Ulmus*). The simulations captured the essentials of the isopoll maps for both sets of taxa. The results thus provide climatic explanations for the patterns, including features that had previously been attributed to migrational lag.

An example is provided by *Quercus* and *Carya*. These two genera have rather similar present distributions but *Quercus* was ahead of *Carya* in its northward migration along the east coast during the early Holocene (Davis, 1976). The inferred climate scenario showed that precipitation along the east seaboard was lower in the early Holocene than present. Lake-level records (Harrison, 1989) and GCM results for 9 ka (Webb *et al.*, 1987) support this. The response surfaces show that *Carya* has a slightly higher precipitation requirement than *Quercus*; hence the lag.

Prentice *et al.* (1990) did not correctly predict the observed high abundances of *Ulmus* at 12 ka in the region south of the Great Lakes. The mixture of abundant *Picea* and hardwoods characteristic of this region then has no analogue today (Webb *et al.*, 1983; Overpeck *et al.*, 1985) and may have been produced by no-analogue climatic conditions around the southern margin of the ice sheet (section 8.3.3; Solomon and Shugart, 1984; Wright, 1984). Response surfaces, being empirical models based on present conditions, cannot accurately simulate assemblages of plants that do not occur today.

Finer scale analyses of a network of accurately dated Holocene pollen records (Davis *et al.*, 1986; Davis, 1987; Woods and Davis, 1989) have focused on the spatial dynamics of the migration of *Fagus grandifolia* and *Tsuga canadensis* into the Midwest, which started ≈7–8 ka and proceeded at an uneven pace towards the present. The northward and westward spread of these taxa through eastern North America after ≈12 ka was driven first by increasing winter temperatures generally, then (after 6 ka) by increasing precipitation in the Midwest (Woods and Davis, 1989; Prentice *et al.*, 1990). The immigration of *Fagus* and *Tsuga* into the Midwest between 7–8 ka and 5.5–6 ka was rapid and spatially incoherent, suggesting limitation by dispersal rather than climate (Davis *et al.*, 1986). Their later migrations were slower and spatially coherent, suggesting that their ranges were then in equilibrium with climate (Davis *et al.*, 1986). Woods and Davis (1989) showed that the range limit of *Fagus* in the Midwest was close to equilibrium with climate from 5 ka and possibly from 8 ka onwards, with maximum migrational lags <1 ka.

Woods and Davis (1989) also observed that climate probably did not become uniformly more favourable to *Fagus* during this continuous period of expansion, yet range *contractions* were not observed. Their interpretation was that the climatic conditions for establishment are more stringent than those for growth and survival. Whereas climate changes in the orbital band may produce both expansions and contractions of species ranges, the response time of advancing range boundaries may be faster than that of retreating boundaries. The vegetational response to climate changes in the millennial band may therefore be expressed more strongly in range advances than retreats.

The apparent migration rates of plant taxa in response to climate change can be estimated from isopoll maps. Some were exceedingly fast, up to $0.4 \, \text{km} \, \text{a}^{-1}$ (Davis, 1981a) or $1.5 \, \text{km} \, \text{a}^{-1}$ (Huntley and Birks, 1983; Huntley, 1988) for early Holocene migrations of *Pinus* species in eastern North America and Europe, respectively, or even $2 \, \text{km} \, \text{a}^{-1}$ for *Picea glauca* in the Western Interior of Canada (Ritchie and MacDonald, 1986). These rates may possibly refer to waves of increase in abundance within the range boundary, rather than range boundary advances (Bennett, 1986b; Gaudreau and Webb, 1985; Dexter *et al.*, 1987), since an advancing range boundary characterized by small, scattered populations may be hard to detect in the pollen record. Nevertheless, plant species may be more mobile than data on normal dispersal distances and generation times suggest.

This paradox focuses attention on the mechanisms of plant migration (Huntley and Webb, 1989). Part of the explanation for fast migration rates may lie in the ubiquity of natural disturbance (Walker, 1982a). In much of the older palynological literature it was assumed that vegetated landscapes not disturbed by human activities provided effective barriers to immigration. This is inconsistent with present understanding of community and landscape dynamics (Chapter 4). A more modern view is that natural disturbance helps to maintain the overall composition of the vegetation close to equilibrium with a changing climate (Bradshaw and Zackrisson, 1990). Furthermore, many shade-tolerant species, like *Fagus grandifolia*, do not require a disturbance in order to colonize previously vegetated land; they simply establish in the shade of other trees, and a proportion ultimately reaches the canopy (Woods and Davis, 1989). A disturbance may speed the migration of such species, but it is not a *sine qua non*.

Migration on to unvegetated land seems to take place at similar rates to migration into vegetated landscapes (Huntley and Webb, 1989). The lag in vegetation development associated with soil formation after deglaciation has been estimated to be 1.5 ka in a case study of the post-glacial spread of *Betula pubescens* in Britain (Pennington, 1986), and may in fact be 0.5 ka or less (Berglund *et al.*, 1984; Prentice, 1986b).

Rapid migration in general may be facilitated by the occurrence of 'advance' populations in topographically or edaphically favourable habitats beyond the main distribution area (Grimm, 1983; Bennett, 1986b; Woods and Davis, 1989). Expansion may then occur outwards from these populations when the climate changes. Occasional long-distance dispersal to more-or-less random locations may also be particularly important in providing new foci from which the species can colonize surrounding areas during rapid range expansion (Bennett, 1985; Ritchie and MacDonald, 1986). A model study by Mack (1985) showed that range expansion by long-distance dispersal and expansion from outlying foci is more rapid than range expansion by gradual advance along a front. This is also consistent with evidence that species' distributions become spatially incoherent during phases of rapid expansion of their potential range (Davis *et al.*, 1986; Ritchie and MacDonald, 1986).

The total response time of regional vegetation patterns to climate change includes both lags due to migration into new landscapes, and population expansion within landscapes (Webb, 1986). The consensus of the studies discussed above is for total response times on the order of 200–1000 years to climate changes during the Holocene, the response time probably varying according to a number of factors including the rate of increase of the potential range of the species, and whether the response involves establishment in new localities or mortality in existing localities.

Whatever the response times and mechanisms of change, the broad patterns of vegetation change on Late Quaternary and Holocene time scales appear to be climatically controlled. Soil changes (Iversen, 1958; Andersen, 1966) and direct CO_2 effects (Solomon, 1984) are not implicated as primary forcing factors, although changes in soil characteristics may well have occurred as a secondary response to changes in the vegetation.

A significant role is, however, ascribed to disease epidemics, and to human impact on vegetation. A general decrease in the abundance of *Tsuga* in eastern North America at ≈4.6 ka (Webb, 1982) has no obvious climatic explanation and may have been caused by an epidemic analogous to the more recent chestnut blight and Dutch elm disease outbreaks; the *Ulmus* decline at ≈5 ka in Europe may have a similar explanation (Davis, 1981b). Vegetation changes in Europe in recent millennia show much evidence for human impact (Behre, 1988). Nonetheless, the large-scale geographic patterns have remained remarkably coherent, implying ultimate control by climate (Huntley, 1988). Presumably, climate change has also influenced the susceptibility of plant populations to disease, and the patterns of prehistoric land use, although the connections are complicated because both plant pathogens and human societies have evolved on time scales comparable with Quaternary climate change.

Lags in the response of vegetation to climate change are long in the perspective of field studies, but they are slight in the Quaternary perspective and negligible in an evolutionary perspective. The ability of plant species to migrate rapidly can be regarded partially as an evolutionary adaptation to climate changes in the orbital band. Such changes have been a feature of climate even during the long ice-free periods of earth history (Bartlein and Prentice, 1989). This idea is consistent with Good's (1931) 'Theory of tolerance', although Good did not tie the concept specifically to the orbital time scale. Good proposed that species migrate rather than evolve new climatic tolerances. Those that cannot find suitable habitat in time go extinct. This fate befell many warm-temperate forest taxa in Europe during the earlier part of the Quaternary (West, 1980; Davis, 1983; Watts, 1988), ultimately as a consequence of Late Cainozoic cooling.

8.4.5   The global scale

Plant species migrate over subcontinental areas on an orbital time scale, but their capacity for long-distance dispersal and establishment is not unlimited. The fact that widely separated regions with similar climates have substantially distinct floras is *prima facie* evidence that the great oceans, at least, are effective dispersal barriers. Global *floristic* patterns are not in equilibrium with climate.

Large-scale geographic patterns in vegetation *physiognomy*, however, are subject to powerful climatic constraints (Woodward, 1987). The world distribution of vegetation formations can be predicted by equilibrium models that assume direct control by climate (Box, 1981; Emanuel *et al.*, 1985; Woodward, 1987; Woodward and Williams, 1987; Prentice, 1990). Like the continental-scale patterns of species abundance, the global patterns of vegetation physiognomy have varied on an orbital time scale. These variations may be predictable from Late Quaternary climate changes by means of equilibrium models, with the Quaternary pollen record providing the data needed for a critical test. The most appropriate models for this purpose are not empirical models based on the present-day distribution of vegetation formations since these can dissolve under climate change (section 8.2.4), but models that incorporate environmental constraints on the occurrence of different morphological and physiological plant types (Woodward, 1987).

Quaternary changes in vegetation at a global scale also imply changes in the total carbon storage in vegetation, soils and peats. These terrestrial carbon pools are a significant item in the global carbon budget (Bolin, 1983), and changes in their size may have impacts on atmospheric CO_2 and climate (section 8.3.2; Fig. 8.3). Experiments in which global vegetation models are forced by past climates (Prentice and Fung, 1990) can therefore contribute to an understanding of the dynamic role of terrestrial vegetation in the variations of the climate system on an orbital time scale. This is an area of great uncertainty at present.

For example, it is not clear whether global terrestrial carbon storage was less than or greater than present at the last glacial maximum. The expected reduction in biomass due to widespread tropical aridity (Grove, 1984) may have been counteracted by an extension of tropical forests on to the exposed continental shelf (Prentice and Fung, 1990). The causes of low atmospheric CH_4 at the glacial maximum are also unclear but may involve changes in the extent of terrestrial CH_4 sources such as peatlands. Such questions cannot be resolved by models or data in isolation. The development of a predictive understanding of global vegetation change and its role in the carbon cycle will require creative model development, well-motivated data acquisition, and rigorous testing of model results against the global Quaternary record.

SUMMARY

1. Climate controls vegetation through direct effects on establishment and growth, indirect effects mediated by soils, and effects on the disturbance regime. These effects modify the internal dynamics of vegetation, producing both geographic patterns and long-term changes in vegetation.
2. The causes of long-term vegetation changes are different from the causes of succession and patch dynamics. To understand the former requires knowledge of species' responses to climate, and of climate change.
3. Analyses of foraminiferal assemblages and stable isotopes from deep-sea sediment cores record the Quaternary history of sea-surface temperatures, global ice volume and atmospheric CO_2. These are important factors in the climate system.
4. Pollen in mor humus under a forest canopy integrates over $\approx 1000\,m^2$ and $\geqslant 10$ years; in lake sediments, $\approx 1000\,km^2$ and $10-100$ years. Pollen percentages are quantitatively related to relative abundances, and maps of pollen percentages give an approximation of spatial patterns in relative abundance.
5. Pollen data from Europe and eastern North America have been mapped at $\approx 1\,ka$ intervals for the Late Quaternary. Coverage of other areas is very uneven. There are interglacial pollen records from glaciated regions, and longer continuous records from unglaciated regions.
6. Quaternary climate changes affected vegetation in all latitudes. The changes were complex in structure and spatial pattern. Plant taxa migrate individualistically, rather than as communities, in response to climate change.
7. Climate variations at different frequencies have different causes and effects. Orbital variations with periods at ≈ 20, 40 and $100\,ka$ have predominated during the Quaternary. Global ice volume responds to

these variations with a 10–40 ka lag. The 100 ka cycle is amplified by an as yet imperfectly understood mechanism, probably involving variations in atmospheric CO_2.

8. Past climates can be simulated by subjecting climate models to changed insolation patterns and surface boundary conditions. Model studies and data–model comparisons have elucidated the major patterns and mechanisms of Late Quaternary climate change.

9. Single 'patches' of vegetation (100–1000 m^2 in forests) undergo changes due to regeneration cycles and/or disturbance. Mor-humus pollen analysis records these phenomena. Climate change can alter the course of succession as it occurs, or more gradually over many cycles.

10. Pollen diagrams from laminated lake sediments record cycles of succession and disturbance in landscapes influenced by fire, and climatically induced changes in fire frequency and post-fire succession. Pollen diagrams also record human impacts on landscapes.

11. Most pollen diagrams do not record succession, only changes in average landscape composition due to climate change or human effects. Landscape composition can respond to climate in ≈100 years; high disturbance frequency accelerates the response. Climate changes on the time scale of the Little Ice Age appear in the pollen record, though possibly damped or lagged. Abrupt responses to gradual climate changes can be caused by transitions between alternative landscape equilibria.

12. Closely spaced pollen flux-density measurements document the transient dynamics of landscape invasions. Estimated population doubling times range from 25 to 400 years; most invasions are completed in 200–1000 years.

13. Response surfaces relating modern pollen abundances to climate describe the realized niche of each taxon in climate space. Response surfaces are multifactorial and unique, reflecting individualistic physiological constraints and competitive effects on species' distribution and abundance.

14. Changes in regional vegetation patterns are displayed by continental-scale isopoll maps. The major changes in the vegetation of eastern North America during the Late Quaternary were a response to continuous climate changes caused by insolation variations and ice-sheet retreat. Migration lags are unimportant at this scale. Lags of up to ≈1 ka have been inferred from higher resolution studies of sub-continental areas.

15. Observed migration rates vary widely, but can reach 2 km a^{-1}. Established vegetation is not a barrier because natural disturbances allow new species in, and because many species do not require disturbance. Primary succession after deglaciation is also rapid. Rapid migration into newly favourable areas may be accelerated by nucleation, resulting in spatially incoherent spread.

16. Species migrate, rather than evolve new climatic tolerances, in the face of climatic changes at orbital frequencies. Extinctions have occurred in response to Late Cainozoic cooling.
17. Global floristic patterns are not in equilibrium with climate, but vegetation physiognomy is constrained by climate. Global vegetation model predictions can be tested against the Quaternary record. Total terrestrial carbon storages change with climate and may have feedback effects on climate.

ACKNOWLEDGEMENTS

I thank Pat Bartlein, Richard Bradshaw, Eric Grimm, Sandy Harrison, Rob Hengeveld, Brian Huntley, Matt McGlone, Hank Shugart, Al Solomon, Cajo ter Braak and Tom Webb for discussion and comment. This research was supported by US Department of Energy (CO_2 Research Division) grants to COHMAP (Co-operative Holocene Mapping Project) and Swedish Natural Science Research Council (NFR) grants to the project 'Simulation modelling of natural forest dynamics'.

REFERENCES

Allen, T. F. H., O'Neill, R. V. and Hoekstra, T. W. (1984) *Interlevel Relations in Ecological Research and Management: Some Working Principles from Hierarchy Theory*. US Department of Agriculture, General Technical Report RM-110.

Andersen, S. T. (1966) Interglacial vegetational succession and lake development in Denmark. *Palaeobotanist*, **15**, 117-27.

Andersen, S. T. (1970) The relative pollen productivity and pollen representation of north European trees, and correction factors for tree pollen spectra. *Danmarks Geol. Undersøg.*, **II-96**.

Atkinson, T. C., Briffa, K. R., Coope, G. R., Joachim, M. J. and Perzy, D. W. (1986) Climatic calibration of coleopteran data, in *Handbook of Holocene Palaeoecology and Palaeohydrology*, (ed. B. E. Berglund), Wiley, Chichester, pp. 851-8.

Austin, M. P. and Smith, T. M. (1989) A new model for the continuum concept. *Vegetatio*, **83**, 35-47.

Barber, K. (1981) *Peat Stratigraphy and Climatic Change: A Palaeo-ecological Test of the Theory of Cyclic Peat Bog Regeneration*, Balkema, Rotterdam.

Barber, K. (1985) Peat stratigraphy and climatic change: some speculations, in *The Climatic Scene: Essays in Honour of Gordon Manley*, (eds M. J. Tooley and G. M. Sheail), Allen and Unwin, London, pp. 175-85.

Barnola, J. M., Raynaud, D., Korotkevich Y. S. and Lorius, C. (1987) Vostok ice core provides 160 000-year record of atmospheric CO_2. *Nature*, **329**, 408-14.

Barnosky, C. W., Anderson, P. M. and Bartlein, P. J. (1987) in *North America and Adjacent Oceans During the Last Deglaciation*, (eds W. F. Ruddiman and H. E. Wright, Jr), Geology of North America K-3, Geological Society of America, Boulder, pp. 289-322.

Barry, R. G. (1985) Snow and ice data, in *Paleoclimate Analysis and Modeling*, (ed. A. D. Hecht), Wiley, New York, pp. 259–90.

Bartlein, P. (1988) Late-Tertiary and Quaternary palaeo-environments, in *Vegetation History*, (eds B. Huntley and T. Webb III), Kluwer, Dordrecht, pp. 113–54.

Bartlein, P. J. and Prentice, I. C. (1989) Orbital variations, climate and palaeoecology. *Trends Ecol. Evol.*, **4**, 195–9.

Bartlein, P. J., Prentice, I. C. and Webb, T., III. (1986) Climatic response surfaces for some eastern North American taxa. *J. Biogeog.*, **13**, 35–57.

Bartlein, P. J., Webb, T., III and Fleri, E. (1984) Holocene climatic change in the northern midwest: pollen-derived estimates. *Quat. Res.*, **22**, 361–74.

Behre, K.-E. (1988) The role of man in European vegetation history, in *Vegetation History*, (eds B. Huntley and T. Webb III), Kluwer, Dordrecht, pp. 633–72.

Behre, K.-E. (1989) Biostratigraphy of the last glacial period in Europe. *Quat. Sci. Rev.*, **8**, 25–44.

Bennett, K. D. (1983) Postglacial population expansion of forest trees in Norfolk, UK. *Nature*, **303**, 164–7.

Bennett, K. D. (1985) The spread of *Fagus grandifolia* across eastern North America during the last 18000 years. *J. Biogeog.*, **12**, 147–64.

Bennett, K. D. (1986a) Competitive interactions among forest tree populations in Norfolk, England, during the last 10000 years. *New Phytol.*, **103**, 603–20.

Bennett, K. D. (1986b) The rate of spread and population increase of forest trees during the postglacial. *Phil. Trans. Roy. Soc. London B*, **314**, 523–31.

Benson, L. and Thompson R. S. (1987) The physical record of lakes in the Great Basin, in *North America and Adjacent Oceans During the Last Deglaciation*, (eds W. F. Ruddiman and H. E. Wright, Jr), Geology of North America K-3, Geological Society of America, Boulder, pp. 241–60.

Berger, A. L. (1978a) Long-term variations of daily insolation and Quaternary climatic changes. *J. Atmos. Sci.*, **35**, 2362–7.

Berger, A. L. (1978b) Long-term variations of caloric solar radiation resulting from the earth's orbital elements. *Quat. Res.*, **9**, 139–67.

Berger, A. L. (1981) Spectrum of climatic variations and possible causes, in *Climatic Variations and Variability: Facts and Theories* (ed. A. Berger), Reidel, Dordrecht, pp. 411–32.

Berglund, B. E. (ed.) (1986) *Handbook of Holocene Palaeoecology and Palaeohydrology*, Wiley, Chichester.

Berglund, B. E., Lemdahl, G., Liedberg-Jönsson, B. and Persson, T. (1984) Biotic response to climatic changes during the time span 13000–10000 B.P. – A case study from SW Sweden, in *Climatic Changes on a Yearly to Millenial Basis* (eds N. A. Mörner and W. Karlén), Reidel, Dordrecht, pp. 25–36.

Bernabo, J. C. (1981) Quantitative estimates of temperature changes over the last 2700 years in Michigan based on pollen data. *Quat. Res.*, **15**, 143–59.

Bernabo, J. C. and Webb, T. III. (1977) Changing patterns of the Holocene pollen record of north-eastern North America: a mapped summary. *Quat. Res.*, **8**, 64–96.

Bolin, B. (1983) The carbon cycle, in *The Major Biogeochemical Cycles and Their Interactions* (ed. B. Bolin), Wiley, Chichester, pp. 41–5.

Bormann, F. H. and Likens, G. E. (1979) *Pattern and Process in a Forested*

Ecosystem, Springer-Verlag, New York.

Box, E. O. (1981) *Macroclimate and Plant Forms: An Introduction to Predictive Modeling in Phytogeography*, Junk, The Hague.

Boyle, E. A. and Keigwin, L. (1987) North Atlantic thermohaline circulation during the past 20 000 years linked to high-latitude surface temperature. *Nature*, **330**, 35–40.

Bradley, R. S. (1985) *Quaternary Paleoclimatology*. Allen and Unwin, Boston.

Bradshaw, R. H. W. (1981) Modern pollen representation factors for woods in south-east England. *J. Ecol.*, **69**, 45–70.

Bradshaw, R. H. W. (1988) Spatially-precise studies of forest dynamics, in *Vegetation History* (eds B. Huntley and T. Webb III), Kluwer, Dordrecht, pp. 725–52.

Bradshaw, R. H. W. and Miller, N. (1988) Recent successional processes investigated by pollen analysis of closed-canopy forest sites. *Vegetatio*, **76**, 45–54.

Bradshaw, R. H. W. and Webb, T. III. (1985) Relationships between contemporary pollen and vegetation data from Wisconsin and Michigan, USA. *Ecology*, **66**, 721–37.

Bradshaw, R. H. W. and Zackrisson, O. (1990) A two thousand year history of a northern Swedish boreal forest stand. *J. Veg. Sci.*, **1**, 519–28.

Broccoli, A. J. and Manabe, S. (1987) The influence of continental ice, atmospheric CO_2, and land albedo on the climate of the last glacial maximum. *Climate Dynamics*, **1**, 87–99.

Broecker, W. S. (1982) Ocean chemistry during glacial time. *Geochim. Cosmochim. Acta*, **46**, 1689–1705.

Broecker, W. S. (1984) Terminations, in *Milankovitch and Climate*, Part 2 (eds A. Berger, J. Imbrie, J. Hays, G. Kukla and B. Saltzman), Reidel, Dordrecht, pp. 687–98.

Broecker, W. S., Peteet, D. and Rind, D. (1985) Does the ocean-atmosphere system have more than one stable mode of operation? *Nature*, **315**, 21.

Brokaw, N. V. L. (1985) Gap-phase regeneration in a tropical forest. *Ecology*, **66**, 682–7.

Brubaker, L. B. (1975) Post-glacial forest patterns associated with till and outwash in north-central Upper Michigan. *Quat. Res.*, **5**, 499–527.

Brubaker, L. B. (1986) Responses of tree populations to climatic change. *Vegetatio*, **67**, 119–30.

Brubaker, L. B. and Cook, E. R. (1983) Tree-ring studies of Holocene environments, in *Late Quaternary Environments of the United States*, Vol. 2. *The Holocene* (ed. H. E. Wright Jr), Univ. of Minnesota Press, Minneapolis, pp. 222–35.

Bryson, R. A. and Goodman, B. M. (1980) Volcanic activity and climatic changes. *Science*, **207**, 1041–4.

Bryson, R. A. and Wendland, W. M. (1967) Tentative climatic patterns for some late glacial and post-glacial episodes in central North America, in *Life, Land and Water: Proceedings of the Conference on Environmental Studies of the Glacial Lake Agassiz Region, 1966* (ed. W. J. Mayer-Oakes), Manitoba Univ., Deparment of Anthropology Occasional Paper **1**, pp. 271–98.

Bush, M. B. and Colinvaux, P. A. (1988) A 7000-year pollen record from the Amazon lowlands, Ecuador. *Vegetatio*, **76**, 141–54.

Chamberlain, A. C. (1975) The movement of particles in plant communities, in

Vegetation and the Atmosphere, 1 (ed. J. L. Monteith), Academic Press, London, pp. 155–203.

Clark, J. S. (1988a) Particle motion and the theory of charcoal analysis: source area, transport, deposition and sampling. *Quat. Res.*, **30**, 67–80.

Clark, J. S. (1988b) Stratigraphic charcoal analysis on petrographic thin sections: application to fire history in northwestern Minnesota. *Quat. Res.*, **30**, 81–91.

Clark, J. S. (1988c) Effect of climate change on fire frequency in northwestern Minnesota, *Nature*, **334**, 233–5.

Clark, J. S. (1990) Fire and climate change during the last 750 years in northwestern Minnesota, *Ecol. Monogr.*, **60**, 135–59.

Clark, J. S., Merkt, J. and Müller, H. (1989) Post-glacial fire, vegetation, and human history on the northern Alpine forelands, south-western Germany. *J. Ecol.*, **77**, 897–925.

CLIMAP Project Members (1976) The surface of the ice-age earth. *Science*, **191**, 1131–7.

CLIMAP Project Members (1981) *Seasonal Reconstructions of the Earth's Surface at the Last Glacial Maximum*, Geological Soc. of America Map and Chart Series, MC-36.

COHMAP Members (1988) Climatic changes of the last 18 000 years: observations and model simulations. *Science*, **241**, 1043–52.

Colinvaux, P. (1987) Amazon diversity in light of the paleoecological record. *Quat. Sci. Rev.*, **6**, 93–114.

Coope, G. R. (1970) Interpretations of Quaternary insect fossils. *Ann. Rev. Entomol.*, **15**, 97–120.

Coope, G. R. (1979) Late Cenozoic fossil Coleoptera: evolution, biogeography, and ecology. *Ann. Rev. Ecol. System.*, **10**, 247–67.

Craig, A. J. (1972) Pollen influx to laminated sediments: a pollen diagram from northeastern Minnesota. *Ecology*, **53**, 46–57.

Cwynar, L. C. (1978) Recent history of fire and vegetation from laminated sediment of Greenleaf Lake, Algonquin Park, Ontario. *Canad. J. Bot.*, **56**, 10–21.

Davis, M. B. (1976) Pleistocene biogeography of temperate deciduous forests. *Geosci. Man*, **13**, 13–26.

Davis, M. B. (1981a) Quaternary history and the stability of forest communities, in *Forest Succession: Concepts and Application* (eds D. C. West, H. H. Shugart, and D. B. Botkin), Springer-Verlag, New York, pp. 132–53.

Davis, M. B. (1981b) Outbreaks of forest pathogens in Quaternary history. *4th Internat. Palynol. Conf., Lucknow*, **3**, 216–27.

Davis, M. B. (1983) Quaternary history of deciduous forests of eastern North America and Europe. *Ann. Missouri Bot. Garden*, **70**, 550–63.

Davis, M. B. (1986) Climatic instability, time lags, and community disequilibrium, in *Community Ecology* (eds J. Diamond and T. J. Case), Harper and Row, New York, pp. 269–84.

Davis, M. B. (1987) Invasion of forest communities during the Holocene: beech and hemlock in the Great Lakes region, in *Colonization, Succession and Stability* (eds A. J. Gray, M. J. Crawley, and P. J. Edwards), Blackwell, Oxford, pp. 373–94.

Davis, M. B. and Botkin, D. B. (1985) Sensitivity of cool-temperate forests and their fossil pollen record to rapid temperature change. *Quat. Res.*, **23**, 327–40.

Davis, M. B., Brubaker, L. B. and Webb, T., III. (1973) Calibration of absolute pollen influx, in *Quaternary Plant Ecology* (eds H. J. B. Birks and R. G. West), Blackwell, Oxford, pp. 9–25.

Davis, M. B., Moeller, R. E. and Ford, J. (1984) Sediment focusing and pollen influx, in *Lake Sediments and Environmental History* (eds E. Y. Haworth and J. W. G. Lund), Univ. of Leicester Press, Leicester, pp. 261–93.

Davis, M. B., Woods, K. D., Webb, S. L. and Futyma, R. P. (1986) Dispersal versus climate: expansion of *Fagus* and *Tsuga* into the Upper Great Lakes region. *Vegetatio*, **67**, 93–103.

Davis, R. B. (1974) Stratigraphic effects of tubificids in profundal lake sediments. *Limnol. Oceanog.*, **19**, 466–88.

de Beaulieu, J.-L. and Reille, M. (1984) A long upper Pleistocene pollen record from Les Echets, near Lyon, France. *Boreas*, **13**, 111–32.

Deacon, J. and Lancaster, N. (1990) *Late Quaternary Palaeo-Environments of Southern Africa*, Clarendon Press, Oxford.

Delcourt, H. R., Delcourt, P. A. and Webb, T., III. (1983) Dynamic plant ecology: the spectrum of vegetational change in space and time. *Quat. Sci. Rev.*, **1**, 153–75.

Delcourt, P. A., Delcourt, H. R. and Webb, T., III. (1984) *Atlas of Mapped Distributions of Dominance and Modern Pollen Percentages for Important Eastern North American Tree Taxa*, American Association of Stratigraphic Palynologists Contributions Series **14**.

Delcourt, P. A. and Delcourt, H. R. (1987) *Long-term Forest Dynamics of the Temperate Zone*, Springer-Verlag, New York.

Denton, G. H. and Hughes, T. J. (1981) *The Last Great Ice Sheets*, Wiley, New York.

Dexter, F., Banks, H. T. and Webb, T., III. (1987) Modeling Holocene changes in the location and abundance of beech populations in eastern North America. *Rev. Palaeobot. Palynol.*, **50**, 273–92.

Dodson, J. (1989) Late Pleistocene vegetation and environmental shifts in Australia and their bearing on faunal extinctions. *J. Archaeol. Sci.*, **16**, 207–17.

Edwards, T. W. D. and Fritz, P. (1986) Assessing meteoric water composition and relative humidity from ^{18}O and ^2H in wood cellulose: paleoclimatic implications for southern Ontario, Canada. *Appl. Geochem.*, **1**, 715–23.

Emanuel, W. R., Shugart, H. H. and Stevenson, M. L. (1985) Climate change and the broad-scale distribution of terrestrial ecosystem complexes. *Climatic Change*, **7**, 29–43.

Engstrom, D. R. and Wright, H. E., Jr (1984) Chemical stratigraphy of lake sediments as a record of environmental change, in *Lake Sediments and Environmental History* (eds E. Y. Haworth and J. W. G. Lund), Leicester Univ. Press, Leicester, pp. 11–67.

Finegan, B. (1984) Forest succession. *Nature*, **312**, 109–14.

Flohn, H. (1979) On time scales and causes of abrupt paleo-climatic events. *Quat. Res.*, **12**, 135–49.

Follieri, M., Magri, D. and Sadori, L. (1988) 250 000-year pollen record from Valle di Castiglione (Roma). *Pollen Spores*, **30**, 329–56.

Forrester, R. M. (1987) Late Quaternary paleoclimate records from lacustrine ostracodes, in *North America and Adjacent Oceans During the Last Deglaciation* (eds W. F. Ruddiman and H. E. Wright, Jr), Geology of North

America K-3, Geological Soc. America, Boulder, pp. 261–76.

Fuji, N. (1984) Pollen analysis, in *Lake Biwa* (ed. S. Norie), Junk, Dordrecht, pp. 497–529.

Gajewski, K. (1987) Climatic impacts on the vegetation of eastern North America during the past 2000 years. *Vegetatio*, **68**, 179–90.

Gasse, F., Téhet, R., Durand, A., Gilbert E. and Fontes, J.-C. (1990) The arid-humid transition in the Sahara and the Sahel during the last deglaciation. *Nature*, **346**, 141–6.

Gates, W. L. (1976) Modeling the ice-age climate. *Science*, **191**, 1138–44.

Gaudreau, D. C. and Webb, T., III. (1985) Late-Quaternary pollen stratigraphy and isochrone maps for the northeastern United States, in *Pollen Records of Late Quaternary North American Sediments* (eds V. M. Bryant and R. G. Holloway), American Association of Stratigraphic Palynologists Foundation, Dallas, pp. 247–80.

Genthon, C., Barnola, J. M., Raynaud, D., Lorius, C., Jouzel, J., Barkov, N. I., Korotkevich, Y. S. and Kotlyakov, V. M. (1987) Vostok ice core: climatic response to CO_2 and orbital-forcing changes over the last climatic cycle. *Nature*, **329**, 414–18.

Gilliland, R. L. and Schneider, S. H. (1984) Volcanic, CO_2 and solar forcing of northern and southern hemisphere surface air temperatures. *Nature*, **310**, 38–41.

Gleason, H. A. (1926) The individualistic concept of the plant association. *Bull. Torrey Bot. Club*, **53**, 7–26.

Good, R. D'O. (1931) A theory of plant geography. *New Phytol.*, **30**, 149–71.

Green, D. G. (1981) Time series and postglacial forest ecology. *Quat. Res.*, **15**, 265–77.

Green, D. G. (1982) Fire and stability in the postglacial forests of southwest Nova Scotia. *J. Biogeog.*, **9**, 29–40.

Grichuk, V. P. (1984) Late Pleistocene vegetation history, in *Late Quaternary Environments of the Soviet Union* (ed. A. A. Velichko), Longman, London, pp. 155–78.

Grichuk, V. P., Gurtovaya, Ye. Ye., Zelikson, E. M. and Borisova, O. K. (1984) Methods and results of Late Pleistocene paleoclimatic reconstructions, in *Late Quaternary Environments of the Soviet Union* (ed. A. A. Velichko), Longman, London, pp. 251–60.

Grimm, E. C. (1983) Chronology and dynamics of vegetation change in the prairie-woodland region of southern Minnesota, U.S.A. *New Phytol.*, **93**, 311–50.

Grimm, E. C. (1984) Fire and other factors controlling the Big Woods vegetation of Minnesota in the mid-nineteenth century. *Ecol. Monogr.*, **54**, 291–311.

Grove, A. T. (1984) Changing climate, changing biomass and changing atmospheric CO_2. *Prog. Biometeorol.*, **3**, 5–10.

Grove, J. M. (1979) The glacial history of the Holocene. *Prog. Phys. Geog.* **3**, 1–54.

Grove, J. M. (1988) *The Little Ice Age*, Methuen, London.

Guiot, J. (1987) Reconstruction of seasonal temperatures in central Canada since A.D. 1700 and detection of the 18.6- and 22-year signals. *Climatic Change*, **10**, 249–68.

Harrison, S. P. (1989) Lake levels and climatic change in eastern North America. *Climate Dynamics*, **3**, 157–67.

Harrison, S. P., Prentice, I. C. and **Bartlein, P. J.** (1992) What climate models can tell us about the Holocene palaeoclimates of Europe, in *Evaluation of Climate Proxydata in Relation to the European Holocene*, European Science Foundation, Strasbourg.

Harvey, L. D. D. (1980) Solar variability as a contributing factor to Holocene climatic change. *Prog. Phys. Geog.*, **4**, 487–530.

Harvey, L. D. D. (1988) Climatic impact of ice-age aerosols. *Nature*, **334**, 333–5.

Harvey, L. D. D. (1989) Modelling the Younger Dryas. *Quat. Sci. Rev.*, **8**, 137–49.

Hays, J. D., Imbrie, J. and Shackleton N. J. (1976) Variations in the earth's orbit: pacemaker of the ice ages. *Science*, **194**, 1121–32.

Hecht, A. D. (ed.) (1985) *Paleoclimate Analysis and Modeling*, Wiley, New York.

Heinselman, M. L. (1973) Fire in the virgin forests of the Boundary Waters Canoe Area, Minnesota, *Quat. Res.*, **3**, 329–82.

Heusser, C. J. (1989) Polar perspective of Late-Quaternary climates in the southern hemisphere. *Quat. Res.*, **32**, 60–71.

Heusser, C. J. and Rabassa, J. (1987) Cold climatic episode of Younger Dryas age in Tierra del Fuego. *Nature*, **328**, 609–11.

Heusser, L. E. and King, J. E. (1988) North America, in *Vegetation History* (eds B. Huntley and T. Webb III), Kluwer, Dordrecht, pp. 193–236.

Hooghiemstra, H. (1984) Vegetational and climatic history of the high plain of Bogotá, Colombia: a continuous record of the last 3.5 million years. *Diss. Botanicae*, **79**.

Hooghiemstra, H. (1988) The orbital-tuned marine oxygen isotope record applied to the Middle and late Pleistocene pollen record of Funza (Colombian Andes). *Palaeogeog., Palaeoclimatol., Palaeoecol.*, **66**, 9–17.

Huntley, B. (1988) Europe, in *Vegetation History* (eds B. Huntley and T. Webb III), Kluwer, Dordrecht, pp. 341–84.

Huntley, B. and Birks, H. J. B. (1983) *Atlas of Past and Present Pollen Maps for Europe 0–13 000 Years Ago*. Cambridge Univ. Press, Cambridge.

Huntley, B. and Prentice, I. C. (1988) July temperatures in Europe from pollen data, 6000 years before present. *Science*, **241**, 687–90.

Huntley, B. and Webb, T., III. (1988) *Vegetation History*, Kluwer, Dordrecht.

Huntley, B. and Webb, T., III. (1989) Migration: species' response to climatic variations caused by changes in the earth's orbit. *J. Biogeog.*, **16**, 5–19.

Huntley, B., Bartlein, P. J. and Prentice, I. C. (1989) Climatic control of the distribution and abundance of beech (*Fagus*) in Europe and North America. *J. Biogeog.*, **16**, 551–60.

Hyvärinen, H. (1976) Flandrian pollen deposition rates and tree-line history in northern Fennoscandia. *Boreas*, **5**, 163–75.

Imbrie, J. (1985) A theoretical framework for the Pleistocene ice ages. *J. Geol. Soc. London*, **142**, 417–32.

Imbrie, J. and Imbrie, J. Z. (1980) Modeling the climatic response to orbital variations. *Science*, **207**, 943–53.

Imbrie, J. and Kipp, N. G. (1971) A new micropaleontological method for quantitative paleoclimatology: application to a late Pleistocene Caribbean core, in *Late Cenozoic Glacial Ages* (ed. K. Turekian), Yale Univ. Press, New Haven, pp. 71–181.

Imbrie, J., Hays, J. D., Martinson, D. G., McIntyre, A., Mix, A. C., Morley,

J. J., Pisias, N. G., Prell, W. L. and Shackleton, N. J. (1984) The orbital theory of Pleistocene climate: support from a revised chronology of the marine $\delta^{18}O$ record, in *Milankovitch and Climate*, Part 2 (eds A. Berger, J. Imbrie, J. Hays, G. Kukla and B. Saltzman), Reidel, Dordrecht, pp. 269–305.

Iversen, J. (1958) The bearing of glacial and interglacial epochs on the formation and extinction of plant taxa. *Uppsala Universitets Årsskrifter*, **6**, 210–15.

Iversen, J. (1960) Problems of the early post-glacial forest development in Denmark. *Danmarks Geol. Undersøg.*, **IV-4** (3), 1–32.

Iversen, J. (1969) Retrogressive development of a forest ecosystem demonstrated by pollen diagrams from fossil mor. *Oikos Suppl.*, **12**, 35–49.

Iversen, J. (1973) The development of Denmark's nature since the last glacial. *Danmarks Geol. Undersøg.*, **V 7-C**.

Jacobson, G. L., Jr (1988) Ancient permanent plots: sampling in paleovegetational studies, in *Vegetation History* (eds B. Huntley and T. Webb III), Kluwer, Dordrecht, pp. 3–16.

Jacobson, G. L., Jr and Bradshaw, R. H. W. (1981) The selection of sites for paleovegetational studies. *Quat. Res.*, **16**, 80–96.

Jacobson, G. L., Jr, Webb, T., III and Grimm, E. C. (1987) Patterns and rates of vegetation change during the deglaciation of eastern North America, in *North America and Adjacent Oceans During the Last Deglaciation* (eds W. F. Ruddiman and H. E. Wright, Jr), Geology of North America K-3, Geological Soc. America, Boulder, pp. 277–88.

Janssen, C. R. (1970) Problems in the recognition of plant communities in pollen diagrams. *Vegetatio*, **20**, 187–98.

Janssen, C. R. (1973) Local and regional pollen deposition, in *Quaternary Plant Ecology* (eds H. J. B. Birks and R. G. West), Blackwell, Oxford, pp. 31–42.

Karlén, W. (1988) Scandinavian glacial and climatic fluctuations during the Holocene. *Quat. Sci. Rev.*, **7**, 199–209.

Khotinskiy, N. A. (1984). Holocene vegetation history, in *Late Quaternary Environments of the Soviet Union* (ed. A. A. Velichko), Longman, London, pp. 179–201.

Knox, F. and McElroy, M. B. (1984) Changes in atmospheric CO_2: influence of the marine biota at high latitude. *J. Geophys. Res.*, **89**, 4629–37.

Knox, J. C. (1983) Responses of river systems to Holocene climates, in *Late Quaternary Environments of the United States*, Vol. 2, *The Holocene* (ed. H. E. Wright, Jr), Univ. of Minnesota Press, Minneapolis, pp. 26–41.

Kullman, L. (1986) Late Holocene reproductional patterns of *Pinus sylvestris* and *Picea abies* at the forest limit in central Sweden. *Canad. J. Bot.*, **64**, 1682–90.

Kutzbach, J. E. (1976) The nature of climate and climatic variations. *Quat. Res.*, **6**, 471–80.

Kutzbach, J. E. (1981) Monsoon climate of the early Holocene: climate experiment with the earth's orbital parameters for 9000 years ago. *Science*, **214**, 59–61.

Kutzbach, J. E. (1985) Modeling of paleoclimates. *Adv. Geophys.*, **28A**, 159–96.

Kutzbach, J. E. (1987) Model simulations of the climatic patterns during the deglaciation of North America, in *North America and Adjacent Oceans During the Last Deglaciation* (eds W. F. Ruddiman and H. E. Wright, Jr), Geology of North America K-3, Geological Soc. America, Boulder, pp. 425–46.

Kutzbach, J. E. and Gallimore, R. G. (1988) Sensitivity of a coupled atmosphere/

mixed layer ocean to changes in orbital forcing at 9000 years B.P. *J. Geophys. Res.*, **93**, 803–21.

Kutzbach, J. E. and Guetter, P. J. (1986) The influence of changing orbital parameters and surface boundary conditions on climate simulations for the past 18 000 years. *J. Atmos. Sci.*, **43**, 1726–59.

Kutzbach, J. E. and Otto-Bliesner, B. L. (1982) The sensitivity of the African–Asian monsoonal climate to orbital parameter changes for 9000 years B.P. in a low-resolution general circulation model. *J. Atmos. Sci.*, **39**, 1177–88.

Kutzbach, J. E. and Street-Perrott, F. A. (1985) Milankovitch forcing of fluctuations in the level of tropical lakes from 18 to 0 kyr BP. *Nature*, **317**, 130–4.

Kutzbach, J. E. and Wright, H. E., Jr (1985) Simulation of the climate of 18 000 years BP: results for the North American/North Atlantic/European sector and comparison with the geologic record of North America. *Quat. Sci. Rev.*, **4**, 147–87.

Lamb, H. H. and Edwards, M. E. (1988) The Arctic, in *Vegetation History* (eds B. Huntley and T. Webb III), Kluwer, Dordrecht, pp. 519–56.

Le Roy Ladurie, E. (1971) *Times of Feast, Times of Famine*, Allen and Unwin, London.

Legrand, M. R., Lorius, C., Barkov, N. I. and Petrov, V. N. (1988) Vostok (Antarctica) ice core: atmospheric chemistry changes over the last climatic cycle (160 000 years). *Atmos. Environ.*, **22**, 317–31.

Larcher, W. (1983) *Physiological Plant Ecology*, Springer-Verlag, Berlin.

Lehman, J. T. (1975) Reconstructing the rate of accumulation of lake sediment: the effect of sediment focusing. *Quat. Res.*, **5**, 541–50.

Levin, S. A. (1976) Spatial patterning and the structure of ecological communities. *Lect. Math. Life Sciences*, **8**, 1–35.

Levins, R. (1974) The qualitative analysis of partially specified systems. *Ann. New York Acad. Sci.*, **231**, 123–38.

Lézine, A.-M. (1989) Late Quaternary vegetation and climate of the Sahel. *Quat. Res.*, **32**, 317–34.

Lorius, C., Merlivat, L., Jouzel, J. and Pourchet, M. (1979) A 30 000-yr isotope climatic record from Antarctic ice. *Nature*, **280**, 644.

Loucks, O. L. (1970) Evolution of diversity, efficiency, and community stability. *Amer. Zool.*, **10**, 17–25.

Lowe, J. J. and Walker, M. J. C. (1984) *Reconstructing Quaternary Environments*, Longman, London.

Mack, R. N. (1985) Evaluating factors that affect the rate of migrations. *Bull. Ecol. Soc. Amer.*, **66**, 222.

Manabe, S. and Hahn, D. G. (1977) Simulation of the tropical climate of an ice age. *J. Geophys. Res.*, **82**, 3889–3911.

Manabe, S. and Stouffer, R. J. (1988) Two stable equilibria of a coupled ocean-atmosphere model. *J. Climate*, **1**, 841–66.

Mangerud, J. (1980) Ice-front variations of different parts of the Scandinavian ice sheet, 13 000–10 000 years BP, in *Studies in the Lateglacial of North-west Europe* (eds J. J. Lowe, J. M. Gray and J. E. Robinson), Pergamon, Oxford, pp. 23–30.

Mannion, A. M. (1982) Diatoms: their use in physical geography. *Prog. Phys. Geog.*, **6**, 233–59.

Markgraf, V. (1989) Palaeoclimates in central and south America since 18 000 BP based on pollen and lake-level records. *Quat. Sci. Rev.*, **8**, 1–24.

Martinson, D. G., Pisias, N. G., Hays, J. D., Imbrie, J., Moore, T. C., Jr, and Shackleton, N. J. (1987) Age dating and the orbital theory of the ice ages: development of a high-resolution 0 to 300 000-year chronostratigraphy. *Quat. Res.*, **27**, 1–29.

McDowell, P. F., Webb, T., III, and Bartlein P. J. (1991) Long-term environmental change, in *The Earth as Transformed by Human Action* (eds B. L. Turner II *et al.*), Cambridge Univ. Press, Cambridge.

McGlone, M. S. (1988) New Zealand, in *Vegetation History* (eds B. Huntley and T. Webb III), Kluwer, Dordrecht, pp. 557–602.

McIntosh, R. P. (1981) Succession and ecological theory, in *Forest Succession: Concepts and Application* (eds D. C. West, H. H. Shugart, and D. B. Botkin), Springer-Verlag, New York, pp. 10–23.

Milankovitch, M. (1930) Mathematische Klimalehre und astronomische Theorie der Klimaschwankungen. In *Handbuch der Klimatologie I (A)* (eds W. Köppen and R. Geiger), Gebrüder Bornträger, Berlin.

Mitchell, J. F. B., Grahame, N. S. and Needham, K. J. (1988) Climate simulations for 9000 years before present: seasonal variations and effects of the Laurentide ice sheet. *J. Geophys. Res.*, **93**, 8283–303.

Mitchell, J. M. (1976) An overview of climatic variability and its causal mechanisms. *Quat. Res.*, **6**, 481–93.

Mix, A. C. and Ruddiman, W. F. (1984) Oxygen-isotope analyses and Pleistocene ice volumes. *Quat. Res.*, **21**, 1–20.

Mix, A. C. (1987) The oxygen-isotope record of glaciation, in *North America and Adjacent Oceans During the Last Deglaciation* (eds W. F. Ruddiman and H. E. Wright, Jr), Geology of North America K-3, Geological Soc. America, Boulder, pp. 111–36.

Neftel, A., Oeschger, H., Schwander, J., Stauffer, B. and Zumbrunn, R. (1982) Ice core sample measurements give atmospheric CO_2 content during the past 40 000 yr. *Nature*, **295**, 220–3.

Neilson, R. P. (1987) Biotic regionalization and climatic controls in western North America. *Vegetatio*, **70**, 135–47.

Neilson, R. P. and Wullstein, L. H. (1983) Biogeography of two southwest American oaks in relation to atmospheric dynamics. *J. Biogeog.*, **10**, 275–97.

Oeschger, H. and Langway, C. C. (eds) (1989) *The Environmental Record in Glaciers and Ice Sheets*, Wiley, Chichester.

Oeschger, H., Beer, J., Siegenthaler, U., Stauffer, B., Dansgaard, W. and Langway, C. C. (1984) Late glacial climate history from ice cores, in *Climate Processes and Climate Sensitivity* (eds J. E. Hansen and T. Takahashi), Geophysical Monograph Series 29, American Geophysical Union, pp. 299–306.

Olsson, I. U. (1986) Radiometric dating, in *Handbook of Holocene Palaeoecology and Palaeohydrology* (ed. B. E. Berglund), Wiley, Chichester, pp. 273–312.

Overpeck, J. T., Webb, T., III and Prentice, I. C. (1985) Quantitative interpretation of fossil pollen spectra: dissimilarity coefficients and the method of modern analogs. *Quat. Res.*, **23**, 87–108.

Overpeck, J. T., Peterson, L. C., Kipp, N., Imbrie, J. and Rind, D. (1989) Climate change in the circum-North Atlantic region during the last deglaciation. *Nature*, **338**, 553–7.

Paterson, W. S. B. and Hammer, C. U. (1987) Ice core and other glaciological data, in *North America and Adjacent Oceans During the Last Deglaciation* (eds W. F. Ruddiman and H. E. Wright, Jr), Geology of North America K-3, Geological Soc. America, Boulder, pp. 91–110.

Patterson, W. A., III, and Backman, A. E. (1988) Fire and disease history of forests, in *Vegetation History* (eds B. Huntley and T. Webb III), Kluwer, Dordrecht, pp. 603–32.

Pennington, W. (1986) Lags in adjustment of vegetation to climate caused by the pace of soil development: evidence from Britain. *Vegetatio*, **67**, 105–18.

Peterson, G. M., Webb, T., III, Kutzbach, J. E., van der Hammen, T., Wijmstra, T. A. and Street, F. A. (1979) The continental record of environmental conditions at 18 000 yr B.P.: an initial evaluation. *Quat. Res.*, **12**, 47–82.

Pickett, S. T. A. (1980) Non-equilibrium coexistence of plants. *Bull. Torrey Bot. Club*, **107**, 238–48.

Pisias, N. G. and Shackleton, N. J. (1984) Modelling the global climate response to orbital forcing and atmospheric carbon dioxide changes. *Nature*, **310**, 757–9.

Pittock, A. B. (1983) Solar variability, weather and climate: an update. *Quat. J. Roy. Meteorol. Soc.*, **109**, 23–55.

Porter, S. C. (1986) Pattern and forcing of northern hemisphere glacier variations during the last millenium. *Quat. Res.*, **26**, 27–48.

Prell, W. L., and Kutzbach, J. E. (1987) Monsoon variability over the past 150 000 years. *J. Geophys. Res.*, **92**, 8411–25.

Prentice, I. C. (1978) Modern pollen spectra from lake sediments in Finland and Finnmark, north Norway. *Boreas*, **7**, 131–53.

Prentice, I. C. (1983) Postglacial climatic change: vegetation dynamics and the pollen record. *Prog. Phys. Geog.*, **7**, 273–86.

Prentice, I. C. (1984) Seasonality and community response: quantitative prediction and calibration. *Amer. Quat. Ass. Abstracts*, **8**, 105.

Prentice, I. C. (1985) Pollen representation, source area, and basin size: toward a unified theory of pollen analysis. *Quat. Res.*, **23**, 76–86.

Prentice, I. C. (1986a) Some concepts and objectives of forest dynamics research, in *Forest Dynamics Research in Western and Central Europe*, (ed. J. Fanta), PUDOC, Wageningen, pp. 32–9.

Prentice, I. C. (1986b) Vegetation responses to past climatic variation. *Vegetatio*, **67**, 131–41.

Prentice, I. C. (1986c) Forest-composition calibration of pollen spectra, in *Handbook of Holocene Palaeoecology and Palaeohydrology*, (ed. B. E. Berglund), Wiley, Chichester, pp. 799–816.

Prentice, I. C. (1988a) Records of vegetation in time and space: the principles of pollen analysis. in *Vegetation History*, (eds B. Huntley and T. Webb III), Junk, Dordrecht, pp. 17–42.

Prentice, I. C. (1988b) Plant population dynamics and palaeoecology. *Trends Ecol. Evol.*, **3**, 343–5.

Prentice, I. C. and Leemans, R. (1991) Pattern and process and the dynamics of forest structure: a simulation approach. *J. Ecol.*, **78**, 340–55.

Prentice, I. C. and Webb, T., III. (1986) Pollen percentages, tree abundances and the Fagerlind effect. *J. Quat. Sci.*, **1**, 35–43.

Prentice, I. C., Bartlein, P. J. and Webb, T., III. (1991) Vegetation and climate change in eastern North America since the last glacial maximum. *Ecology* (1972), pp. 2038–56.

Prentice, I. C., Berglund, B. E. and Olsson, T. (1987) Quantitative forest-composition sensing characteristics of pollen samples from Swedish lakes. *Boreas*, **16**, 43–54.

Prentice, K. C. (1990) Bioclimatic distribution of vegetation for general circulation model studies. *J. Geophys. Res.*, **95**, 11811–30.

Prentice, K. C. and Fung, I. Y. (1990) The sensitivity of terrestrial carbon storage to climate change. *Nature*, **346**, 48–50.

Rasmusson, E. M. (1985) El Niño and variations in climate. *Amer. Scient.*, **73**, 168–77.

Raynaud, D., Chappellaz, J. Barnola, J. M., Korotkevich, Y. S. and Lorius, C. (1988) Climatic and CH_4 cycle implications of glacial–interglacial CH_4 change in the Vostok ice core. *Nature*, **333**, 655–7.

Rind, D., Peteet, D., Broecker, W., McIntyre, A. and Ruddiman, W. (1986) The impact of cold North Atlantic sea surface temperatures on climate: implications for the Younger Dryas cooling (11–10k). *Climate Dynamics*, **1**, 3–34.

Ritchie, J. C. (1984) *Past and Present Vegetation of the Far Northwest of Canada*, Univ. of Toronto Press, Toronto.

Ritchie, J. C. (1986) Climate change and vegetation response. *Vegetatio*, **67**, 65–74.

Ritchie, J. C. (1987) *Postglacial Vegetation of Canada*, Cambridge Univ. Press, Cambridge.

Ritchie, J. C. and MacDonald, G. M. (1986) The patterns of post-glacial spread of white spruce. *J. Biogeog.*, **13**, 527–40.

Ritchie, J. C., Cwynar, L. C. and Spear, R. W. (1983) Evidence from northwest Canada for an early Holocene Milankovitch thermal maximum. *Nature*, **305**, 126–8.

Roberts, D. W. (1987) A dynamical systems perspective on vegetation theory. *Vegetatio*, **69**, 27–33.

Robin, G. de Q. (ed.) (1983) *The Climate Record in Polar Ice Sheets*, Cambridge Univ. Press, Cambridge.

Rognon, P. (1987) Late Quaternary climatic reconstruction for the Maghreb (North Africa). *Palaeogeog., Palaeoclimatol., Palaeoecol.*, **58**, 11–34.

Romme, W. H. and Knight, D. H. (1981) Fire frequency and subalpine forest succession along a topographic gradient in Wyoming. *Ecology*, **62**, 319–26.

Rossignol-Strick, M. (1983) African monsoons, an immediate climate response to orbital insolation. *Nature*, **304**, 46–9.

Ruddiman, W. F. (1985) Climate studies in ocean cores, in *Paleoclimate Analysis and Modeling* (ed. A. D. Hecht), Wiley, New York, pp. 197–258.

Ruddiman, W. F. and McIntyre, A. (1981) The mode and mechanism of the last deglaciation: oceanic evidence. *Quat. Res.*, **16**, 125–34.

Ruddiman, W. F., Prell, W. L. and Raymo, M. E. (1989) Late Cenozoic uplift in southern Asia and the American West: rationale for general circulation modelling experiments. *J. Geophys. Res.*, **94D**, 18379–91.

Runkle, J. R. (1985) Disturbance regimes in temperate forests, in *The Ecology of Natural Disturbance and Patch Dynamics* (eds S. T. A. Pickett and P. S. White), Academic Press, Orlando, pp. 17–33.

Saarnisto, M. (1986) Annually laminated lake sediments, in *Handbook of Holocene Palaeoecology and Palaeohydrology* (ed. B. E. Berglund), Wiley, Chichester, pp. 343–70.

Saarnisto, M. (1988) Time-scales and dating, in *Vegetation History* (eds B. Huntley and T. Webb III), Kluwer, Dordrecht, pp. 77–112.

Saltzman, B. (1985) Paleoclimatic modeling, in *Paleoclimate Analysis and Modeling* (ed. A. D. Hecht), Wiley, New York, pp. 341–96.

Sarmiento, J. L. and Toggweiler, J. R. (1984) A new model for the role of the oceans in determining atmospheric pCO_2. *Nature*, **308**, 621–4.

Schneider, S. H., Peteet, D. M. and North, G. R. (1987) A climate model intercomparison for the Younger Dryas and its implications for paleoclimatic data collection, in *Abrupt Climatic Change* (eds W. H. Berger and L. D. Labeyrie), Reidel, Dordrecht, pp. 399–417.

Schumm, S. A. and Brakenridge, G. R. (1987) River responses, in *North America and Adjacent Oceans During the Last Deglaciation* (eds W. F. Ruddiman and H. E. Wright, Jr), Geology of North America K-3, Geological Soc. America, Boulder, pp. 221–40.

Schuurmans, C. (1981) Climate of the last 1000 years, in *Climatic Variations and Variability: Facts and Theories* (ed. A. L. Berger), Reidel, Dordrecht, pp. 245–58.

Shackleton, N. J. (1977) Carbon 13 in *Uvigerina*: tropical rain forest history and the Equatorial Pacific carbonate dissolution cycles, in *The Fate of Fossil Fuel CO_2 in the Oceans* (eds N. R. Anderson and A. Malahoff), Plenum, New York, pp. 401–27.

Shackleton, N. J. and Opdyke, N. D. (1973) Oxygen isotope and paleomagnetic stratigraphy of equatorial Pacific core V28–238 oxygen isotope temperatures and ice volumes on a 10^5 and 10^6 year scale. *Quat. Res.*, **3**, 39–55.

Shackleton, N. J. and Pisias, N. G. (1985) Atmospheric CO_2, orbital forcing and climate, in *The Carbon Cycle and Atmospheric CO_2: Natural Variations Archean to Present* (eds E. T. Sundquist and W. S. Broecker), American Geophysical Union Monograph **32**, American Geophysical Union, Washington, D.C., pp. 303–18.

Shackleton, N. J., Hall, M. A., Line, J. and Cang Shuxi. (1983) Carbon isotope data in core V19–30 confirm reduced carbon dioxide concentrations in the ice age atmosphere. *Nature*, **306**, 319–22.

Shugart, H. H. (1984) *A Theory of Forest Dynamics*, Springer, New York.

Shugart, H. H., Emanuel, W. R., West, D. C. and DeAngelis, D. L. (1980) Environmental gradients in a simulation model of a beech-yellow-poplar stand. *Math. Biosciences*, **50**, 163–70.

Siegenthaler, U. and Eicher, U. (1986) Stable oxygen and carbon isotope analyses, in *Handbook of Holocene Palaeoecology and Palaeohydrology* (ed. B. E. Berglund), Wiley, Chichester, pp. 407–22.

Siegenthaler, U. and Wenk, T. (1984) Rapid atmospheric CO_2 variations and ocean circulation. *Nature*, **308**, 624–5.

Solomon, A. M. (1984) Forest responses to complex interacting full glacial environmental conditions. *Amer. Quat. Ass. Abstracts*, **8**, 120.

Solomon, A. M. and Shugart, H. H. (1984) Integrating forest-stand simulations with paleoecological records to examine long-term forest dynamics, in *State and Change of Forest Ecosystems – Indicators in Current Research* (ed. G. I. Ågren),

Swedish University of Agricultural Science, Uppsala, pp. 333–56.

Solomon, A. M. and Webb, T., III. (1985) Computer-aided reconstruction of Late-Quaternary landscape dynamics. *Ann. Rev. Ecol. System.*, **16**, 63–84.

Solomon, A. M., Delcourt, H. R., West, D. C. and Blasing, T. J. (1980) Testing a simulation model for reconstruction of prehistoric forest stand dynamics. *Quat. Res.*, **14**, 275–93.

Solomon, A. M., West, D. C. and Solomon, J. A. (1981) Simulating the role of climate change and species immigration in forest succession, in *Forest Succession: Concepts and Application* (eds D. C. West, H. H. Shugart and D. B. Botkin), Springer-Verlag, New York, pp. 154–79.

Sousa, W. P. (1984) The role of disturbance in natural communities. *Ann. Rev. Ecol. System.*, **15**, 353–91.

Spaulding, W. G. (1984) The last glacial–interglacial climatic cycle: its effects on woodlands and forests in the American West, in *Proc. 8th North American Forest Biology Workshop* (ed. R. M. Lanner), Utah State University, Logan, Utah, pp. 42–69.

Spaulding, W. G., Leopold, E. B. and van Devender, T. R. (1983) Late Wisconsin paleoecology of the American southwest, in *Late-Quaternary Environments of the United States*, Vol. 1. *The Late Pleistocene* (ed. S. C. Porter), Univ. of Minnesota Press, Minneapolis, pp. 259–93.

Stauffer, B., Lochbronner, E., Oeschger, H. and Schwander, J. (1988) Methane concentration in the glacial atmosphere was only half that of the preindustrial Holocene. *Nature*, **332**, 812–14.

Stockton, C. W., Boggess, W. R. and Meko, D. M. (1985) Climate and tree rings, in *Paleoclimate Analysis and Modeling* (ed. A. D. Hecht), Wiley, New York, pp. 71–151.

Street, F. A. and Grove, A. T. (1979) Global maps of lake-level fluctuations since 30 000 yr B.P. *Quat. Res.*, **12**, 83–118.

Street-Perrott, F. A. and Harrison, S. P. (1985) Lake levels and climate reconstruction, in *Paleoclimate Analysis and Modeling* (ed. A. D. Hecht), Wiley, New York, pp. 291–340.

Street-Perrott, F. A. and Perrott, R. A. (1990) Abrupt climate fluctuations in the tropics: the influence of Atlantic Ocean circulation. *Nature*, **343**, 607–12.

Stuiver, M. (1980) Solar variability and climatic change during the current millenium. *Nature*, **286**, 868–71.

Stuiver, M. (1989) Dating proxy data, in *Climate and Geo-Sciences* (eds A. Berger *et al.*,), Kluwer, Dordrecht, pp. 39–45.

Stuiver, M. and Braziunas, T. F. (1987) Tree cellulose $^{13}C/^{12}C$ isotope ratios and climatic change. *Nature*, **327**, 58–60.

Stuiver, M. and Burk, R. L. (1985) Appendix: Paleoclimatic studies using isotopes in trees, in *Paleoclimate Analysis and Modeling* (ed. A. D. Hecht), Wiley, New York, pp. 151–61.

Swain, A. M. (1973) A history of fire and vegetation in northeastern Minnesota as recorded by lake sediments. *Quat. Res.*, **3**, 383–97.

Swain, A. M. (1978) Environmental changes during the past 2000 years in north-central Wisconsin: analysis of pollen, charcoal and seeds from varved lake sediments. *Quat. Res.*, **10**, 55–68.

Tauber, H. (1965) Differential pollen dispersion and the interpretation of pollen diagrams. *Danmarks Geol. Undersøg.*, **II-89**.

ter Braak, C. J. F. and Prentice, I. C. (1988) A theory of gradient analysis. *Adv. Ecol. Res.*, **18**, 271–317.

Thompson, L. G. and Mosley-Thompson, E. (1987) Evidence of abrupt climatic changes during the last 1500 years recorded in ice cores from the tropical Quelccaya ice cap, Peru, in *Abrupt Climatic Change* (eds W. H. Berger and L. D. Labeyrie), Reidel, Dordrecht, pp. 99–110.

Thompson, L. G., Mosley-Thompson, E. and Arnao, B. M. (1984) El Niño–Southern Oscillation events recorded in the stratigraphy of the tropical Quelccaya ice cap, Peru. *Science*, **226**, 50–3.

Thompson, R. (1980) Use of the word 'influx' in palaeo-limnological studies. *Quat. Res.*, **14**, 269–70.

Thompson, R. S. (1985) Palynology and *Neotoma* middens. *Amer. Ass. Strat. Palynol. Cont. Series*, **16**, 89–112.

Thompson, R. S. (1988) Western North America, in *Vegetation History* (eds B. Huntley and T. Webb III), Kluwer, Dordrecht, pp. 415–58.

Tsukada, M. (1988) Japan, in *Vegetation History* (eds B. Huntley and T. Webb III), Kluwer, Dordrecht, pp. 459–518.

Tsukada, M. and Sugita, S. (1982) Late Quaternary dynamics of pollen influx at Mineral Lake, Washington. *Bot. Mag.*, *Tokyo*, **95**, 401–18.

Turner, J. and Peglar, S. M. (1988) Temporally-precise studies of vegetation history, in *Vegetation History* (eds B. Huntley and T. Webb III), Kluwer, Dordrecht, pp. 753–78.

van Campo, E., Duplessy, J. C. and Rossignol-Strick, M. (1982) Climatic conditions deduced from a 150-kyr oxygen isotope-pollen record from the Arabian Sea. *Nature*, **296**, 56–9.

van der Hammen, T. (1988) South America, in *Vegetation History* (eds B. Huntley and T. Webb III), Kluwer, Dordrecht, pp. 307–40.

van Devender, T. R., Thompson, R. S. and Betancourt, J. L. (1987) Vegetation history of the deserts of southwestern North America: the nature and timing of the Late Wisconsin-Holocene transition, in *North America and Adjacent Oceans During the Last Deglaciation* (eds W. F. Ruddiman and H. E. Wright, Jr), Geology of North America K-3, Geological Soc. America, Boulder, pp. 323–52.

Velichko, A. A. (1984) Late Pleistocene spatial paleoclimatic reconstructions, in *Late Quaternary Environments of the Soviet Union* (ed. A. A. Velichko), Longman, London, pp. 261–85.

Velichko, A. A. (1987) Relationship of climatic changes in high and low latitudes of the earth during the Late Pleistocene and Holocene, in *Palaeogeography and Loess*, (eds M. Pécsi and A. A. Velichko), Akadémiai Kladó, Budapest, pp. 9–26.

Walker, D. (1982a) Vegetation's fourth dimension. *New Phytol.*, **90**, 419–29.

Walker, D. (1982b) The development of resilience in burned vegetation, in *The Plant Community as a Working Mechanism* (ed. E. I. Newman), Blackwell, Oxford, pp. 27–43.

Walker, D. and Chen, Y. (1987) Palynological light on tropical rainforest dynamics. *Quat. Sci. Rev.*, **6**, 77–92.

Walker, D. and Flenley, J. R. (1979) Late Quaternary vegetational history of the Enga Province of upland Papua New Guinea. *Philos. Trans. Roy. Soc. London B*, **286**, 265–344.

Watt, A. S. (1947) Pattern and process in the plant community. *J. Ecol.*, **35**, 1–22.

Watts, W. A. (1973) Rates of change and stability of vegetation in the perspective of long periods of time, in *Quaternary Plant Ecology*, (eds H. J. B. Birks and R. G. West), Blackwell, Oxford, pp. 195–206.

Watts, W. A. (1978) Plant macrofossils and Quaternary palaeoecology, in *Biology and Quaternary Environments* (eds D. Walker and J. C. Guppy), Australian Academy of Science, Canberra, pp. 53–67.

Watts, W. A. (1980) Regional variations in the response of vegetation to late glacial climatic events in Europe, in *Studies in the Late glacial of North-West Europe* (eds J. J. Lowe, J. M. Gray and J. E. Robinson), Pergamon, Oxford, pp. 1–22.

Watts, W. A. (1983) Vegetational history of the eastern United States 25 000 to 10 000 years ago, in *Late-Quaternary Environments of the United States*, Vol. 1. *The Late Pleistocene* (ed. S. C. Porter), Univ. of Minnesota Press, Minneapolis, pp. 294–310.

Watts, W. A. (1988) Europe, in *Vegetation History* (eds B. Huntley and T. Webb III), Kluwer, Dordrecht, pp. 155–92.

Webb, T., III. (1974) Corresponding patterns of pollen and vegetation in Lower Michigan: a comparison of quantitative data. *Ecology*, **55**, 17–28.

Webb, T., III. (1982) Temporal resolution in Holocene pollen data. *Third N. Amer. Paleontological Conv., Proc.*, **2**, 569–72.

Webb, T., III (1985) Holocene palynology and climate, in *Paleoclimate Analysis and Modeling* (ed. A. D. Hecht), Wiley, New York, pp. 163–95.

Webb, T., III (1986) Is vegetation in equilibrium with climate? How to interpret Late-Quaternary pollen data. *Vegetatio*, **67**, 75–91.

Webb, T., III (1987) The appearance and disappearance of major vegetation assemblages: long-term vegetational dynamics in eastern North America. *Vegetatio*, **69**, 177–87.

Webb, T., III (1988) Eastern North America, in *Vegetation History* (eds B. Huntley and T. Webb III), Kluwer, Dordrecht, pp. 385–414.

Webb, T., III, and Bryson, R. A. (1972) Late- and postglacial climatic change in the northern Midwest, U.S.A.: quantitative estimates derived from fossil pollen spectra by multivariate statistical analysis. *Quat. Res.*, **2**, 70–115.

Webb, T., III, Bartlein, P. J. and Kutzbach, J. E. (1987) Climatic change in eastern North America during the past 18 000 years: comparisons of pollen data with model results, in *North America and Adjacent Oceans During the Last Deglaciation* (eds W. F. Ruddiman and H. E. Wright, Jr), Geological Soc. America, Boulder, pp. 447–62.

Webb, T., III, Cushing, E. J. and Wright, H. E. Jr (1983) Holocene changes in the vegetation of the Midwest, in *Late Quaternary Environments of the United States*, Vol. 2. *The Holocene* (ed. H. E. Wright Jr), Univ. of Minnesota Press, Minneapolis, pp. 142–65.

Webb, T., III, Laseski, R. A. and Bernabo, J. C. (1978) Sensing vegetational patterns with pollen data: choosing the data. *Ecology*, **59**, 1151–63.

Webb, T., III, Howe, S. E., Bradshaw, R. H. W. and Heide, K. M. (1981) Estimating plant abundances from pollen percentages: the use of regression analysis. *Rev. Palaeobot. Palynol.*, **34**, 269–300.

Wells, P. V. (1976) Macrofossil analysis of woodrat (*Neotoma*) middens as a key to the Quaternary vegetational history of arid America. *Quat. Res.*, **6**, 223–48.

West, R. G. (1964) Inter-relations of ecology and Quaternary palaeobotany. *J. Ecol.*, **52** (Suppl.), 47–57.

West, R. G. (1980) Pleistocene forest history in East Anglia. *New Phytol.*, **85**, 571–622.

White, P. S. (1979) Pattern, process, and natural disturbance in vegetation. *Bot. Rev.*, **45**, 229–99.

Whitmore, T. C. (1982) On pattern and process in forests, in *The Plant Community as a Working Mechanism* (ed. E. I. Newman), Blackwell, Oxford, pp. 45–60.

Whittaker, R. H. and Levin, S. A. (1977) The role of mosaic phenomena in natural communities. *Theor. Pop. Biol.*, **12**, 117–39.

Winkler, M. G. (1985a) Charcoal analysis for paleoenvironmental interpretation: a chemical assay. *Quat. Res.*, **23**, 313–26.

Winkler, M. G. (1985b) A 12 000-year history of vegetation and climate for Cape Cod, Massachussets. *Quat. Res.*, **23**, 301–12.

Woillard, G. (1978) Grande Pile Peat Bog: a continuous pollen record for the last 140 000 years. *Quat. Res.*, **9**, 1–24.

Woods, K. and Davis, M. B. (1989) Paleoecology of range limits: beech in the Upper Peninsula of Michigan. *Ecology*, **70**, 681–96.

Woodward, F. I. (1987) *Climate and Plant Distribution*, Cambridge Univ. Press, Cambridge.

Woodward, F. I. and Williams, B. G. (1987) Climate and plant distribution at global and local scales. *Vegetatio*, **69**, 189–97.

Wright, H. E., Jr (1984) Sensitivity and response time of natural systems to climatic change in the Late Quaternary. *Quat. Sci. Rev.*, **3**, 91–131.

Epilogue

Within the chapters of this book we have seen several recurring themes. These might be taken as a consensus view of the research directions and theoretical issues that have broad application and interest. In this section we identify five of these recurrent themes and use them to project the most important and productive directions for future research in vegetation dynamics.

THE NEED FOR EMPIRICAL STUDIES

Biology, and especially ecology, is the study of diversity. The millions of extant species and the infinite variation in possible environmental conditions means that every example of vegetation dynamics will be unique. Vegetation composition and dynamics are inseparably linked to initial conditions of site characteristics and species availability. Thus, vegetation change cannot be predicted from first principles alone. Instead, any attempt at a general understanding of vegetation dynamics must build from numerous case studies of vegetation change. While adherence to deductive theory has intellectual appeal, the field of vegetation dynamics has advanced through empirical generalization based on numerous case studies. An enormous amount of natural variation remains unstudied. Carefully executed case studies aimed towards conceptual generalization will continue to provide the basic ingredients for increased understanding of vegetation dynamics.

While no chapter in this book focuses exclusively on empirical data, several provide summaries of types of empirical data. Glenn-Lewin and van der Maarel (Chapter 1) provide an overview of the types of data often employed in succession studies. Van der Valk (Chapter 2) describes the attributes of plants that become established at different times during succession. The diversity of reported patterns in such community and ecosystem properties as species diversity, stability, biomass and production are reported and analysed by Peet (Chapter 3). Veblen (Chapter 4) examines the processes of regeneration of tree species in several types of near-equilibrium forest. Finally, Prentice (Chapter 8) describes trends in vegetation change over geologic time scales. Each of these chapters provides clear evidence of the fundamental role that empirical studies have played in the development of vegetation dynamics theory.

Model systems are often employed for studying complex biological systems; take, for example, the pivotal roles played by *Drosophila melanogaster* and *Escherichia coli* in genetics and molecular biology. Just as there is a need for ecologists to understand a broad array of systems, there is an equally compelling need to understand the vegetation dynamics of a few model systems very deeply. As yet there are many unanswered questions about even the most thoroughly studied vegetation types. For example, very few studies link population dynamics to changes in ecosystem properties. Continued, reductionist-oriented studies of the best-known vegetation types need to be encouraged.

IDENTIFICATION OF CRITICAL MECHANISMS

Predictions of vegetation change based simply on extrapolation from observations over short time periods or small areas are likely to be of limited value. Most of the more important and more difficult predictions require longer intervals or extrapolation to distant areas. To make these kinds of predictions with any accuracy requires a detailed understanding of the mechanisms of vegetation change. At the simplest level, the mechanisms of change outlined by Clements in 1916 remain valid today: nudation, migration, ecesis, competition, reaction, stabilization (Prologue and Chapter 1). But we need a deeper understanding. Starting with Clements' general scheme, Pickett *et al.* (1987) compiled a long list of mechanisms potentially responsible for vegetation change (e.g. Chapter 1, Table 1.1). Connell and Slatyer (1977), Noble and Slatyer (1980) and Walker and Chapin (1987) have all made compelling cases for the need to understand vegetation dynamics through study of the underlying mechanisms. Connell and Slatyer in particular caught the attention of vegetation scientists, and their three models of vegetation change (each a composite of several mechanisms) have provided a focus for numerous studies of vegetation dynamics. Tilman's work at Cedar Creek in Minnesota (Chapters 1, 3 and 5) is exemplary in the emphasis it places on experimental separation of critical component mechanisms of vegetation change. In this book, van der Valk illustrates many facets of colonization and establishment, and Veblen extends the regeneration theme to maintenance of species in relatively mature, stable forest systems.

As mechanisms of vegetation change become better understood, we can anticipate comparative studies that examine the shifting roles of various mechanisms of vegetation change as investigators move between vegetation types or geographic regions. Walker and Chapin's (1987) paper, with its eloquent plea for studies of vegetation dynamics that partition the relative importance of the various potential mechanisms of change that operate across a particular succession sequence, could well serve as a guide for future studies.

MODELS OF VEGETATION DYNAMICS

Models provide a way of organizing and extending our understanding of vegetation dynamics to make predictions. Both qualitative (conceptual) and quantitative (mathematical) models have proved valuable. Advances in our understanding of the mechanisms and processes of vegetation change (e.g. Connell and Slatyer, 1977; Noble and Slatyer, 1980; Peet and Christensen, 1980) have permitted the development of more realistic conceptual models (Chapter 3; Pickett *et al.*, 1987; Walker and Chapin, 1987). Theoretical advances in vegetation dynamics have been based on the iterative process of conceptual generalization, leading to empirical testing, leading to revision of concepts.

More and more, advances in the development of a theoretical framework for vegetation dynamics are based on combinations of empirical research and quantitative modelling. As mechanisms have become better understood, quantitative population modelling has become more realistic. This, too, is an iterative process, here involving development and parametrization of quantitative models, leading to testing of predictions, leading in turn back to revision of the quantitative and underlying qualitative models.

The individual-based gap models (Chapter 7) have demonstrated the value of computer-intensive simulators for predicting the impact of environmental change on vegetation. As computing power increases, particularly the availability of machines with massively parallel architecture, there will almost certainly be an increased emphasis on modelling spatial processes, both at the stand level and across large landscapes.

NON-EQUILIBRIUM PERSPECTIVES

The 1970s saw a rejection of equilibrium-based models in many areas of ecology. The seductive allure of the niche partitioning approach to understanding community structure developed by Robert MacArthur and his associates during the 1960s (Schoener, 1974) was largely abandoned as disturbance and extreme events came to be seen as at least equally critical for understanding species coexistence (e.g. Connell, 1971; Weins, 1977). The concept of climax became less important in studies of vegetation dynamics as the patch dynamics paradigm, first fully articulated by Watt (1947), achieved widespread acceptance (Pickett and White, 1985). As Veblen explains in Chapter 4, the seeming equilibrium of the mature 'climax' forest is at best a dynamic equilibrium resembling a mosaic with each of the component cells going through cycles of recovery, crash and recovery. The regeneration niche concept (Chapters 1, 2 and 4; Grubb, 1977), which explains coexistence based on fine-scale environmental variation, was a logical extension of this perspective in that it applies non-equilibrium concepts to the smaller spatial and temporal scales experienced by seeds and seedlings. At the other extreme of scale, global climatic instabilities lead to long-term and large-scale variations in vegetation (Chapter 8), which provide even greater challenges for predictive modelling of vegetation change.

If models of vegetation dynamics are to be broadly predictive, they must be compatible with the largely non-equilibrium character of ecological systems. While the Markov models described by Usher in Chapter 6 assume an equilibrium in the sense of stationary transition matrices, tests are provided to assure the user that the transition matrices are sufficiently stable. More complex, near-equilibrium formulations are possible, but have yet to be applied to vegetation dynamics. The now widely used JABOWA–FORET–ZELIG-style gap models described by Urban and Shugart (Chapter 7) can readily accommodate changing environmental conditions, and have, in fact, been used to predict the consequences of broad-scale environmental changes (e.g. Solomon, 1986; Pastor and Post, 1988).

LANDSCAPE AND GLOBAL-SCALE VEGETATION DYNAMICS

In Chapter 8, Prentice refers to the work of Margaret Davis (e.g. Davis 1976, 1981) and her reconstructions of post-glacial migration patterns of tree species in eastern North America. She found that not only are these species individualistic in their behaviour, but that the communities they formed as recently as 10 000 years ago lack modern analogues. While the plant communities of eastern North America have changed in a predictable fashion over the last 18 000 years in response to climatic variation (Prentice, Bartlein and Webb 1991), species have exhibited noticeable time-lags in approaching equilibrium with climate. As a result, climate alone is often insufficient as a predictor of vegetational composition.

Davis's findings have major implications for our ability to make predictions about large-scale changes in the earth's biota as a consequence of global climatic change (Davis, 1986). If each species has a unique migrational behaviour that reflects the peculiarities of its biology and the landscape that it must traverse, we shall need theory and models that incorporate the details of large-scale spatial pattern and initial conditions. The climatic changes that will result from human activities are anticipated to be much more rapid than those of the past 18 000 years. Certainly, we will not be dealing with an equilibrium situation.

Vegetation science has reached the point where the predictive ability of our models at the stand scale (e.g. $0.1 - 10$ ha) is very good, but in relation to global change issues we are now asked to make predictions at scales of $10^5 - 10^7$ km^2. Application of current theory at such large scales takes us well beyond the conditions for which that theory was designed. A more fruitful approach would be a stepwise increment of spatial scales as models are broadened and refined, first to the landscape level (e.g. $10-100 \, \text{km}^2$), and then the small region level (e.g. $1000-10000 \, \text{km}^2$). Botkin et al.'s (1991) examination of the possible impacts of climate change on the survival of the endangered Kirtland's warbler serves as a useful example of meso-scale applications of current models.

Species availability (migration) will become critical at different scales for different species, depending on dispersal abilities and initial conditions. Due to the lack of an empirical basis for realistic formulation of population-based models at regional scales, there is a danger of an inappropriate return to equilibrium assumptions in modelling region- and continent-scale climate-induced vegetation change. Experience with stand-level gap models (Chapter 7) has made clear the importance of such non-equilibrium phenomena as vegetational inertia and timelags in migration. Regional-scale models of vegetation dynamics that incorporate spatial processes must be a major focus of future modelling efforts if we are to address the implications of global-scale environmental change effectively.

REFERENCES

Botkin, D. B., Woodbury, D. A. and Nisbet, R. A. (1991) Kirtland's warbler habitats: a possible early indicator of climatic warming. *Biol. Conserv.*, **56**, 63–78.

Clements, F. E. (1916) *Plant Succession: An Analysis of the Development of Vegetation*, Carnegie Institution of Washington, Publication 242.

Connell, J. H. (1978) Diversity in tropical rain forests and coral reefs. *Science*, **199**, 1302–10.

Connell, J. H. and Slatyer, R. O. (1977) Mechanisms of succession in natural communities and their roles in community stability and organization. *Amer. Nat.*, **111**, 1119–44.

Davis, M. B. (1976) Pleistocene biogeography of temperate deciduous forests. *Geosci. Man*, **13**, 13–26.

Davis, M. B. (1981) Quaternary history and the stability of forest communities, in *Forest Succession: Concepts and Application* (eds D. C. West, H. H. Shugart and D. B. Botkin), Springer-Verlag, New York, pp. 132–53.

Davis, M. B. (1986) Climatic instability, time lags, and community disequilibrium, in *Community Ecology* (eds J. Diamond and T. J. Case), Harper and Row, New York, pp. 269–84.

Grubb, P. J. (1977) The maintenance of species-richness in plant communities: the importance of the regeneration niche. *Biol. Rev.*, **52**, 107–45.

Noble, I. R. and Slatyer, R. O. (1980) The use of vital attributes to predict successional changes in plant communities subject to recurrent disturbances. *Vegetatio*, **43**, 5–21.

Pastor, J. and Post, W. M. (1988) Response of northern forests to CO_2-induced climate change. *Nature*, **334**, 55–8.

Peet, R. K. and Christensen, N. L. (1980) Succession: a population process. *Vegetatio*, **43**, 131–40.

Pickett, S. T. A., Collins, S. L. and Armesto, J. J. (1987) Models, mechanisms and pathways of succession. *Bot. Rev.*, **53**, 335–71.

Pickett, S. T. A. and White, P. S. (1985) *The Ecology of Natural Disturbance and Patch Dynamics*, Academic Press, Orlando.

Prentice, I. C., Bartlein, P. J. and Webb, T., III (1991) Vegetation and climate change in eastern North America since the last glacial maximum. *Ecology*, **72**, 2038–56.

Schoener, T. W. (1974) Resource partitioning in ecological communities. *Science*, **185**, 27–39.

Solomon, A. M. (1986) Transient response of forests to CO_2-induced climate change: simulation modeling experiments in eastern North America. *Oecologia*, **68**, 567–79.

Walker, L. R. and Chapin, F. S. III. (1987) Interactions among processes controlling successional change. *Oikos*, **50**, 131–5.

Watt, A. S. (1947) Pattern and process in the plant community. *J. Ecol.*, **35**, 1–22.

Wiens, J. A. (1977) On competition and variable environments. *Amer. Scient.*, **65**, 590–7.

Index